STUDENT ATLAS OF

World Geography

Seventh Edition

John L. Allen
University of Wyoming

Christopher J. Sutton
Western Illinois University

STUDENT ATLAS OF WORLD GEOGRAPHY, SEVENTH EDITION

Published by McGraw-Hill, a business unit of The McGraw-Hill Companies, Inc., 1221 Avenue of the Americas, New York, NY 10020. Copyright © 2012 by The McGraw-Hill Companies, Inc. All rights reserved. Previous edition(s) © 2010, 2008, and 2005. No part of this publication may be reproduced or distributed in any form or by any means, or stored in a database or retrieval system, without the prior written consent of The McGraw-Hill Companies, Inc., including, but not limited to, in any network or other electronic storage or transmission, or broadcast for distance learning.

Some ancillaries, including electronic and print components, may not be available to customers outside the United States.

Student Atlas® is a registered trademark of the McGraw-Hill Companies, Inc.
Student Atlas is published by the **Contemporary Learning Series** group within the McGraw-Hill Higher Education division.

1 2 3 4 5 6 7 8 9 0 QDB / QDB 1 0 9 8 7 6 5 4 3 2 1

MHID 0-07-352762-9
ISBN 978-0-07-352762-8
ISSN 1531-0221

Managing Editor: *Larry Loeppke*
Developmental Editor II: *Debra A. Henricks*
Senior Permissions Coordinator: *Shirley Lanners*
Marketing Specialist: *Alice Link*
Senior Project Manager: *Jane Mohr*
Design Coordinator: *Brenda A. Rolwes*
Cover Graphics: *Rick D. Noel*
Buyer: *Nicole Baumgartner*
Media Project Manager: *Sridevi Palani*

Compositor: Lachina Publishing Services
Cover Image: © Doug Sherman/Geofile

Library of Congress Cataloging-in-Publication Data

Allen, John Logan, 1941–
   Student atlas of world geography / John L. Allen. — 7th ed.
      p. cm.
   "Contemporary Learning Series."
   "ISSN 1531-0221."
   Includes bibliographical references and index.
   ISBN 978-0-07-352762-8 (alk. paper)
   1. Atlases. 2. Geography—Maps. 3. Atlases. I. Title.
   G1021.A48 2012
   912—dc22

                                        2010054497

www.mhhe.com

A Note to the Student

The study of geography has become an increasingly important part of the curriculum in secondary schools and institutions of higher education over the last decade. This trend, a most welcome one from the standpoint of geographers, has begun to address the massive problem of "geographic illiteracy" that has characterized the United States, almost alone among the world's developed nations. When a number of international comparative studies on world geography were undertaken, beginning in the 1970s, it became apparent that most American students fell far short of their counterparts in Europe, Russia, Canada, Australia, and Japan in their abilities to recognize geographic locations, to identify countries or regions on maps, or to explain the significance of such key geographic phenomena such as population distribution, economic or urban location, or the availability of natural resources. Indeed, many American students could not even locate the United States on world maps, let alone countries like France, or Indonesia, or Nigeria. This atlas, and the texts it is intended to accompany, is a small part of the process of attempting to increase the geographic literacy of American students. As the true meaning of "the global community" becomes more apparent, such an increase in geographic awareness is not only important but necessary. If the United States has learned any lesson from the tragic events at the World Trade Center and the Pentagon on September 11, 2001, these lessons would surely include the considerations that we are not isolated from events that transpire in other parts of the world; our boundaries do not make us secure; and we ignore the conditions of political, economic, cultural, and physical geography outside those boundaries at our great peril.

The maps in the *Student Atlas of World Geography* are designed to introduce you to the patterns or "spatial distribution" of the wide variety of human and physical features of the earth's surface and to help you understand the relationships between these patterns. We call such relationships "spatial correlation" and whenever you compare the patterns made by two or more phenomena that exist at or near the earth's surface—the distribution of human population and the types of climate, for example—you are engaging in spatial correlation. At the very outset of your study of this atlas, you should be aware of some limitations of the data used to create the maps. In some instances, there may be data missing. In such cases, the cause may represent the failure of a country to report information to a central international body (like the United Nations or the World Bank), or it may mean that the shifting of political boundaries and changed responsibility for reporting data have caused some countries (for example, those countries that made up the former Soviet Union or the former Yugoslavia) to delay their reports. It is always our aim to use the most up-to-date data that is possible. Subsequent editions of this atlas will have increased data on countries like Serbia, Montenegro, or Kosovo when it becomes available. In the meantime, as events continue to restructure our world, it's an exciting time to be a student of world geography!

You will find your study of this atlas more productive if you study the maps and tables on the following pages in the context of the five distinct themes that have been developed as part of the increasing awareness of the importance of geographic education:

1. *Location: Where Is It?* This theme offers a starting point from which you discover the precise location of places in both absolute terms (the latitude and longitude of a place) and in relative terms (the location of a place in relation to the location of other places). When you think of location, you should automatically think of both forms. Knowing something about absolute location will help you to understand a variety of features of physical geography, since such key elements are so closely related to their position on the earth. But it is equally important to think of location in relative terms. The location of places in relation to other places is often more important as a determinant of social, economic, and cultural characteristics than the factors of physical geography.

2. *Place: What Is It Like?* This theme investigates the political, economic, cultural, environmental, and other characteristics that give a place its identity. You should seek to understand the similarities and differences of places by exploring their basic characteristics. Why are some places with similar environmental characteristics so very different in economic, cultural, social, and political ways? Why are other places with such different environmental characteristics so seemingly alike in terms of their institutions, their economies, and their cultures?

3. *Human/Environment Interactions: How Is the Landscape Shaped?* This theme illustrates the ways in which people respond to and modify their environments. Certainly the environment is an important factor in influencing human activities and behavior. But the characteristics of the environment do not exert a controlling influence over human activities; they only provide a set of alternatives from which different cultures, in different times, make their choices. Observe the relationship between the basic elements of physical geography such as climate and terrain and the host of ways in which humans have used the land surfaces of the world.

4. *Movement: How Do People Stay in Touch?* This theme examines the transportation and communications systems that link people and places. Movement or "spatial interaction" is the chief mechanism for the spread of ideas and innovations from one place to another. It is spatial interaction that validates the old cliché, "the world is getting smaller." We find McDonald's restaurants in Tokyo and Honda automobiles in New York City because of spatial interaction. Advanced transportation and communications systems

have transformed the world into which your parents were born. And the world your children will be born into will be very different from your world. None of this would happen without the force of movement or spatial interaction.

5. *Regions: Worlds Within a World.* This theme helps to organize knowledge about the land and its people. The world consists of a mosaic of "regions" or areas that are somehow different and distinctive from other areas. The region of Anglo-America (the United States and Canada) is, for example, different enough from the region of Western Europe that geographers clearly identify them as two unique and separate areas. Yet despite their differences, Anglo-Americans and Europeans share a number of similarities: common cultural backgrounds, comparable economic patterns, shared religious traditions, and even some shared physical environmental characteristics. Conversely, although the regions of Anglo-America and Eastern Asia are also easily distinguished as distinctive units of the earth's surface, they have a greater number of shared physical environmental characteristics. But those who live in Anglo-America and Eastern Asia have fewer similarities and more differences between them than is the case with Anglo-America and Western Europe: different cultural traditions, different institutions, different linguistic and religious patterns. An understanding of both the differences and similarities between regions like Anglo-America and Europe on the one hand, or Anglo-America and Eastern Asia on the other, will help you to understand the world around you. At the very least, an understanding of regional similarities and differences will help you to interpret what you read on the front page of your daily newspaper or view on the evening news report on your television set.

Not all of these themes will be immediately apparent on each of the maps and tables in this atlas. But if you study the contents of *Student Atlas of World Geography*, along with the reading of your text and think about the five themes, maps and tables and text will complement one another and improve your understanding of global geography.

John L. Allen
Christopher J. Sutton

About the Authors

John L. Allen is professor and chair of geography, emeritus, at the University of Wyoming where he taught from 2000 to 2007, and emeritus professor at the University of Connecticut, where he taught from 1967 to 2000. He is a native of Wyoming. He received his bachelor's degree in 1963 and his M.A. in 1964 from the University of Wyoming, and in 1969 his Ph.D. from Clark University. His special areas of interest are perceptions of the environment and the impact of human societies on environmental systems. Dr. Allen is the author and editor of many books and articles as well as several other student atlases, including the best-selling *Student Atlas of World Politics*.

Christopher J. Sutton is professor of geography at Western Illinois University. Born in Virginia and raised in Illinois, he received his bachelor's degree (1988) and master's degree (1991) in Geography from Western Illinois University. In 1995 he earned his Ph.D. in Geography from the University of Denver. He is the author of numerous research articles and educational materials. A broadly trained geographer, his areas of interest include cartographic design, cultural geography, and urban transportation. After teaching at Northwestern State University of Louisiana for three years, Dr. Sutton returned to Western Illinois University in 1998, serving as chair of the Department of Geography from 2002 to 2007. Additionally, Dr. Sutton has served as president of the Illinois Geographical Society.

Academic Advisory Board

Members of the Academic Advisory Board review maps for content, accuracy, and usefulness. We think you will find their careful consideration reflected in this edition.

Courtney Donovan
San Francisco State University

Stephen Herring
Edgecombe Community College

Frank Ibe
Wayne County Community College District

Richard Inscore
Charleston Southern University

Jean-Gabriel Jolivet
Southwestern College

Vishnu Khade
Manchester Community College

Taylor Mack
Louisiana Tech University

Monica Milburn
Lone Star College–Kingwood

Bryant Mullen
Newport Business Institute

Linda Murphy
Blinn Community College

Mahendra Singh
Grambling State University

Acknowledgments

Nozar Alaolmolki
Hiram College

Barbara Batterson-Rossi
Palomar College

A. Steele Becker
University of Nebraska at Kearney

Koop Berry
Walsh University

Daniel A. Bunye
South Plains College

Winifred F. Caponigri
Holy Cross College

Femi Ferreira
Hutchinson Community College

Eric J. Fournier
Samford University

William J. Frazier
Columbus State College

Hari P. Garbharran
Middle Tennessee State University

Baher Gosheh
Edinboro University of Pennsylvania

Donald Hagan
Northwest Missouri State University

Robert Janiskee
University of South Carolina

David C. Johnson
University of Louisiana

Effie Jones
Crichton College

Cub Kahn
Marythurst University

Artimus Keiffer
Franklin College

Leonard E. Lancette
Mercer University

Donald W. Lovejoy
Palm Beach Atlantic College

Mark Maschhoff
Harris-Stowe State College

Richard Matthews
University of South Carolina

Madolia Mills
University of Colorado–Colorado Springs

Robert Mulcahy
Providence College

Otto H. Muller
Alfred University

J. Henry Owusu
University of Northern Iowa

Steven Parkansky
Morehead State University

William Preston
California Polytechnic State University, San Luis Obispo

Neil Reid
The University of Toledo

A. L. Rydant
Keene State College

Deborah Berman Santana
Mills College

Steven Slakey
University of La Verne

Rolf Sternberg
Montclair State University

Richard Ulack
University of Kentucky

David Woo
California State University, Haywood

Donald J. Zeigler
Old Dominion University

Table of Contents

Unit IV   Global Economic Patterns   67

Unit V   Global Patterns of Environmental Disturbance   95

Unit VI   Global Political Patterns   127

Unit VII   World Regions   165

Introduction: How to Read an Atlas

An atlas is a book containing maps which are "models" of the real world. By the term "model" we mean exactly what you think of when you think of a model: a representation of reality that is generalized, usually considerably smaller than the original, and with certain features emphasized, depending on the purpose of the model. A model of a car does not contain all of the parts of the original but it may contain enough parts that it is recognizable as a car and can be used to study principles of automotive design or maintenance. A car model designed for racing, on the other hand, may contain fewer parts but would have the mobility of a real automobile. Car models come in a wide variety of types containing almost anything you can think of relative to automobiles that doesn't require the presence of a full-size car. Since geographers deal with the real world, virtually all of the printed or published studies of that world require models. Unlike a mechanic in an automotive shop, we can't roll our study subject into the shop, take it apart, put it back together. We must use models. In other words, we must generalize our subject, and the way we do that is by using maps. Some maps are designed to show specific geographic phenomena, such as the climates of the world or the relative rates of population growth for the world's countries. We call these maps "thematic maps" and Units I through VI of this atlas contain maps of this type. Other maps are designed to show the geographic location of towns and cities and rivers and lakes and mountain ranges and so on. These are called "reference maps" and they make up many of the maps in Unit VII. All of these maps, whether thematic or reference, are models of the real world that selectively emphasize the features that we want to show on the map.

In order to read maps effectively—in other words, in order to understand the models of the world presented in the following pages—it is important for you to know certain things about maps: how they are made using what are called *projections*; how the level of mathematical proportion of the map or what geographers call *scale* affects what you see; and how geographers use *generalization* techniques such as simplification and symbols where it would be impossible to draw a small version of the real world feature. In this brief introduction, then, we'll explain to you three of the most important elements of map interpretation: projection, scale, and generalization.

The Coordinate System

Map Projections

Perhaps the most basic problem in *cartography*, or the art and science of map-making, is the fact that the subject of maps—the earth's surface—is what is called by mathematicians "a non-developable surface." Since the world is a sphere (or nearly so—it's actually slightly flattened at the poles and bulges a tiny bit at the equator), it is impossible to flatten out the world or any part of its curved surface without producing some kind of distortion. This "near sphere" is represented by a geographic grid or coordinate system of lines of latitude or *parallels* that run east and west and are used to measure distance north and south on the globe, and lines of longitude or *meridians* that run north and south and are used to measure distance east and west. All the lines of longitude are half circles of equal length and they all converge at the poles. These meridians are numbered from 0 degrees (Prime or Greenwich Meridian) east and west to 180 degrees. The meridian of 0 degrees and the meridian of 180 degrees are halves of the same "great circle" or line representing a plane that bisects the globe into two equal hemispheres. All lines of longitude are halves of great circles. All the lines of latitude are complete circles that are parallel to one another and are spaced equidistant on the meridians. The circumference of these circles lessens as you move north or south from the equator. Parallels of latitude are numbered from 0 degrees at the equator north and south to 90 degrees at the North

and South poles. The only line of latitude that is a great circle is the equator, which equally divides the world into a northern and southern hemisphere. In the real world, all these grid lines of latitude and longitude intersect at right angles. The problem for cartographers is to convert this spherical or curved grid into a geometrical shape that is "developable"; that is, it can be flattened (such as a cylinder or cone) or is already flat (a plane). The reason the results of the conversion process are called "projections" is that we imagine a world globe (or some part of it) that is made up of wires running north-south and east-west to represent the grid lines of latitude and longitude and other wires or even solid curved plates to represent the coastlines of continents or the continents themselves. We then imagine a light source at some location inside or outside the wire globe that can "project" or cast shadows of the wires representing grid lines onto a developable surface. Sometimes the basic geometric principles of projection may be modified by other mathematical principles to yield projections that are not truly geometric but have certain desirable features. We call these types of projections "arbitrary." The three most basic types of projections are named according to the type of developable surface: cylindrical, conic, or azimuthal (plane). Each type has certain characteristic features: they may be *equal area* projections in which the size of each area on the map is a direct proportional representation of that same area in the real world but shapes are distorted; they may be *conformal* projections in which area may be distorted but shapes are shown correctly; or they may be *compromise* projections in which both shape and area are distorted but the overall picture presented is fairly close to reality. It is important to remember that all maps distort the geographic grid and continental outlines in characteristic ways. The only representation of the world that does not distort either shape or area is a globe. You can see why we must use projections—can you imagine an atlas that you would have to carry back and forth across campus that would be made up entirely of globes?

Cylindrical Projections

Cylindrical projections are drawn as if the geographic grid were projected onto a cylinder. Cylindrical projections have the advantage of having all lines of latitude as true parallels or straight lines. This makes these projections quite useful for showing geographic relationships in which latitude or distance north-south is important (many physical features, such as climate, are influenced by latitude). Unfortunately, most cylindrical-type projections distort area significantly. One of the most famous is the Mercator projection shown on the next page.

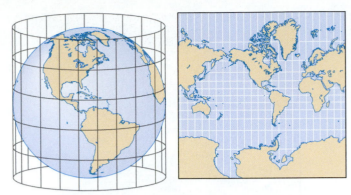

Cylindrical Projection: The Mercator Projection

This projection makes areas disproportionately large as you move toward the pole, making Greenland, which is actually about one-seventh the size of South America, appear to be as large as the southern continent. But the Mercator projection has the quality of conformality: landmasses on the map are true in shape and thus all coastlines on the map intersect lines of latitude and longitude at the proper angles. This makes the Mercator projection, named after its inventor, a sixteenth-century Dutch cartographer, ideal for its original purpose as a tool for navigation—but not a good projection for attempting to show some geographical feature in which areal relationship is important. Unfortunately, the Mercator projection has often been used for wall maps for schoolrooms and the consequence is that generations of American school children have been "tricked" into thinking that Greenland is actually larger than South America. Much better cylindrical-type projections are those like the Robinson projection used in this atlas that is neither equal area nor conformal but a compromise that portrays the real world much as it actually looks, enough so that we can use it for areal comparisons.

The Robinson Projection

Conic Projections

Conic projections are those that are imagined as being projected onto a cone that is tangent to the globe along a standard parallel, or a series of cones tangent along several parallels or even intersecting the

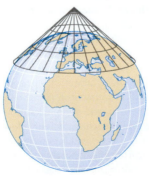

Conic Projection of Europe

globe. Conic projections usually show latitude as curved lines and longitude as straight lines. They are good projections for areas with north-south extent, like the preceding map of Europe, and may be either conformal, equal area, or compromise, depending on how they are constructed. Many of the regional maps in the last map section of this atlas are conic projections.

Azimuthal Projections

Azimuthal projections are those that are imagined as being projected onto a plane or flat surface. They are named for one of their essential properties. An "azimuth" is a line of compass bearing and azimuthal projections have the property of yielding true compass directions from the center of the map. This makes azimuthal maps useful for navigation purposes, particularly air navigation. But, because they distort area and shape so greatly, they are seldom used for maps designed to show geographic relationships. When they are used as illustrative rather than navigation maps, it is often in the "polar case" projection shown above where the plane has been made tangent to the globe at the North Pole.

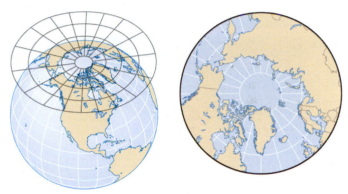

Azimuthal Projection of the North Polar Region

Map Scale

Since maps are models of the real world, it follows that they are not the same size as the real world or any portion of it. Every map, then, is subject to generalization, which is another way of saying that maps are drawn to certain scales. The term *scale* refers to the mathematical quality of *proportional representation,* and is expressed as a ratio between an area of the real world or the distance between places on the real world and the same area or distance on the map. We show map scale on maps in three different ways. Sometimes we simply use the proportion and write what is called a *natural scale* or representative fraction: for example, we might show on a map the mathematical proportion of 1:62,500. A map at this scale is one that is one sixty-two thousand five-hundredth the size of the same area in the real world. Other times we convert the proportion to a written description that approximates the relationship between distance on the map and distance in the real world. Since there are nearly 62,500 inches in a mile, we would refer to a map having a natural scale of 1:62,500 as having an "inch-mile" scale of "1 inch represents 1 mile." If we draw a line one inch long on this map, that line represents a distance of approximately one mile in the real world. Finally, we usually use a graphic or linear scale: a bar or line, often graduated into miles or kilometers, that shows graphically the proportional representation. A graphic scale for our 1:62,500 map might be about five inches long, divided into five equal units clearly labeled as "1 mile," "2 miles," and so on. Our examples on the following page show all three kinds of scales.

The most important thing to keep in mind about scale, and the reason why knowing map scale is important to being able to read a map correctly, is the relationship between proportional representation and generalization. A map that fills a page but shows the whole world is

Map 1 Small Scale Map of the United States

Map 2 Map of the Northeast

Map 3 Map of Southeastern New England

Map 4 Large Scale Map of Boston, MA

much more highly generalized than a map that fills a page but shows a single city. On the world map, the city may appear as a dot. On the city map, streets and other features may be clearly seen. We call the first map, the world map, a *small scale* map because the proportional representation is a small number. A page-size map showing the whole world may be drawn at a scale of 1:150,000,000. That is a very small number indeed—hence the term *small scale* map even though the area shown is large. Conversely, the second map, a city map, may be drawn at a scale of 1:250,000. That is still a very small number but it is a great deal larger than 1:150,000,000! And so we'd refer to the city map as a *large scale* map, even though it shows only a small area. On our world map, geographical features are generalized greatly and many features can't even be shown at all. On the city map, much less generalization occurs—we can show specific features that we couldn't on the world map—but generalization still takes place. The general rule is that the smaller the map scale, the greater the degree of generalization; the larger the map scale, the less the degree of generalization. The only map that would not generalize would be a map at a scale of 1:1 and that map wouldn't be very handy to use. Examine the relationship between scale and generalization in the four maps on this page.

Generalization on Maps

A review of the four maps on this page should give you some indication of how cartographers generalize on maps. One thing that you should have noticed is that the first map, that of the United States, is

much simpler than the other three and that the level of *simplification* decreases with each map. When a cartographer simplifies map data, information that is not important for the purposes of the map is just left off. For example, on Map 1 the objective may have been to show cities over 1 million in population. To do that clearly and effectively, it is not necessary to show and label rivers and lakes. The map has been simplified by leaving those items out. Map 4, on the other hand, is more complex and shows and labels geographic features that are important to the character of the city of Boston; therefore, the Charles River is clearly indicated on the map.

Another type of generalization is *classification*. Map 1 shows cities over 1 million in population. Map 2 shows cities of several different sizes and a different symbol is used for each size classification or category. Many of the thematic maps used in this atlas rely on classification to show data. A thematic map showing population growth rates (see Map 26) will use different colors to show growth rates in different classification levels or what are sometimes called *class intervals*. Thus, there will be one color applied to all countries with population growth rates between 1.0 percent and 1.4 percent, another color applied to all countries with population growth rates between 1.5 percent and 2.1 percent, and so on. Classification is necessary because it is impossible to find enough symbols or colors to represent precise values. Classification may also be used for qualitative data, such as the national or regional origin of migrating populations. Cartographers show both quantitative and qualitative classification levels or class intervals in important sections of maps called *legends*.

These legends, as in the samples that follow, make it possible for the reader of the map to interpret the patterns shown.

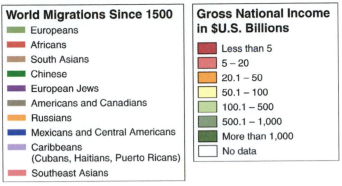

Map Legends from Maps 25 and 52

A third technique of generalization is *symbolization* and we've already noted several different kinds of symbols: those used to represent cities on the preceding maps, or the colors used to indicate population growth levels on Map 26. One general category of map symbols is quantitative in nature and this category can further be divided into a number of different types. For example, the symbols showing city size on Maps 1 and 2 on the preceding page can be categorized as *ordinal* in that they show relative differences in quantities (the size of cities). A cartographer might also use lines of different widths to express the quantities of movement of people or goods between two or more points as on Map 25.

The color symbols used to show rates of population growth can be categorized as *interval* in that they express certain levels of a mathematical quantity (the percentage of population growth). Interval symbols are often used to show physical geographic characteristics such as inches of precipitation, degrees of temperature, or elevation above sea level. The following sample, for example, shows precipitation (from Map 3a).

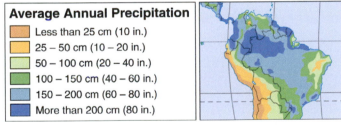

Interval Symbols

Still another type of mathematical symbolization is the *ratio* in which sets of mathematical quantities are compared: the number of persons per square mile (population density) or the growth in gross national product per capita (per person). The map below shows Gross National Income per capita (from Map 53).

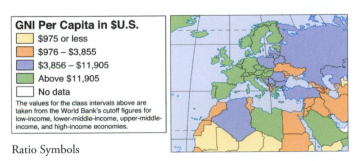

Ratio Symbols

Finally, there are a vast number of cartographic symbols that are not mathematical but show differences in the kind of information being portrayed. These symbols are called *nominal* and they range from the simplest differences such as land and water to more complex differences such as those between different types of vegetation. Shapes or patterns or colors or iconographic drawings may all be used as nominal symbols on maps. The following map (from Map 8) uses color to show the distribution of soil types.

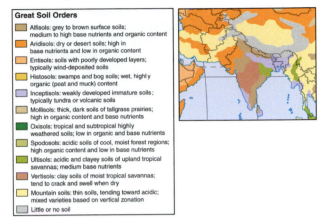

Nominal Symbols

The final technique of generalization is what cartographers refer to as *interpolation*. Here, the maker of a map may actually show more information on the map than is actually supplied by the original data. In understanding the process of interpolation it is necessary for you to visualize the quantitative data shown on maps as being three dimensional: x values provide geographic location along a north-south axis of the map; y values provide geographic location along the east-west axis of the map; and z values are those values of whatever data (for example, temperature) are being shown on the map at specific points. We all can imagine a real three-dimensional surface in which the x and y values are directions and the z values are the heights of mountains and the depths of valleys. On a topographic map showing a real three-dimensional surface, contour lines are used to connect points of equal elevation above sea level. These contour lines are not measured directly; they are estimated by interpolation on the basis of the elevation points that are provided.

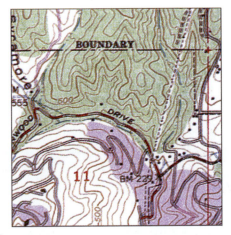

Interpolation

It is harder to imagine the statistical surface of a temperature map in which the x and y values are directions and the z values represent degrees of temperature at precise points. But that is just what cartographers do. And to obtain the values between two or more specific points where z values exist, they interpolate based on a class interval they have decided is appropriate and use *isolines* (which are statistical equivalents of a contour line) to show increases or decreases in value. The diagram to the right shows an example of an interpolation process. Occasionally interpolation is referred to as *induction*. By whatever name, it is one of the most difficult parts of the cartographic process.

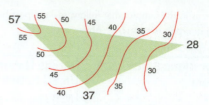

Degrees of Temperature (Celsius)
Interval = 5 degrees

And you thought all you had to do to read an atlas was look at the maps! You've now learned that it is a bit more involved than that. As you read and study this atlas, keep in mind the principles of projection and scale and generalization (including simplification, classification, symbolization, and interpolation) and you'll do just fine. Good luck and enjoy your study of the world of maps as well as maps of the world!

Unit I

Global Physical Patterns

Map 1   World Political Boundaries

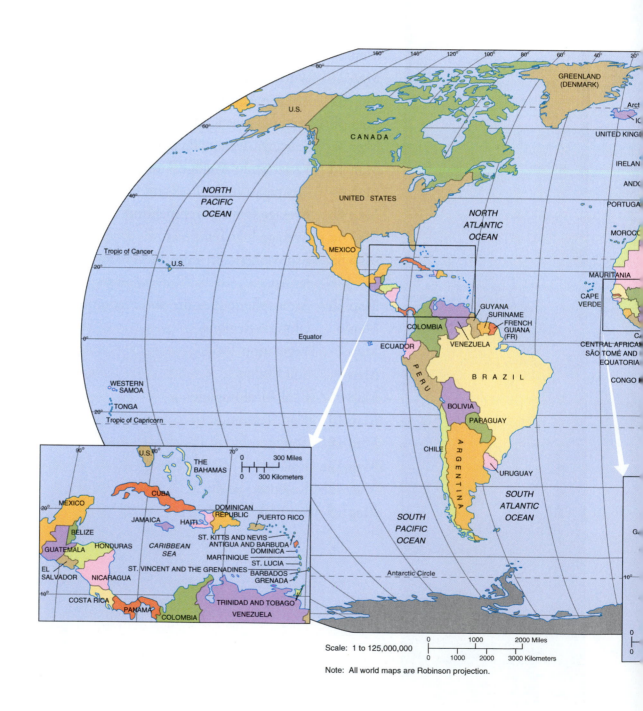

Scale: 1 to 125,000,000

Note: All world maps are Robinson projection.

Map 2 · World Physical Features

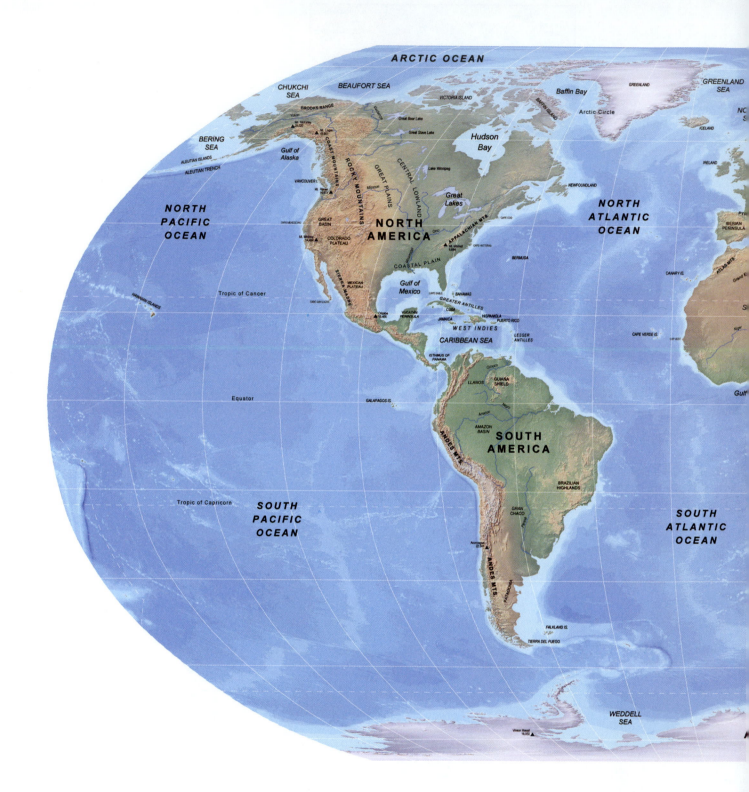

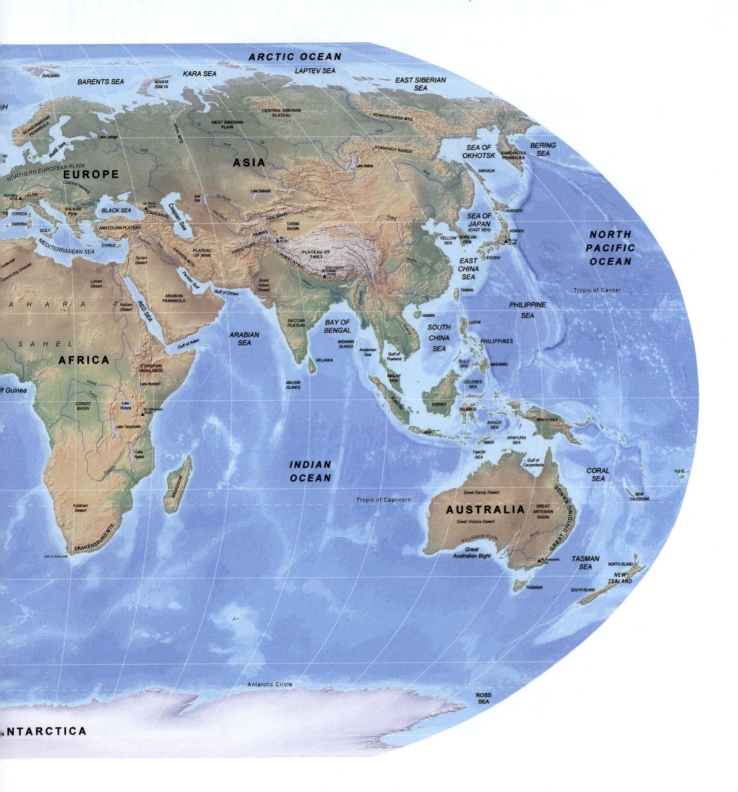

Map 3a   Average Annual Precipitation

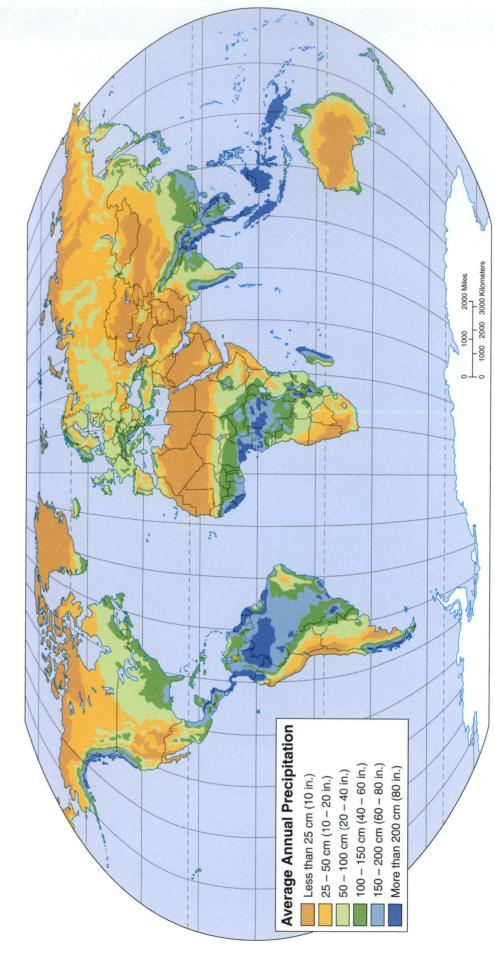

Average Annual Precipitation

- Less than 25 cm (10 in.)
- 25 – 50 cm (10 – 20 in.)
- 50 – 100 cm (20 – 40 in.)
- 100 – 150 cm (40 – 60 in.)
- 150 – 200 cm (60 – 80 in.)
- More than 200 cm (80 in.)

The two most important physical geographic variables are precipitation and temperature, the essential elements of weather and climate. Precipitation is a conditioner of both soil type and vegetation. More than any other single environmental element, it influences where people do or do not live. Water is the most precious resource available to humans, and water availability is largely a function of precipitation. Water availability is also a function of several precipitation variables that do not appear on this map: the seasonal distribution of precipitation (is precipitation or drought concentrated in a particular season?), the ratio between precipitation and temperature (how much of the water that comes to the earth in the form of precipitation is lost through mechanisms such as evaporation and transpiration that are a function of temperature?), and the annual variability of precipitation (how much do annual precipitation totals for a place or region tend to vary from the "normal" or average precipitation?). In order to obtain a complete understanding of precipitation, these variables should be examined along with the more general data presented on this map. The study of precipitation and other climatic elements is the concern of the branch of physical geography called "climatology."

Map 3b Seasonal Average Precipitation, November through April

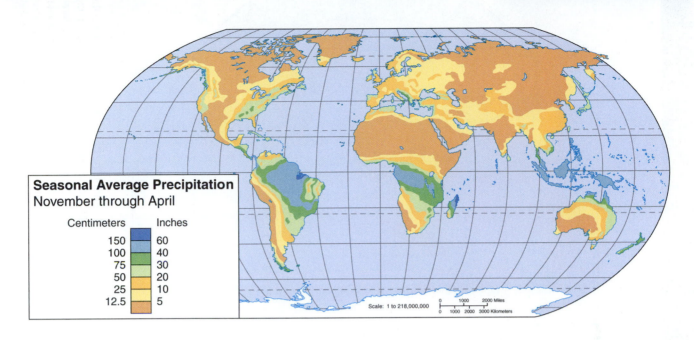

Seasonal Average Precipitation
November through April

Centimeters	Inches
150	60
100	40
75	30
50	20
25	10
12.5	5

Scale: 1 to 218,000,000

Map 3c Seasonal Average Precipitation, May through October

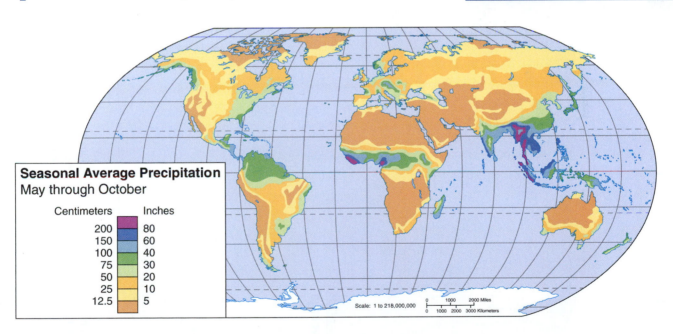

Seasonal Average Precipitation
May through October

Centimeters	Inches
200	80
150	60
100	40
75	30
50	20
25	10
12.5	5

Scale: 1 to 218,000,000

Seasonal average precipitation is nearly as important as annual precipitation totals in determining the habitability of an area. Critical factors are such things as whether precipitation coincides with the growing season and thus facilitates agriculture or during the winter when it is less effective in aiding plant growth, and whether precipitation occurs during summer with its higher water loss through evaporation and transpiration or during the winter when more of it can go into storage. Several of the world's great climate zones have pronounced seasonal precipitation rhythms. The tropical and subtropical savanna grasslands have a long winter dry season and abundant precipitation in the summer. The Mediterranean climate is the only major climate with a marked dry season during the summer, making agriculture possible only through irrigation or other adjustments to cope with drought during the period of plant growth. And the great monsoon climates of South and Southeast Asia have their winter dry season and summer rain that have conditioned the development of Asian agriculture and the rhythms of Asian life.

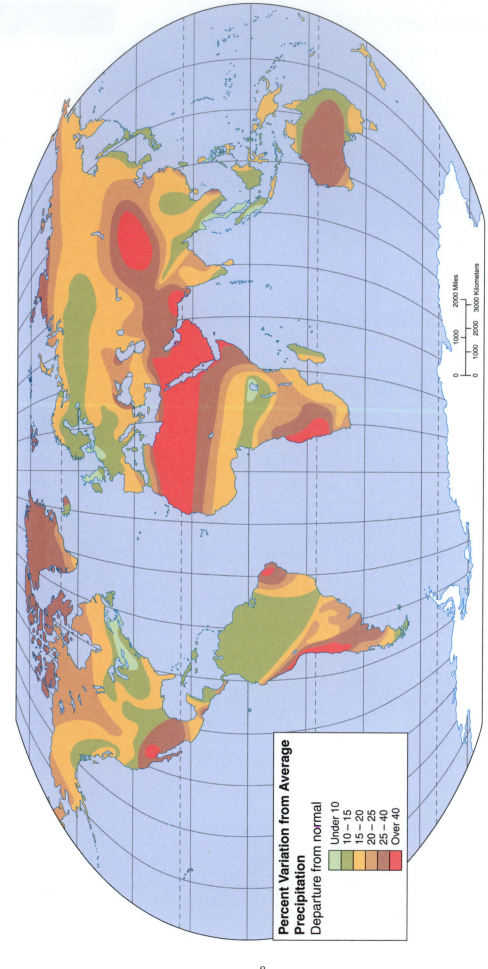

Percent Variation from Average Precipitation

Departure from normal

- Under 10
- 10 – 15
- 15 – 20
- 20 – 25
- 25 – 40
- Over 40

While annual precipitation totals and seasonal distribution of precipitation are important variables, the variability of precipitation from one year to the next may be even more critical. You will note from the map that there is a general spatial correlation between the world's drylands and the amount of annual variation in precipitation. Generally, the drier the climate, the more likely it is that there will be considerable differences in rainfall and/or snowfall from one year to the next. We might determine that the average precipitation of the mid-Sahara is 2 inches per year. What this really means is that a particular location in the

Sahara during one year might receive .5", during the next year 3.5", and during a third year 2". If you add these together and divide by the number of years, the "average" precipitation is 2" per year. The significance of this is that much of the world's crucial agricultural output of cereals (grains) comes from dryland climates (the Great Plains of the United States, the Pampas of Argentina, the steppes of Ukraine and Russia, for example), and variations in annual rainfall totals can have significant impacts on levels of grain production and, therefore, important consequences for both economic and political processes.

Map 4a Temperature Regions and Ocean Currents

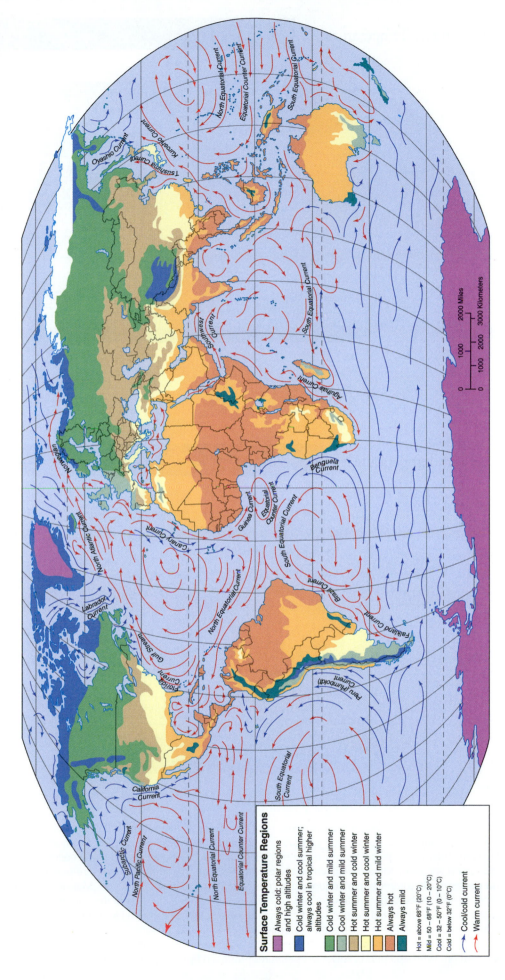

Surface Temperature Regions

- Always cold: polar regions and high altitudes
- Cold winter and cool summer; always cool in tropical higher altitudes
- Cold winter and mild summer
- Cool winter and mild summer
- Hot summer and cold winter
- Hot summer and cool winter
- Hot summer and mild winter
- Always hot
- Always mild

Hot = above 68°F (20°C)
Mild = 50 – 68°F (10 – 20°C)
Cool = 32 – 50°F (0 – 10°C)
Cold = below 32°F (0°C)

↗ Cool/cold current
↗ Warm current

Along with precipitation, temperature is one of the two most important environmental variables, defining the climate conditions so essential for the distribution of such human activities as agriculture and the distribution of the human population. The seasonal rhythm of temperature, including such measures as the average annual temperature range (difference between the average temperature of the warmest month and that of the coldest month), is an additional variable not shown on the map but,

like the seasonality of precipitation, should be a part of any comprehensive study of climate. The ocean currents illustrated exert a significant influence over the climate of adjacent regions and are the most important mechanism for redistributing surplus heat from the equatorial region into middle and high latitudes. Physical geographers known as "climatologists" study the phenomenon of temperature and related climatic characteristics.

Map 4b Average January Temperature

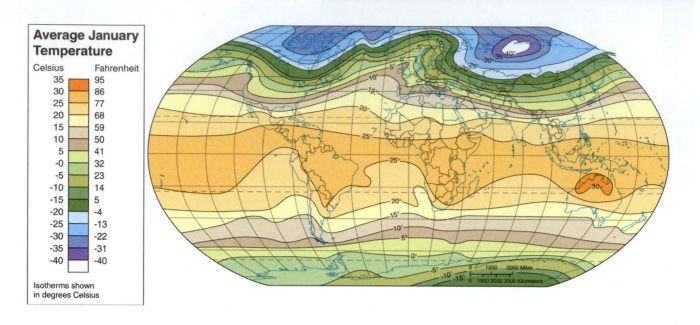

Average January Temperature

Celsius	Fahrenheit
35	95
30	86
25	77
20	68
15	59
10	50
5	41
-0	32
-5	23
-10	14
-15	5
-20	-4
-25	-13
-30	-22
-35	-31
-40	-40

Isotherms shown
in degrees Celsius

Map 4c Average July Temperature

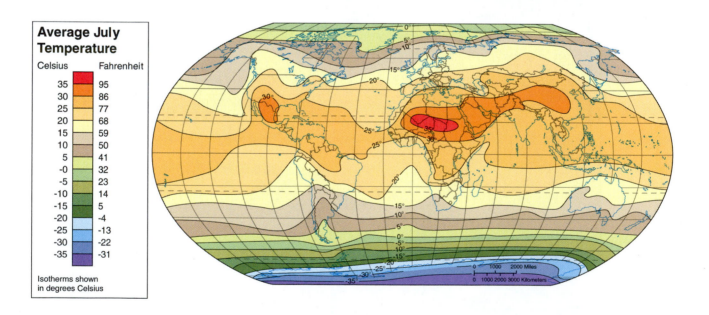

Average July Temperature

Celsius	Fahrenheit
35	95
30	86
25	77
20	68
15	59
10	50
5	41
-0	32
-5	23
-10	14
-15	5
-20	-4
-25	-13
-30	-22
-35	-31

Isotherms shown
in degrees Celsius

Where moisture availability tends to mark the seasons in the tropics and subtropics, in the mid-latitudes, seasons are marked by temperature. Temperature is determined by latitudinal transition, by altitude or elevation above sea level, and by location of a place relative to the world's landmasses and oceans. The most important of these controls is latitude, and temperatures generally become lower with increasing latitude. Proximity to water, however, tends to moderate temperature extremes, and "maritime" climates influenced by the oceans will be warmer in the winter and cooler in the summer than continental climates in the same general latitude. Maritime climates will also show smaller temperature ranges, the difference between January and July temperatures, while climates of the continental interiors, far from the moderating influences of the oceans, will tend to have greater temperature ranges. In the Northern Hemisphere, where there are both large landmasses and oceans, the range is great. But in the Southern Hemisphere, dominated by water and, hence, by the more moderate maritime air masses, the temperature range is comparatively small. Significant temperature departures from the "normal" produced by latitude may also be the result of elevation. With exceptions, lower temperatures produced by topography are difficult to see on maps of this scale.

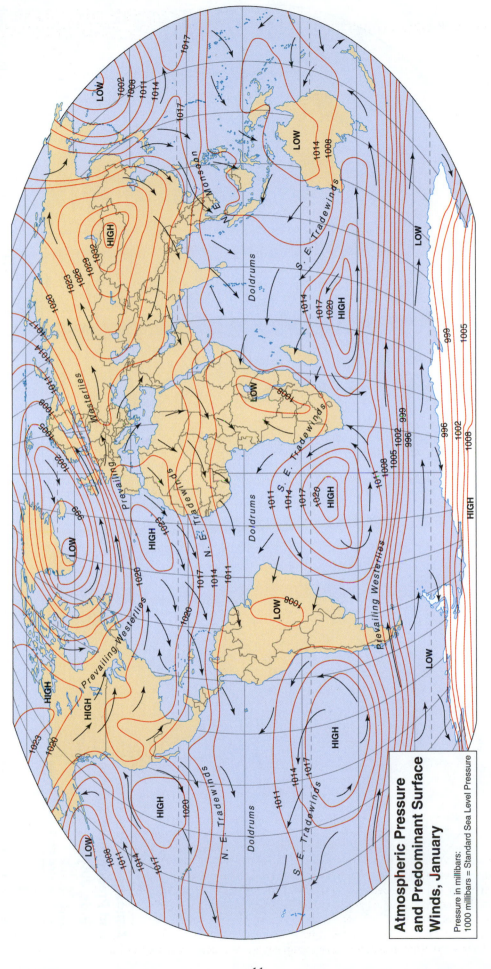

**Atmospheric Pressure
and Predominant Surface
Winds, January**

Pressure in millibars:
1000 millibars = Standard Sea Level Pressure

Map 5b   Atmospheric Pressure and Predominant Surface Winds, July

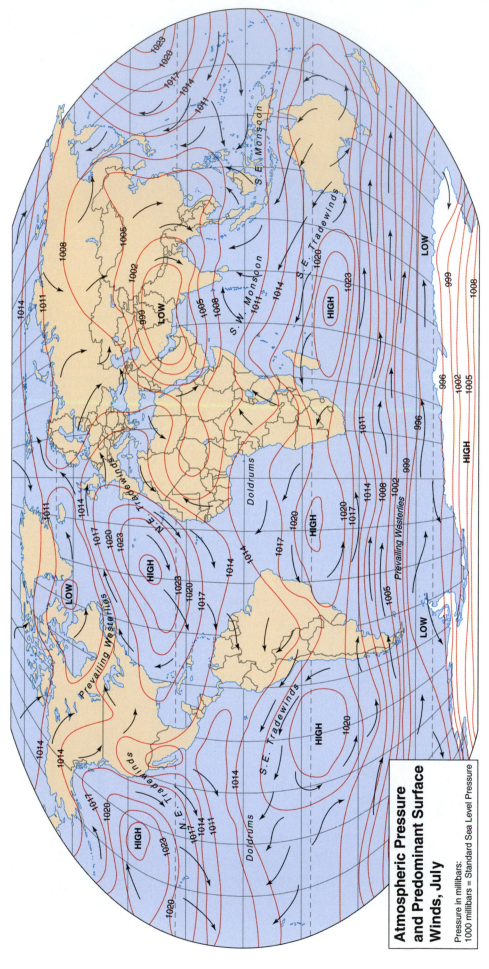

Atmospheric Pressure and Predominant Surface Winds, July

Pressure in millibars:
1000 millibars = Standard Sea Level Pressure

Atmospheric pressure, or the density of air, is a function largely of air temperature: the colder the air, the denser and heavier it is, hence the higher its pressure; the warmer the air, the lighter and less stable it is, hence the lower its pressure. Global pressure systems are the alternating low and high pressure systems that, from the equator north and south, include: the equatorial low (sometimes called the intertropical convergence) centered on the equator for much of the year; the subtropical highs with their centers near the 30th parallel of north and south latitude; the subpolar lows or polar fronts centered near the

60th parallel of north and south latitude; and the polar highs near the north and south poles. Air flows from high pressure to low pressure regions, and this air flow constitutes the earth's major surface winds such as the tropical tradewinds and the prevailing westerlies. This flow of air is one of the chief mechanisms by which surplus heat energy from the equatorial region is redistributed to higher latitudes. It is also the primary conditioner of the world's major precipitation belts, with rainfall and snowfall associated primarily with lower atmospheric pressure conditions.

Map 6 Climate Regions

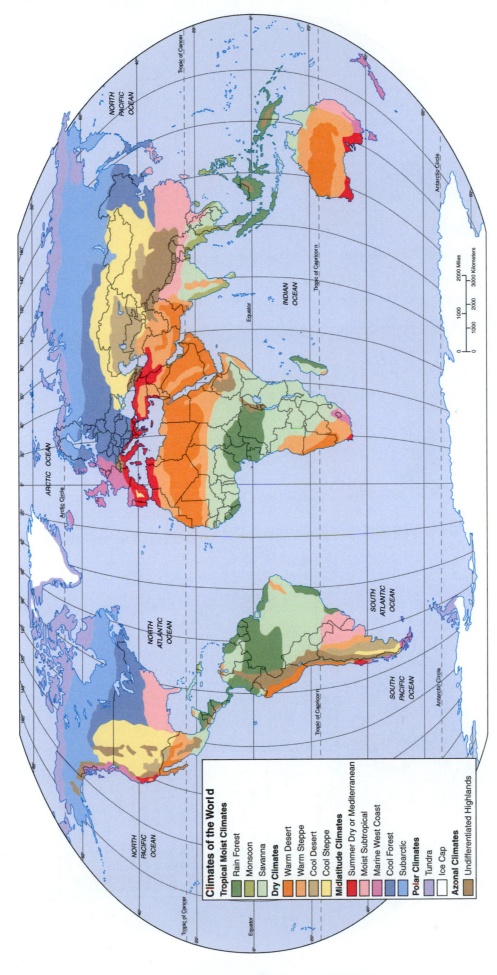

Of the world's many patterns of physical geography, climate or the long-term average of weather conditions such as temperature and precipitation is the most important. It is climate that conditions the distribution of natural vegetation and the types of soils that will exist in an area. Climate also influences the availability of our most crucial resource: water. From an economic standpoint, the world's most important activity is agriculture; no other element of physical geography is more important for agriculture than climate. Ultimately, it is agricultural production that determines where the bulk of human beings live, and therefore, climate

is a basic determinant of the distribution of human populations as well. The study of climates or "climatology" is one of the most important branches of physical geography.

The climate classification system shown on this map is based on that developed by Wladimir Köppen. To establish his climate regions, Köppen used the climatic parameters of *precipitation, temperature,* and *evapotranspiration* as they impacted certain kinds of major vegetative associations. Hence the names for many of the climate regions are also the names of vegetative regions.

Climates of the World

Tropical Moist Climates
Rain Forest
Monsoon
Savanna

Dry Climates
Warm Desert
Warm Steppe
Cool Desert
Cool Steppe

Midlatitude Climates
Summer Dry or Mediterranean
Moist Subtropical
Marine West Coast
Cool Forest
Subarctic

Polar Climates
Tundra
Ice Cap

Azonal Climates
Undifferentiated Highlands

Map 7 Vegetation Types

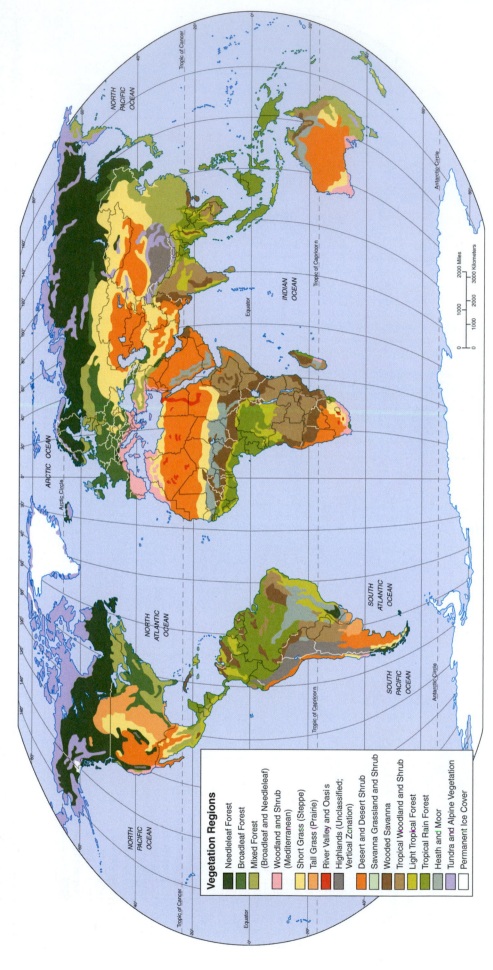

Vegetation Regions

- Needleleaf Forest
- Broadleaf Forest
- Mixed Forest (Broadleaf and Needleleaf)
- Woodland and Shrub (Mediterranean)
- Short Grass (Steppe)
- Tall Grass (Prairie)
- River Valley and Oasis
- Highlands (Unclassified; Vertical Zonation)
- Desert and Desert Shrub
- Savanna Grassland and Shrub
- Wooded Savanna
- Tropical Woodland and Shrub
- Light Tropical Forest
- Tropical Rain Forest
- Heath and Moor
- Tundra and Alpine Vegetation
- Permanent Ice Cover

Vegetation is the most visible consequence of the distribution of temperature and precipitation. The global pattern of vegetative types or "habitat classes" and the global pattern of climate are closely related and make up one of the great global spatial correlations. But not all vegetation types are the consequence of temperature and precipitation or other climatic variables. Many types of vegetation in many areas of the world are the consequence of human activities, particularly the grazing of domesticated livestock, burning, and forest clearance. This map shows the pattern of natural or "potential" vegetation, or vegetation as it might be expected to exist without significant human influences, rather than the actual vegetation that results from a combination of environmental and human factors. Physical geographers who are interested in the distribution and geographic patterns of vegetation are called "biogeographers."

Map 8 Soil Orders

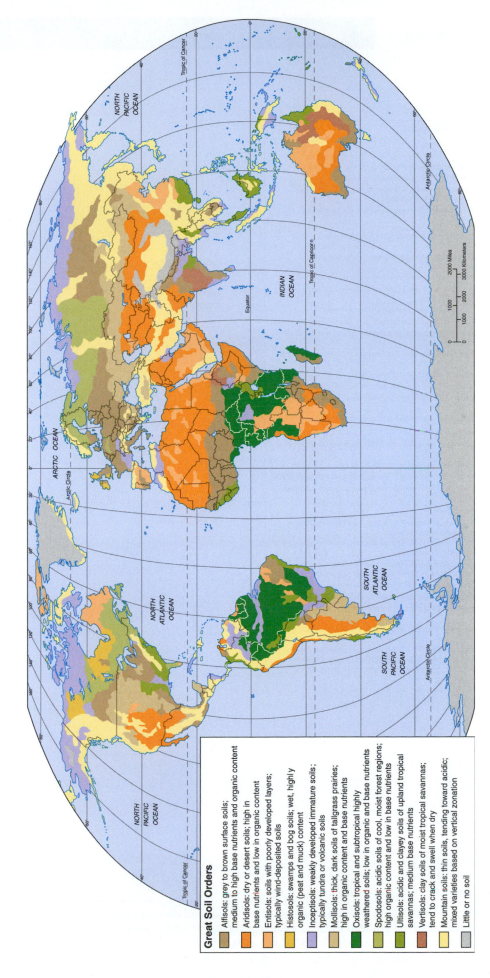

Great Soil Orders

- Affisols: grey to brown surface soils; medium to high base nutrients and organic content
- Aridisols: dry or desert soils; high in base nutrients and low in organic content
- Entisols: soils with poorly developed layers; typically wind-deposited soils
- Histosols: swamps and bog soils; wet, highly organic (peat and muck) content
- Inceptisols: weakly developed immature soils; typically tundra or volcanic soils
- Mollisols: thick, dark soils of tallgrass prairies; high in organic content and base nutrients
- Oxisols: tropical and subtropical highly weathered soils; low in organic and base nutrients
- Spodosols: acidic soils of cool, moist forest regions; high organic content and low in base nutrients
- Ultisols: acidic and clayey soils of upland tropical savannas; medium base nutrients
- Vertisols: clay soils of moist tropical savannas; tend to crack and swell when dry
- Mountain soils: thin soils, tending toward acidic; mixed varieties based on vertical zonation
- Little or no soil

The characteristics of soil are one of the three primary physical geographic factors, along with climate and vegetation, that determine the habitability of regions for humans. In particular, soils influence the kinds of agricultural uses to which land is put. Since soils support the plants that are the primary producers of all food in the terrestrial food chain, their characteristics are crucial to the health and stability of ecosystems. Two types of soil are shown on this map: zonal soils, the characteristics of which are based on climatic patterns; and azonal soils, such as alluvial (water-deposited) or aeolian (wind-deposited)

soils, the characteristics of which are derived from forces other than climate. However, many of the azonal soils, particularly those dependent upon drainage conditions, appear over areas too small to be readily shown on a map of this scale. Thus, almost none of the world's swamp or bog soils appear on this map. People who study the geographic characteristics of soils are most often called "soil scientists," a discipline closely related to that branch of physical geography called "geomorphology."

Map 9 Ecological Regions

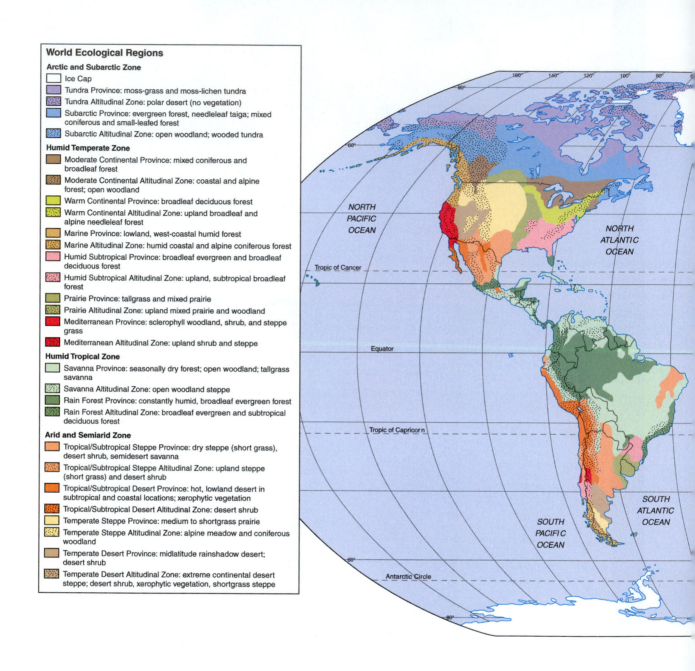

World Ecological Regions

Arctic and Subarctic Zone

- Ice Cap
- Tundra Province: moss-grass and moss-lichen tundra
- Tundra Altitudinal Zone: polar desert (no vegetation)
- Subarctic Province: evergreen forest, needleleaf taiga; mixed coniferous and small-leafed forest
- Subarctic Altitudinal Zone: open woodland; wooded tundra

Humid Temperate Zone

- Moderate Continental Province: mixed coniferous and broadleaf forest
- Moderate Continental Altitudinal Zone: coastal and alpine forest; open woodland
- Warm Continental Province: broadleaf deciduous forest
- Warm Continental Altitudinal Zone: upland broadleaf and alpine needleleaf forest
- Marine Province: lowland, west-coastal humid forest
- Marine Altitudinal Zone: humid coastal and alpine coniferous forest
- Humid Subtropical Province: broadleaf evergreen and broadleaf deciduous forest
- Humid Subtropical Altitudinal Zone: upland, subtropical broadleaf forest
- Prairie Province: tallgrass and mixed prairie
- Prairie Altitudinal Zone: upland mixed prairie and woodland
- Mediterranean Province: sclerophyll woodland, shrub, and steppe grass
- Mediterranean Altitudinal Zone: upland shrub and steppe

Humid Tropical Zone

- Savanna Province: seasonally dry forest; open woodland; tallgrass savanna
- Savanna Altitudinal Zone: open woodland steppe
- Rain Forest Province: constantly humid, broadleaf evergreen forest
- Rain Forest Altitudinal Zone: broadleaf evergreen and subtropical deciduous forest

Arid and Semiarid Zone

- Tropical/Subtropical Steppe Province: dry steppe (short grass), desert shrub, semidesert savanna
- Tropical/Subtropical Steppe Altitudinal Zone: upland steppe (short grass) and desert shrub
- Tropical/Subtropical Desert Province: hot, lowland desert in subtropical and coastal locations; xerophytic vegetation
- Tropical/Subtropical Desert Altitudinal Zone: desert shrub
- Temperate Steppe Province: medium to shortgrass prairie
- Temperate Steppe Altitudinal Zone: alpine meadow and coniferous woodland
- Temperate Desert Province: midlatitude rainshadow desert; desert shrub
- Temperate Desert Altitudinal Zone: extreme continental desert steppe; desert shrub, xerophytic vegetation, shortgrass steppe

Ecological regions are distinctive areas within which unique sets of organisms and environments are found. We call the study of the relationships between organisms and their environmental surroundings "ecology." Within each of the ecological regions portrayed on the map, a particular combination of vegetation, wildlife, soil, water, climate, and terrain defines that region's habitability, or ability to support life, including human life. Like climate and landforms, ecological relationships are crucial to the existence of agriculture, the

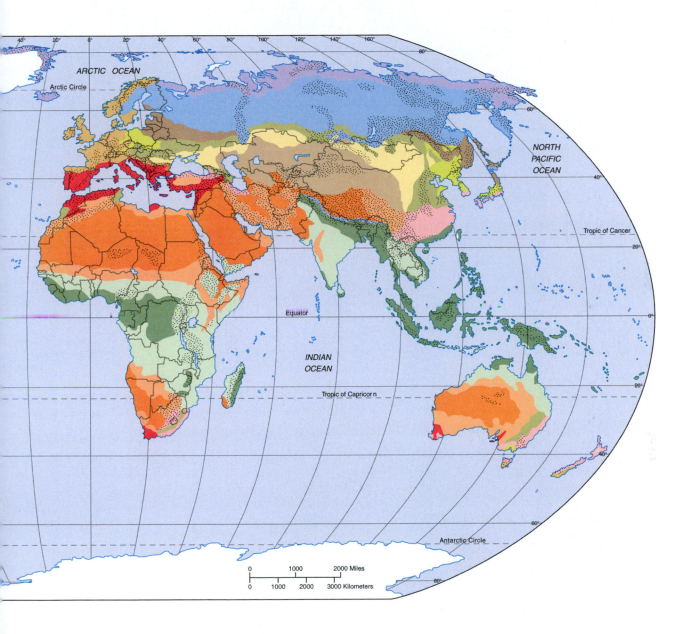

most basic of our economic activities, and important for many other kinds of economic activity as well. Biogeographers are especially concerned with the concept of ecological regions since such regions so clearly depend upon the geographic distribution of plants and animals in their environmental settings.

-17-

Map 10 Plate Tectonics

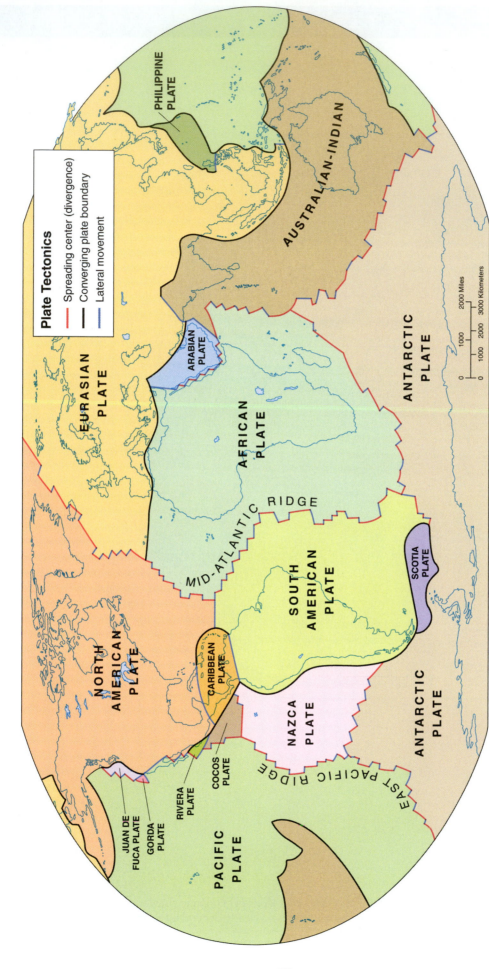

An understanding of the forces that shape the primary features of the earth's surface—the continents and ocean basins—requires a view of the earth's crust as fragments or "lithospheric plates" that shift position relative to one another. There are three dominant types of plate movement: *convergence*, in which plates move together, compressing former ocean floor or continental rocks together to produce mountain ranges, or producing mountain ranges through volcanic activity if one plate slides beneath another; *divergence*, in which the plates move away from one another, producing rifts in the earth's crust through which molten material wells up to produce new sea floors

and mid-oceanic ridges; and *lateral shift*, in which plates move horizontally relative to one another, causing significant earthquake activity. All the major forms of these types of shifts are extremely slow and take place over long periods of geologic time. The movement of crustal plates, or what is known as "plate tectonics," is responsible for the present shape and location of the continents but is also the driving force behind some much shorter-term earth phenomena like earthquakes and volcanoes. A comparison of the map of plates with maps of hazards and terrain will reveal some interesting relationships.

-18-

Map 11 Topography

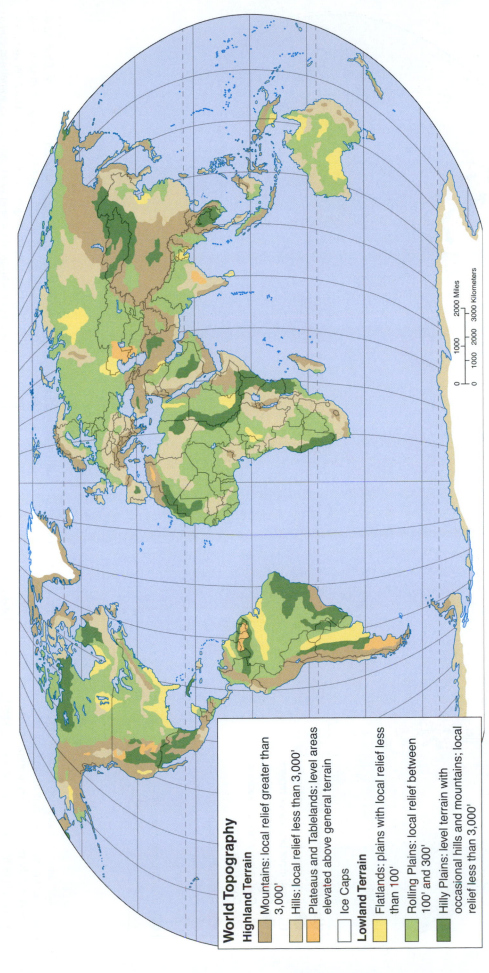

World Topography

Highland Terrain

- Mountains: local relief greater than 3,000'
- Hills: local relief less than 3,000'
- Plateaus and Tablelands: level areas elevated above general terrain
- Ice Caps

Lowland Terrain

- Flatlands: plains with local relief less than 100'
- Rolling Plains: local relief between 100' and 300'
- Hilly Plains: level terrain with occasional hills and mountains; local relief less than 3,000'

0   1000   2000   3000 Kilometers
0   1000   2000 Miles

Topography or terrain, also called "landforms," is second only to climate as a conditioner of human activity, particularly agriculture but also the location of cities and industry. A comparison of this map of mountains, valleys, plains, plateaus, and other features of the earth's surface with a map of land use (Map 16) shows that most of the world's productive agricultural zones are located in lowland and relatively level regions. Where large regions of agricultural productivity are found, we also tend to find urban concentrations and, with cities, we find industry. There is also a good spatial correlation between the map of topography and the map showing the distribution and density of the human population (Map 15). Normally the world's major landforms

are the result of extremely gradual primary geologic activity such as the long-term movement of crustal plates. This activity occurs over hundreds of millions of years. Also important is the more rapid (but still slow by human standards) geomorphological or erosional activity of water, wind, glacial ice, and waves, tides, and currents. Some landforms may be produced by abrupt or "cataclysmic" events such as a major volcanic eruption or a meteor strike, but such events are relatively rare and their effects are usually too minor to show up on a map of this scale. The study of the processes that shape topography is known as "geomorphology" and is an important branch of physical geography.

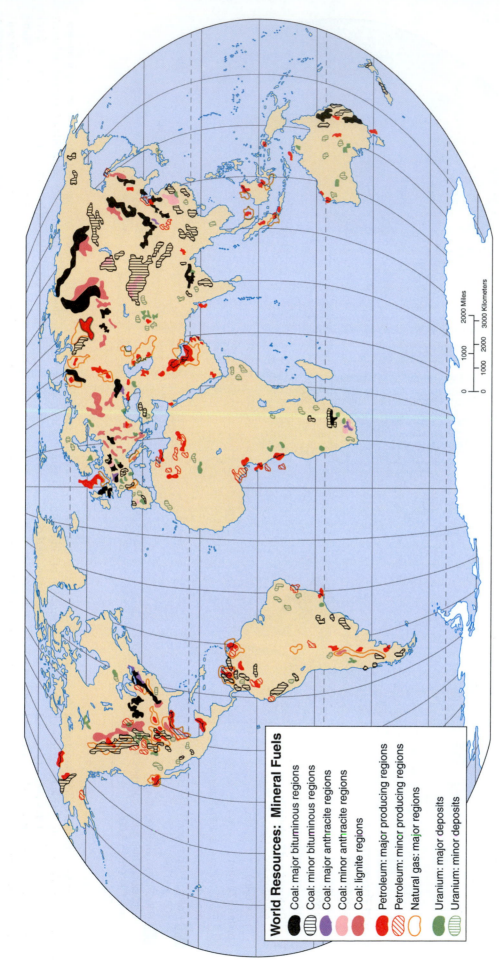

World Resources: Mineral Fuels

- Coal: major bituminous regions
- Coal: minor bituminous regions
- Coal: major anthracite regions
- Coal: minor anthracite regions
- Coal: lignite regions
- Petroleum: major producing regions
- Petroleum: minor producing regions
- Natural gas: major regions
- Uranium: major deposits
- Uranium: minor deposits

The extraction and transportation of mineral fuels rank with agriculture and forestry as "primary" human activities that impact the environment on a global scale. Nearly all of the most highly publicized environmental disasters of recent decades—the Gulf of Mexico oil spill or the Chernobyl nuclear accident, for example—have involved mineral fuels that were being stored, transported, or used. And the continuing extraction of mineral fuels like oil, natural gas, coal, and uranium produces high levels of atmospheric, soil, and water pollution. The location of mineral fuels tells us a great deal about where

environmental degradation is likely to be occurring or to occur in the future. One need only look at the levels of atmospheric pollution and vegetative disruption in central and eastern Europe to recognize the damaging consequences of heavy reliance on coal as a domestic and industrial fuel. The location of mineral fuels also tells us something about existing or potential levels of economic development with those countries possessing abundant reserves of mineral fuels having more of a chance to maintain or attain higher levels of prosperity.

0   1000   2000 Miles
0   1000   2000   3000 Kilometers

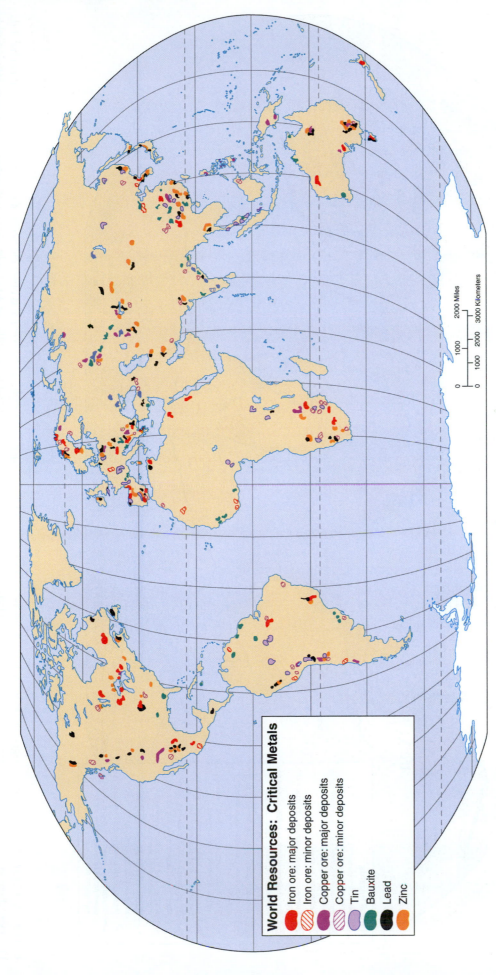

World Resources: Critical Metals

- Iron ore: major deposits
- Iron ore: minor deposits
- Copper ore: major deposits
- Copper ore: minor deposits
- Tin
- Bauxite
- Lead
- Zinc

The location of deposits of critical metals such as iron, copper, tin, and others is an important determinant of the location of mining activities. Also like mineral fuel extraction, mining for critical metallic ores makes significant environmental impact, particularly on vegetation, soils, and water resources. Some of the world's most dramatic examples of human modification of environments are located in areas of metallic ore extraction: the open pit copper mining areas of Arizona and Utah, for example. Environmental impact aside, those countries with significant critical metal deposits tend to stand a better chance of reaching higher levels of economic development, as long as they can extract and market the ores themselves rather than having the extraction process controlled by outside concerns. The average Bolivian, for example, does not benefit greatly from the fact that his/her country is an important producer of tin and other metals. Bolivia is a "colonial dependency" country and the wealth generated by metallic ore production there tends to flow out of the country to Europe and North America. On the other hand, another South American country, Brazil, is paying for much of its own current economic development by utilizing its reserves of iron and other metals and more of the wealth from the extraction of those resources stays within the country.

Map 13 Natural Hazards

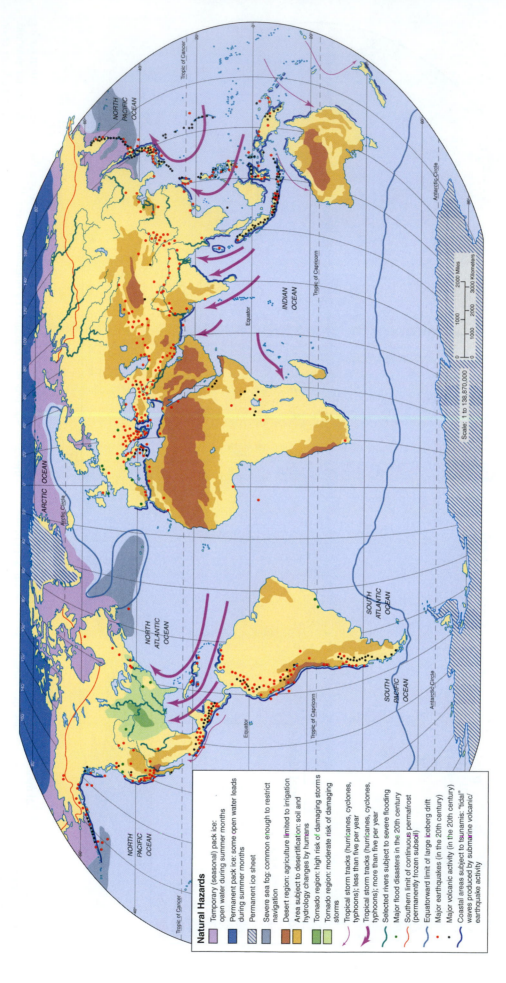

Natural Hazards

- Temporary (seasonal) pack ice: open water during summer months
- Permanent pack ice: some open water leads during summer months
- Permanent ice sheet
- Severe sea fog: common enough to restrict navigation
- Desert region: agriculture limited to irrigation
- Area subject to desertification: soil and hydrology changes by humans
- Tornado region: high risk of damaging storms
- Tornado region: moderate risk of damaging storms
- Tropical storm tracks (hurricanes, cyclones, typhoons); less than five per year
- Tropical storm tracks (hurricanes, cyclones, typhoons); more than five per year
- Selected rivers subject to severe flooding
- Major flood disasters in the 20th century
- Southern limit of continuous permafrost (permanently frozen subsoil)
- Equatorward limit of large iceberg drift
- Major earthquakes (in the 20th century)
- Major volcanic activity (in the 20th century)
- Coastal areas subject to tsunamis: "tidal" waves produced by submarine volcanic/ earthquake activity

Scale: 1 to 138,870,000

Unlike other elements of physical geography, most natural hazards are unpredictable. However, there are certain regions where the probability of the occurrence of a particular natural hazard is high. This map shows regions affected by major natural hazards at rates that are higher than the global norm. The presence of persistent natural hazards may influence the types of modifications that people make in the environment and certainly influence the styles of housing and other elements of cultural geography. Natural hazards may also undermine the utility of an area for economic purposes and some scholars suggest that regions of environmental instability may be regions of political instability as well. The study of natural hazards has become an important activity for "resource geographers" whose areas of interest overlap both human and physical fields of geography.

Unit II

Global Human Patterns

Map 14 Past Population Distributions and Densities

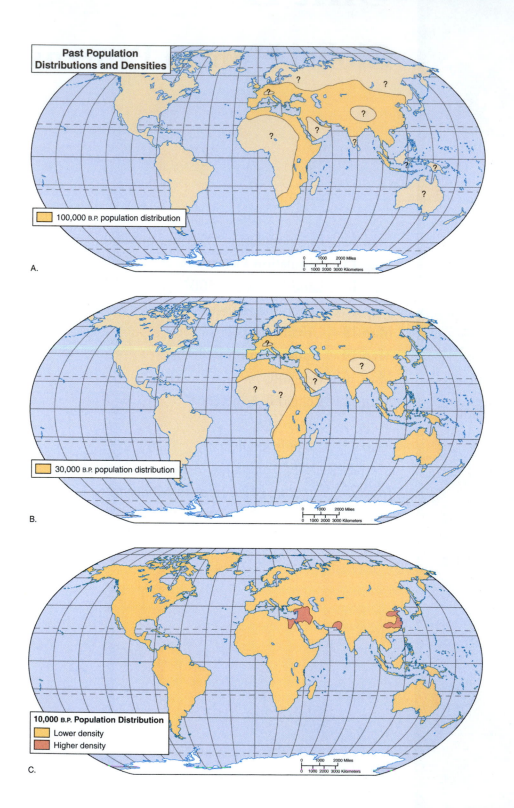

Past Population Distributions and Densities

100,000 B.P. population distribution

A.

30,000 B.P. population distribution

B.

10,000 B.P. Population Distribution
Lower density
Higher density

C.

The map of the world at 100,000 B.P. (Before Present) shows the distributions of hominids who at that time had spread from their probable origin in Africa into parts of the Old World. At 30,000 B.P. few places in the Old World remained uninhabited. All the people then were hunters and gatherers who subsisted on wild foods. The environment probably was not at its carrying capacity for human populations until about 15,000 B.P.

By 10,000 B.P. hunting and gathering people had spread throughout the world. Plant and animal domestication (farming and pastoralism) had begun in some parts of the Old World perhaps as a response to behavioral changes necessitated by the population, which

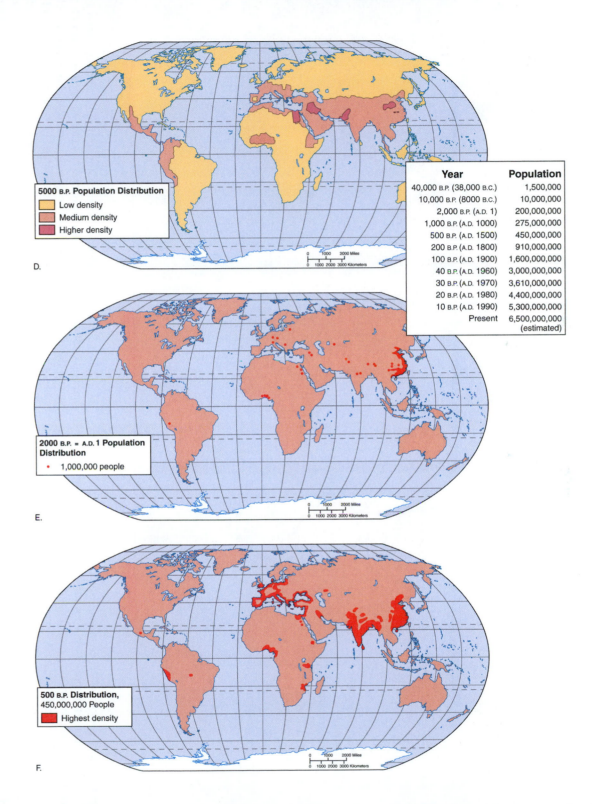

Year	Population
40,000 B.P. (38,000 B.C.)	1,500,000
10,000 B.P. (8000 B.C.)	10,000,000
2,000 B.P. (A.D. 1)	200,000,000
1,000 B.P. (A.D. 1000)	275,000,000
500 B.P. (A.D. 1500)	450,000,000
200 B.P. (A.D. 1800)	910,000,000
100 B.P. (A.D. 1900)	1,600,000,000
40 B.P. (A.D. 1960)	3,000,000,000
30 B.P. (A.D. 1970)	3,610,000,000
20 B.P. (A.D. 1980)	4,400,000,000
10 B.P. (A.D. 1990)	5,300,000,000
Present	6,500,000,000 (estimated)

5000 B.P. Population Distribution
- Low density
- Medium density
- Higher density

D.

2000 B.P. = A.D. 1 Population Distribution
- • 1,000,000 people

E.

500 B.P. Distribution, 450,000,000 People
- Highest density

F.

now exceeded the environmental carrying capacity. Farming supports higher population densities than hunting and gathering. Urban civilization with cities dependent on their hinterlands had developed by 5000 B.P. The maps of A.D. 1 and A.D. 1500 approximate actual population density on a scale of one dot to every million people. A chart provides total world population figures for different time periods. Compare these maps to the contemporary population density map on the following page.

Map 15 Population Density

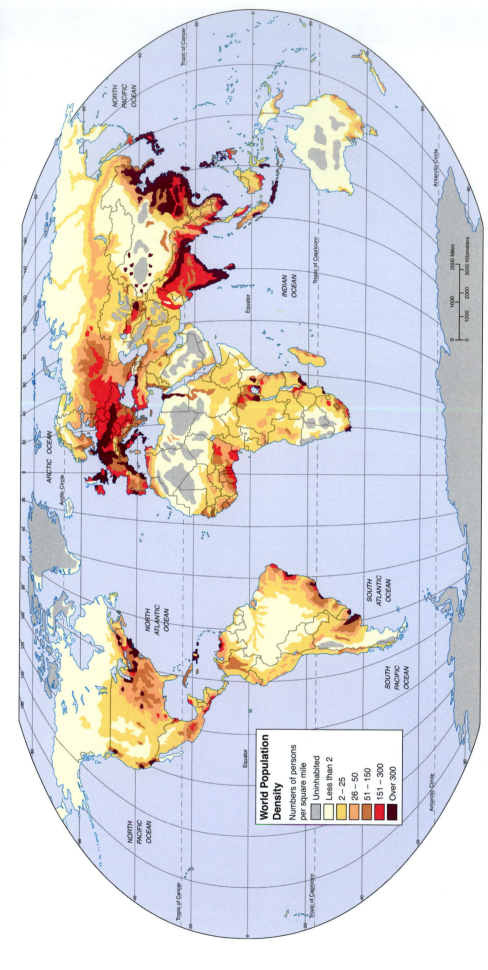

World Population Density

Numbers of persons per square mile

- Uninhabited
- Less than 2
- 2 – 25
- 26 – 50
- 51 – 150
- 151 – 300
- Over 300

No feature of human activity is more reflective of geographic relationships than where people live. In the areas of densest populations, a mixture of natural and human factors has combined to allow maximum food production, maximum urbanization, and maximum centralization of economic activities. Three great concentrations of human population appear on the map—East Asia, South Asia, and Europe—with a fourth, lesser concentration in eastern North America. While population growth is relatively slow in three of these population clusters, in the fourth—South Asia—growth is still rapid and South Asia is expected to become even more densely populated in the early years of the twenty-first century, while density of the other regions is expected to remain about as it

now appears. In Europe and North America, the relatively stable population growth rates are the result of economic development that has caused population growth to level off within the last century. In East Asia, the growth rates have also begun to decline. In the case of Japan, Taiwan, the Koreas, and other more highly developed nations of the Pacific Rim, the reduced growth is the result of economic development. In China, at least until recently, lowered population growth rates have resulted from strict family planning. The areas of future high density of population, in addition to those already existing, are likely to be in Middle and South America and in Central Africa, where population growth rates are well above the world average.

Map 16 Land Use, A.D. 1500

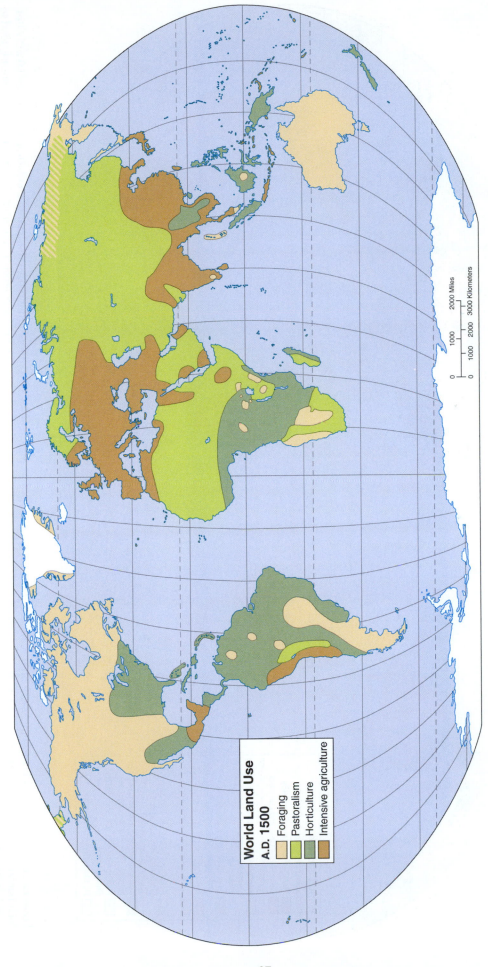

World Land Use
A.D. 1500

- Foraging
- Pastoralism
- Horticulture
- Intensive agriculture

Europeans began to explore the world in the late 1400s. They encountered many independent people with self-sustaining economies at that time. Foraging people practiced hunting and gathering, utilizing the wild forms of plants and animals in their environments. Horticultural people practiced a simple form of agriculture using hoes or digging sticks as their basic tools. They sometimes cleared their land by burning and then planted crops. Pastoralists herded animals as their basic subsistence pattern. Complex state-level societies, such as the Mongols, had pastoralism as their base. Intensive agriculturalists based their societies on complicated irrigation systems and/or the plow and draft animals. Wheat and rice were two kinds of crops that supported large populations. In many of the areas of intensive agriculture—particularly in MesoAmerica, Europe, Southwest Asia, South Asia, and East Asia—complex patterns of market economies had begun to develop well before the fifteenth century and the beginnings of European expansion.

Map 17 Economic Activities

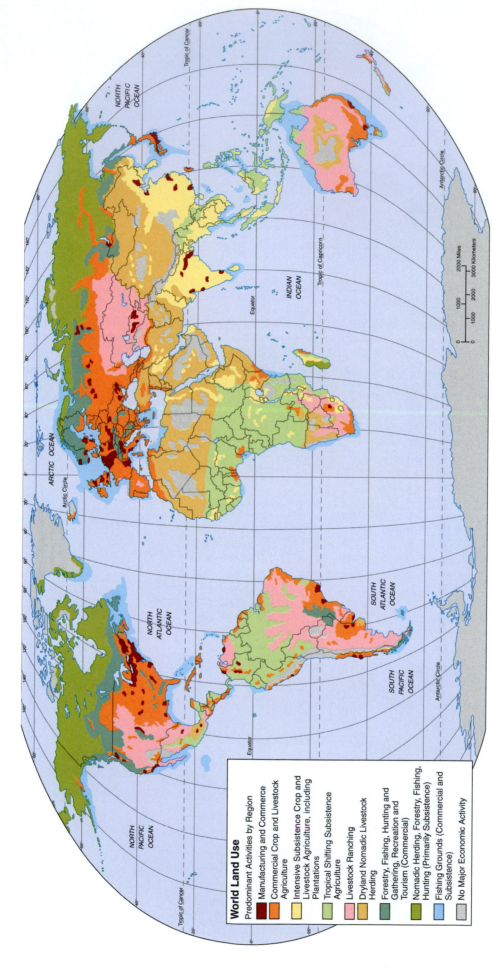

World Land Use

Predominant Activities by Region

- Manufacturing and Commerce
- Commercial Crop and Livestock Agriculture
- Intensive Subsistence Crop and Livestock Agriculture, including Plantations
- Tropical Shifting Subsistence Agriculture
- Livestock Ranching
- Dryland Nomadic Livestock Herding
- Forestry, Fishing, Hunting and Gathering, Recreation and Tourism (Commercial)
- Nomadic Herding, Forestry, Fishing, Hunting (Primarily Subsistence)
- Fishing Grounds (Commercial and Subsistence)
- No Major Economic Activity

Land uses can be categorized as lying somewhere on a scale between extensive uses, in which human activities are dispersed over relatively large areas, and intensive uses, in which human activities are concentrated in relatively small areas. Many of the most important land use patterns of the world (such as urbanization, industry, mining, or transportation) are intensive and therefore relatively small in area and not easily seen on maps of this scale. Hence even in the areas identified as "Manufacturing and Commerce" on the map there are many land uses that are not strictly industrial or commercial in nature, and, in fact, more extensive land uses (farming, residential, open space) may actually cover more ground than the intensive industrial or commercial activities. On the other hand, the more extensive land uses like agriculture and forestry, tend to dominate the areas in which they are found. Thus, primary economic activities such as agriculture and forestry tend to dominate the world map of land use because of their extensive character. Much of this map is, therefore, a map that shows the global variations in agricultural patterns. Note, among other things, the differences between land use patterns in the more developed countries of the temperate zones and the less developed countries of the tropics.

Map 18 Urbanization

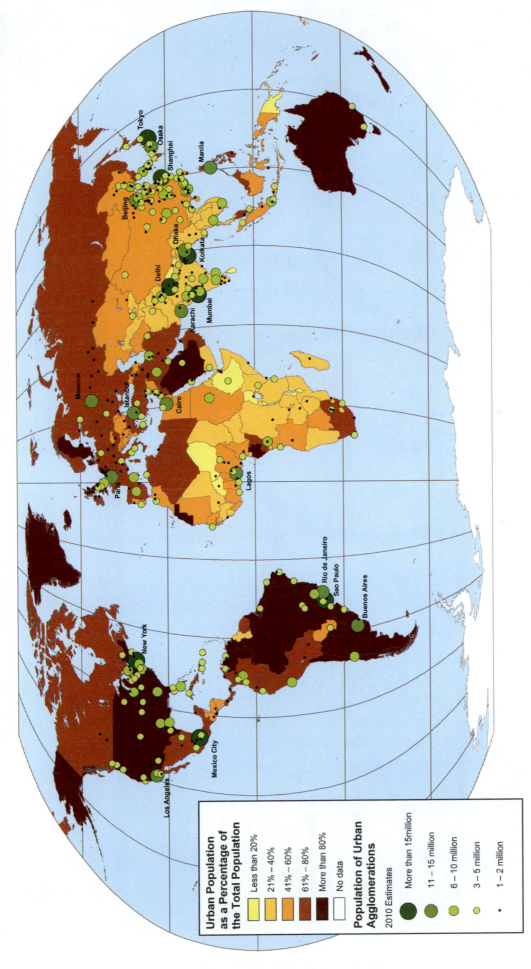

Urban Population as a Percentage of the Total Population

- Less than 20%
- 21% – 40%
- 41% – 60%
- 61% – 80%
- More than 80%
- No data

Population of Urban Agglomerations

2010 Estimates

- More than 15 million
- 11 – 15 million
- 6 – 10 million
- 3 – 5 million
- 1 – 2 million

Map labels: Tokyo, Osaka, Shanghai, Manila, Beijing, Dhaka, Kolkata, Delhi, Mumbai, Karachi, Moscow, Istanbul, Cairo, Lagos, Paris, Rio de Janeiro, Sao Paulo, Buenos Aires, New York, Mexico City, Los Angeles

The degree to which a region's population is concentrated in urban areas is a major indicator of a number of things: the level of economic development, and the problems associated with human concentrations. Urban dwellers are rapidly becoming the norm among the world's people and rates of urbanization are increasing worldwide, with the greatest increases in urbanization taking place in developing regions. Whether in developed or developing countries, those who live in cities exert an influence on the environment, politics, economics, and social systems that goes far beyond the confines of the city itself. Acting as the focal points for

the flow of goods and ideas, cities draw resources and people not just from their immediate hinterland but from the entire world. This process creates far-reaching impacts as resources are extracted, converted through industrial processes, and transported over great distances to metropolitan regions, and as ideas spread or *diffuse* along with the movements of people to cities and the flow of communication from them. The significance of urbanization can be most clearly seen, perhaps, in North America where, in spite of vast areas of relatively unpopulated land, well over 90 percent of the population lives in urban areas.

-29-

Map 19 Transportation Patterns

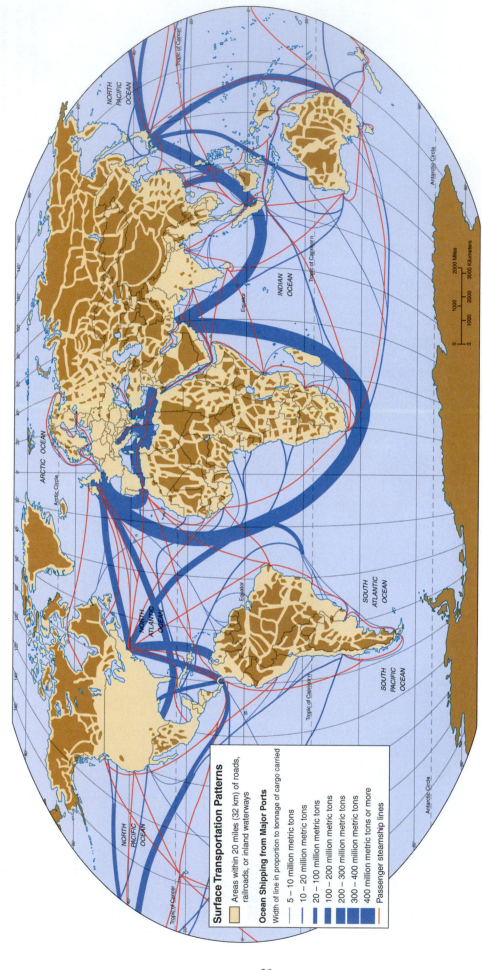

Surface Transportation Patterns

Areas within 20 miles (32 km) of roads, railroads, or inland waterways

Ocean Shipping from Major Ports

Width of line in proportion to tonnage of cargo carried

- 5 – 10 million metric tons
- 10 – 20 million metric tons
- 20 – 100 million metric tons
- 100 – 200 million metric tons
- 200 – 300 million metric tons
- 300 – 400 million metric tons
- 400 million metric tons or more
- Passenger steamship lines

As a form of land use, transportation is second only to agriculture in its coverage of the earth's surface and is one of the clearest examples in the human world of a *network*, a linked system of lines allowing flows from one place to another. The global transportation network and its related communication web is responsible for most of the *spatial interaction*, or movement of goods, people, and ideas between places. As the chief mechanism of spatial interaction, transportation is linked firmly with the concept of a shrinking world and the development of a global community and economy. Because transportation systems require significant modification of the earth's surface, transportation is also responsible for massive alterations in the quantity and quality of water, for major soil degradations and erosion, and (indirectly) for the air pollution that emanates from vehicles utilizing the transportation system. In addition, as improved transportation technology draws together places on the earth that were formerly remote, it allows people to impact environments a great distance away from where they live.

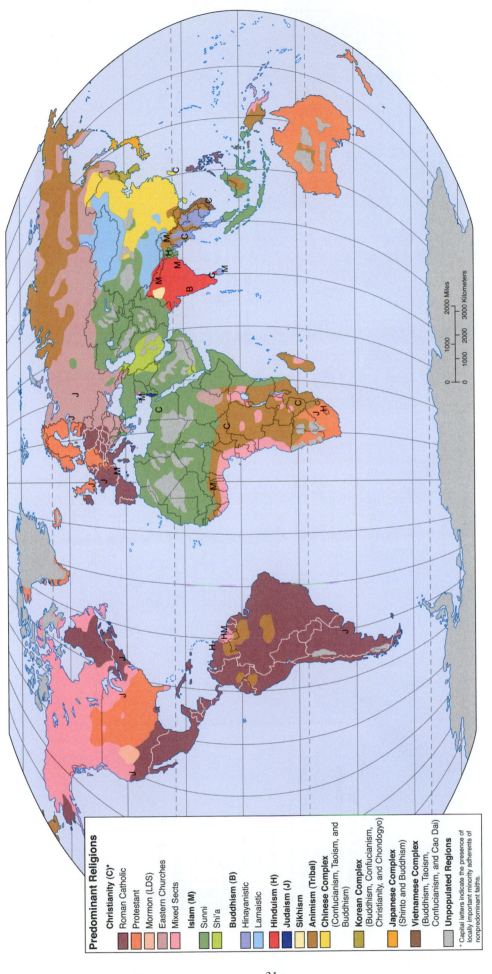

Map 20 Religions

Predominant Religions

Christianity (C)*
- Roman Catholic
- Protestant
- Mormon (LDS)
- Eastern Churches
- Mixed Sects

Islam (M)
- Sunni
- Shi'a

Buddhism (B)
- Hinayanistic
- Lamaistic

Hinduism (H)

Judaism (J)

Sikhism

Animism (Tribal)

Chinese Complex
(Confucianism, Taoism, and Buddhism)

Korean Complex
(Buddhism, Confucianism, Christianity, and Chondogyo)

Japanese Complex
(Shinto and Buddhism)

Vietnamese Complex
(Buddhism, Taoism, Confucianism, and Cao Dai)

Unpopulated Regions

* Capital letters indicate the presence of locally important minority adherents of nonpredominant faiths.

Religious adherence is one of the fundamental defining characteristics of human culture, the style of life adopted by a people and passed from one generation to the next. Because of the importance of religion for culture, a depiction of the spatial distribution of religions is as close as we can come to a map of cultural patterns. More than just a set of behavioral patterns having to do with worship and ceremony, religion is a vital conditioner of the ways that people deal with one another, with their institutions, and with the environments they occupy. In many areas of the world, the ways in which people make a living, the patterns of occupation that they create on the land, and the impacts that they make on ecosystems are the direct consequences of their adherence to a religious faith. An examination of the map in the context of international and intranational conflict will also show that tension between countries and the internal stability of states is also a function of the spatial distribution of religion.

Map **21** Religious Adherence

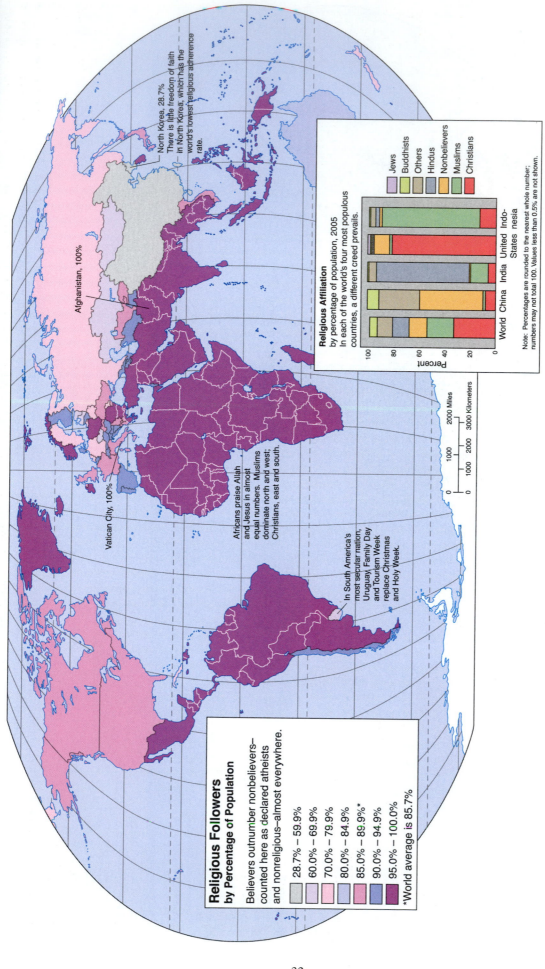

North Korea, 28.7%
There is little freedom of faith in North Korea, which has the world's lowest religious adherence rate.

Afghanistan, 100%

Vatican City, 100%

Africans praise Allah and Jesus in almost equal numbers. Muslims dominate north and west; Christians, east and south.

In South America's most secular nation, Uruguay Family Day and Tourism Week replace Christmas and Holy Week.

Religious Affiliation
by percentage of population, 2005
In each of the world's four most populous countries, a different creed prevails.

Jews
Buddhists
Others
Hindus
Nonbelievers
Muslims
Christians

Percent

World   China   India   United   Indo-
States   nesia

Note: Percentages are rounded to the nearest whole number; numbers may not total 100. Values less than 0.5% are not shown.

Religious Followers
by Percentage of Population

Believers outnumber nonbelievers—counted here as declared atheists and nonreligious—almost everywhere.

	28.7% – 59.9%
	60.0% – 69.9%
	70.0% – 79.9%
	80.0% – 84.9%
	85.0% – 89.9%*
	90.0% – 94.9%
	95.0% – 100.0%

*World average is 85.7%

0   1000   2000 Miles
0   1000   2000   3000 Kilometers

Throughout much of the past century, the numbers of proclaimed religious adherents declined as a percentage of populations as the consequence of the emergence of Communism in what became the U.S.S.R. and in China. Communism enforced atheism as a state "religion," and it is impossible to know how many Orthodox Christians actually existed in Russia or how many Christians, Muslims, Buddhists, Taoists, Confucianists, etc., existed in China. With the downfall of the old Soviet Union and the reemergence of an Orthodox Christian Russia, percentages of religious adherents in that country have increased tremendously. And as the European Union has opened its doors to "guest workers" and others from Southwest Asia and Africa, the number of Muslims has grown considerably in most of western Europe while Islam has reestablished itself as a religion among many inhabitants of the Balkan peninsula. It is worth noting that while past shifts in regional percentages of religious adherents had often been the result of missionary activity (the growing number of Christians in Africa during the twentieth century, for example), the recent changes have come about through political and economic change.

Map 22 Languages

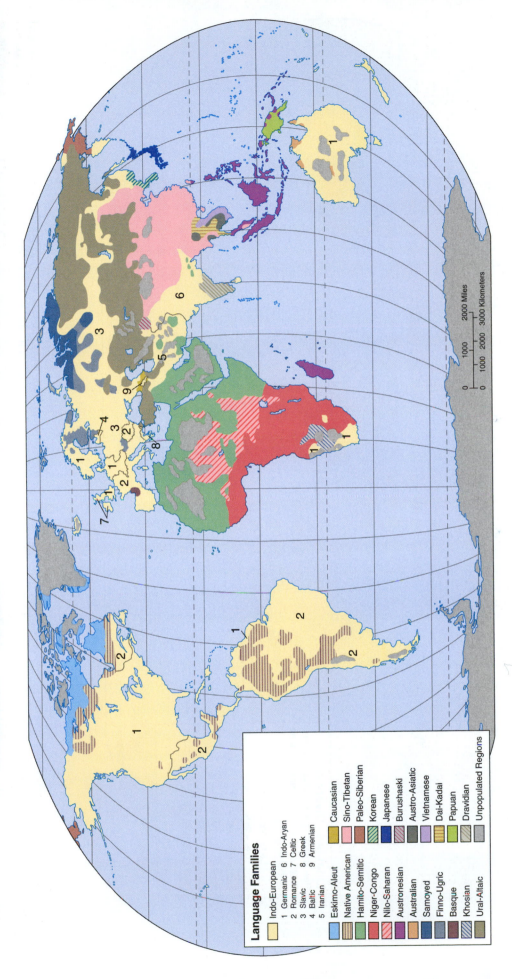

Language, like religion, is an important identifying characteristic of culture. Indeed, it is perhaps the most durable of all those identifying characteristics or *cultural traits*: language, religion, institutions, material technologies, and ways of making a living. After centuries of exposure to other languages or even conquest by speakers of other languages, the speakers of a specific tongue will often retain their own linguistic identity. Language helps us to locate areas of potential conflict, particularly in regions where two or more languages overlap. Many, if not most, of the world's conflict zones are also areas of linguistic diversity. Knowing the distribution of languages helps us to understand some of the reasons behind important current events: for example, linguistic

identity differences played an important part in the disintegration of the Soviet Union in the early 1990s; and in areas emerging from recent colonial rule, such as Africa, the participants in conflicts over territory and power are often defined in terms of linguistic groups. Language distributions also help us to comprehend the nature of the human past by providing clues that enable us to chart the course of human migrations, as shown in the distribution of Indo-European, Austronesian, or Hamito-Semitic languages. Finally, because languages have a great deal to do with the way people perceive and understand the world around them, linguistic patterns help to explain the global variations in the ways that people interact.

Language Families

Indo-European
1 Germanic   6 Indo-Aryan
2 Romance   7 Celtic
3 Slavic   8 Greek
4 Baltic   9 Armenian
5 Iranian

Eskimo-Aleut
Native American
Hamito-Semitic
Niger-Congo
Nilo-Saharan
Austronesian
Australian
Samoyed
Finno-Ugric
Basque
Khosian
Ural-Altaic

Caucasian
Sino-Tibetan
Paleo-Siberian
Korean
Japanese
Burushaski
Austro-Asiatic
Vietnamese
Dai-Kadai
Papuan
Dravidian
Unpopulated Regions

Map **23**   Linguistic Diversity

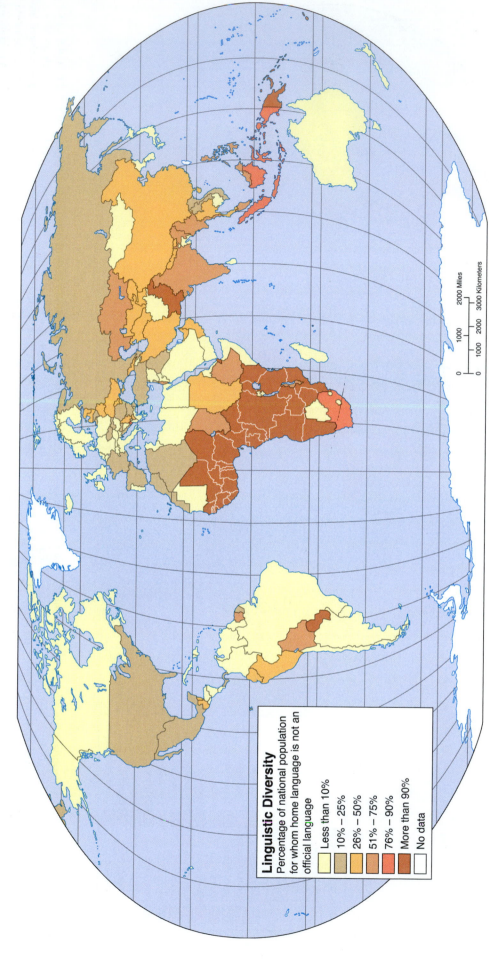

Linguistic Diversity

Percentage of national population for whom home language is not an official language

- Less than 10%
- 10% – 25%
- 26% – 50%
- 51% – 75%
- 76% – 90%
- More than 90%
- No data

Of the world's approximately 6,000 languages, fewer than 100 are official languages, those designated by a country as the language of government, commerce, education, and information. This means that for much of the world's population, the language that is spoken in the home is different from the official language of the country of residence. The world's former colonial areas in Middle and South America, Africa, and South and Southeast Asia stand out on the map as regions in which there is significant disparity between home languages and official languages. To complicate matters further, for most of the world's population, the primary international languages of trade and tourism (French and English) are neither home nor official languages. China is a special case as the official language is the written form of Chinese while several spoken Chinese dialects such as Mandarin and Cantonese, most of them mutually unintelligible, are recognized as official languages. The formal language of government and business is Mandarin.

Map 24 Dying Languages: A Loss of Cultural Wealth

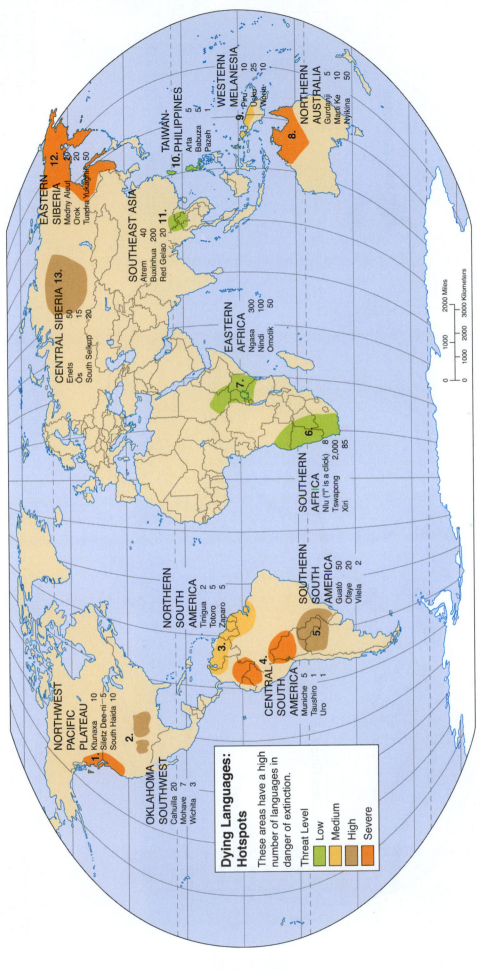

Dying Languages: Hotspots

These areas have a high number of languages in danger of extinction.

Threat Level	
	Low
	Medium
	High
	Severe

NORTHWEST PACIFIC PLATEAU **1.**
Ktunaxa 10
Siletz Dee-ni 5
South Haida 10

2.

OKLAHOMA SOUTHWEST
Cahuilla 20
Mohave 7
Wichita 3

NORTHERN SOUTH AMERICA
Tinigua 2
Totoro 5
Zaparo 5

CENTRAL SOUTH AMERICA **4.**
Muniche 5
Taushiro 1
Uro 1

3.

SOUTHERN SOUTH AMERICA
Guató 50
Ofayé 20
Vilela 2

5.

EASTERN SIBERIA **12.**
Medny Aleut 20
Orok 20
Tundra Yukaghir 50

CENTRAL SIBERIA **13.**
Enets 50
Ös 15
South Selkup 20

SOUTHEAST ASIA **11.**
Atrem 40
Buxinhua 200
Red Gelao 20

TAIWAN-PHILIPPINES **10.**
Arta 5
Babuza 5
Pazeh 1

WESTERN MELANESIA **9.**
Piru 10
Usku 25
Woria 10

NORTHERN AUSTRALIA **8.**
Gurdanji 5
Marti Ke 10
Nyikina 50

EASTERN AFRICA **7.**
Ngasa 300
Nindi 100
Omotik 50

SOUTHERN AFRICA **6.**
Nǀu ("ǀ" is a click) 8
Tswapong 2,000
Xiri 85

0   1000   2000 Miles
0   1000   2000   3000 Kilometers

The number of languages spoken in the world of the sixteenth century was probably in excess of 10,000, possibly half of those in Africa. After four centuries of colonialism, migration, and economic and political change, less than half of the languages spoken in 1500 still exist—and many of these are in jeopardy of becoming extinct. Language is an essential marker of culture and when languages disappear, the cultures that used them tend to disappear as well, becoming assimilated—at least linguistically—into the dominant cultures of their regions. Thus, in North America, languages such as Mojave or Wichita are currently used by fewer than 10 native speakers. If those speakers die before passing their languages on, the languages become "dead" and so do many significant elements of the cultures associated with them. On many American Indian reservations there are currently important efforts to regain or restore native tongues. But these movements are active among the larger American Indian populations; while Northern Arapahoe may well be preserved, the South Haida of the Pacific Northwest may not.

-35-

Map 25   External Migrations in Modern Times

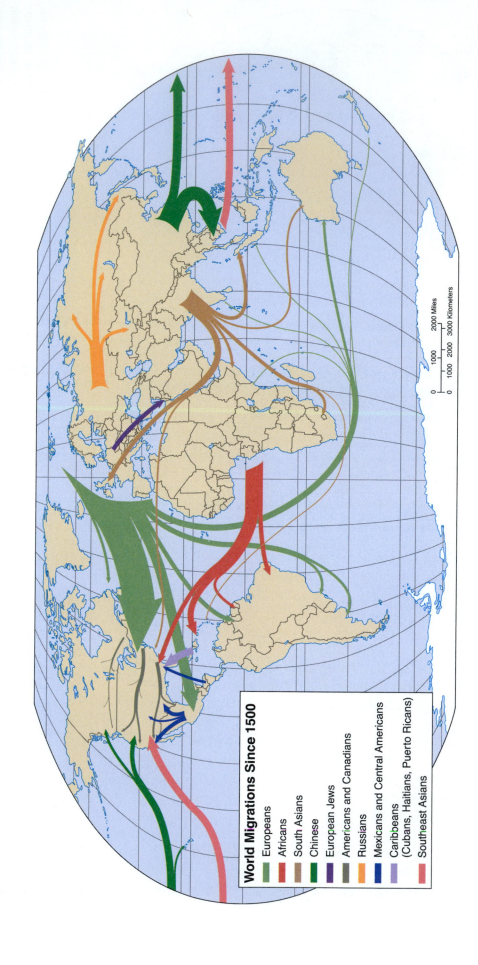

World Migrations Since 1500

- Europeans
- Africans
- South Asians
- Chinese
- European Jews
- Americans and Canadians
- Russians
- Mexicans and Central Americans
- Caribbeans (Cubans, Haitians, Puerto Ricans)
- Southeast Asians

0   1000   2000 Miles

0 1000 2000 3000 Kilometers

Migration has had a significant effect on world geography, contributing to cultural change and development, to the diffusion of ideas and innovations, and to the complex mixture of people and cultures found in the world today. *Internal migration* occurs within the boundaries of a country; *external migration* is movement from one country or region to another. Over the last 50 years, the most important migrations in the world have been internal, largely the rural-to-urban migration that has been responsible for the recent rise of global urbanization. Prior to the mid-twentieth century, three types of external migrations were most important: *voluntary*, most often in search of better economic conditions and opportunities; *involuntary* or *forced*, involving people who have been driven from their homelands by war, political unrest, or environmental disasters, or who have been transported as slaves or prisoners; and *imposed*, not entirely forced but which conditions make highly advisable. Human migrations in recorded history have been responsible for major changes in the patterns of languages, religions, ethnic composition, and economies. Particularly during the last 500 years, migrations of both the voluntary and involuntary or forced type have literally reshaped the human face of the earth.

Unit III

Global Demographic Patterns

Map 26 Population Growth Rates

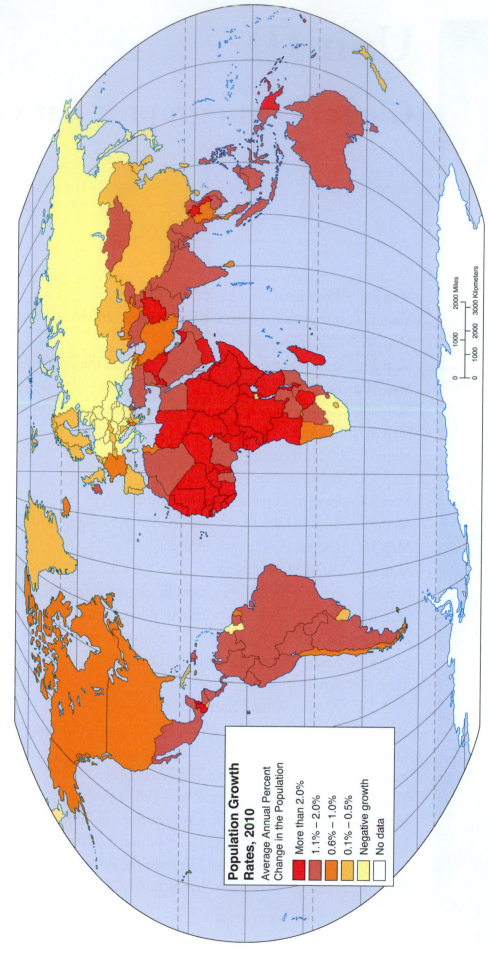

Population Growth Rates, 2010

Average Annual Percent Change in the Population

- More than 2.0%
- 1.1% – 2.0%
- 0.6% – 1.0%
- 0.1% – 0.5%
- Negative growth
- No data

Of all the statistical measurements of human population, that of the rate of population growth is the most important. The growth rate of a population is a combination of natural change (births and deaths), in-migration, and out-migration; it is obtained by adding the number of births to the number of immigrants during a year and subtracting from that total the sum of deaths and emigrants for the same year. For a specific country, this figure will determine many things about the country's future ability to feed, house, educate, and provide medical services to its citizens. Some of the countries with the largest populations (such as India) also have high growth rates. Since these countries tend to be in developing regions, the combination of high population and high growth rates poses special problems for political stability and continuing economic development; the combination also carries heightened risks for environmental degradation. Many people believe that the rapidly expanding world population is a potential crisis that may cause environmental and human disaster by the middle of the twenty-first century.

Map 27   International Migrant Populations, 2009

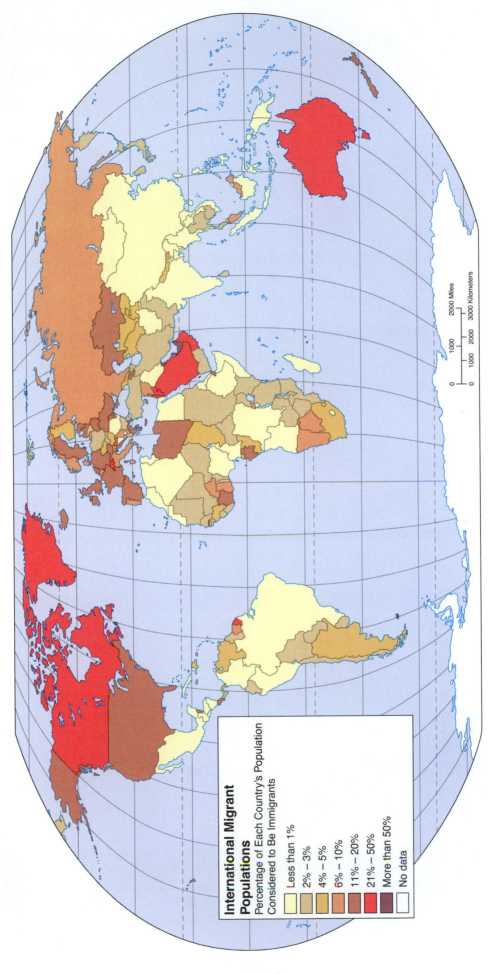

International Migrant Populations

Percentage of Each Country's Population Considered to Be Immigrants

- Less than 1%
- 2% – 3%
- 4% – 5%
- 6% – 10%
- 11% – 20%
- 21% – 50%
- More than 50%
- No data

Migration—the movement from one place to another—takes a number of different forms. It may be a move within a country from an old job to a new one. It may also mean a migration, either forced or voluntary, from one country to another. The map here depicts international migration: migration between countries. Migration is distinguished from a refugee movement in that migrants are not defined as refugees granted a humanitarian and temporary protection status under international law. The Middle Americans who leave the

Central American countries, Mexico, or the Caribbean for the United States plan to live in the United States permanently, although still retaining cultural and family ties to their native country. This map clearly shows that those countries viewed as having the most favorable opportunities for improvement in personal living conditions are those with the highest numbers of in-migrants; those countries that are overcrowded, with little economically upward mobility, or international conflict tend to be those with the greatest number of out-migrants.

Map 28 Migration Rates

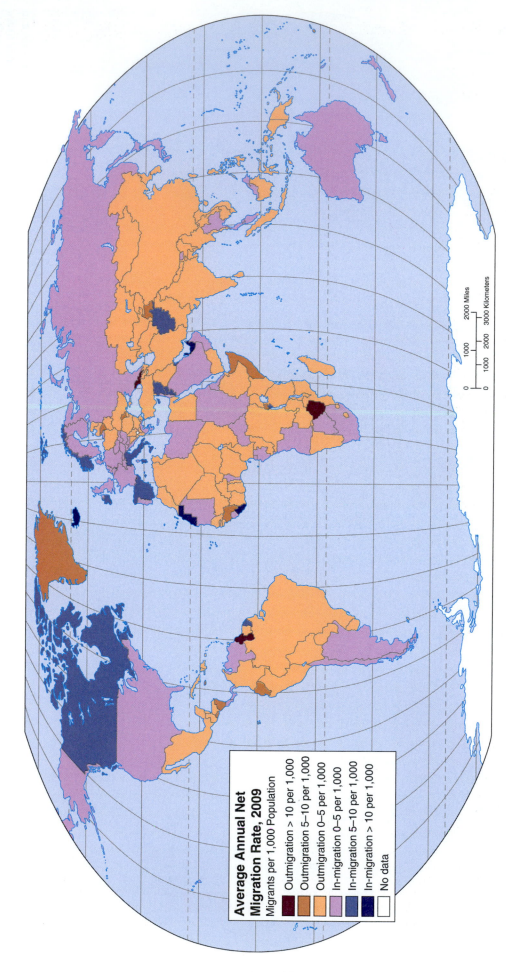

Average Annual Net Migration Rate, 2009
Migrants per 1,000 Population

- Outmigration > 10 per 1,000
- Outmigration 5–10 per 1,000
- Outmigration 0–5 per 1,000
- In-migration 0–5 per 1,000
- In-migration 5–10 per 1,000
- In-migration > 10 per 1,000
- No data

At the most fundamental level, international migration occurs because of two factors: that a *push* factors and *pull* factors. Push factors are those things that compel a person to leave his or her country. Push factors are many, varied, and are perceived in a negative light. Frequently cited push factors include political stresses (e.g., fear of persecution, war, or internal conflict), economic stresses (e.g., lack of employment opportunities), social stresses (e.g., lack of educational opportunities or adequate health care), and environmental stresses (e.g., natural disasters like a major earthquake or hurricane/typhoon,

desertification). Conversely, pull factors are those attributes of another country that a person finds attractive. As might be expected, they are the opposite of push factors. Thus, international migration is spurred by the perception that things are *bad* in one's own country and *better* in another. Both push and pull factors need be present. A person likely will not migrate to another country if it is perceived that conditions in that country are as bad or worse than in that person's home country.

Map 29

Map 29   Europe's Population Bust: A Fizzle in the Population Bomb

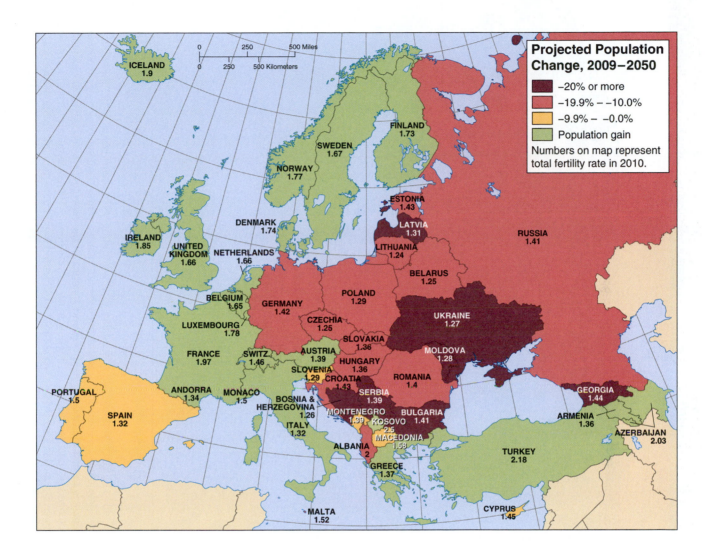

Projected Population Change, 2009–2050

- ■ –20% or more
- ■ –19.9% – –10.0%
- ■ –9.9% – –0.0%
- ■ Population gain

Numbers on map represent total fertility rate in 2010.

ICELAND 1.9

FINLAND 1.73

SWEDEN 1.67

NORWAY 1.77

ESTONIA 1.43

DENMARK 1.74

LATVIA 1.31

RUSSIA 1.41

IRELAND 1.85

UNITED KINGDOM 1.66

NETHERLANDS 1.66

LITHUANIA 1.24

BELARUS 1.25

BELGIUM 1.65

GERMANY 1.42

POLAND 1.29

LUXEMBOURG 1.78

CZECHIA 1.25

UKRAINE 1.27

SLOVAKIA 1.36

MOLDOVA 1.28

FRANCE 1.97

SWITZ. 1.46

AUSTRIA 1.39

HUNGARY 1.36

SLOVENIA 1.29

CROATIA 1.43

ROMANIA 1.4

PORTUGAL 1.5

ANDORRA 1.34

MONACO 1.5

SERBIA 1.39

GEORGIA 1.44

BOSNIA & HERZEGOVINA 1.26

MONTENEGRO 1.39

KOSOVO 2.5

BULGARIA 1.41

ARMENIA 1.36

SPAIN 1.32

ITALY 1.32

MACEDONIA 1.58

AZERBAIJAN 2.03

ALBANIA 2

TURKEY 2.18

GREECE 1.37

MALTA 1.52

CYPRUS 1.45

By the mid-1990s, most countries in Europe had begun to experience a phenomenon that was the logical consequence of the final stages of the "demographic transition" in which small families and increased longevity are the result of economic development, the recognition that fewer numbers of children cost less money, and the emergence of pension plans and other social mechanisms that make larger families unnecessary to protect parents in their old age. The chief manifestation of this last stage of the demographic transition was decreased fertility; indeed, in some countries (such as France) fertility rates or the numbers of children borne by the average woman were so low as to be below "replacement" levels—in other words, men and women were not producing enough offspring to replace themselves. The normal replacement fertility rate level for a wife and husband is 2.1. Throughout Europe, fertility rates are appreciably below that. This does not necessarily mean that all of the countries of Europe will experience shrinking populations. But those that do not will probably experience stable or even growing populations as the result of in-migration rather than natural increase. Even more critically, many of the in-migrants to countries like the United Kingdom are Muslims whose fertility rates are traditionally higher than their Christian neighbors. While stable or even shrinking populations may be beneficial in terms of environmental stress, etc., there are severe market and other pressures that result from aging populations: in a shrinking labor force, for example, will there be a sufficient tax base to support the national pension plans that are the norm in European countries?

Map 30a Total Fertility Rates, 1975

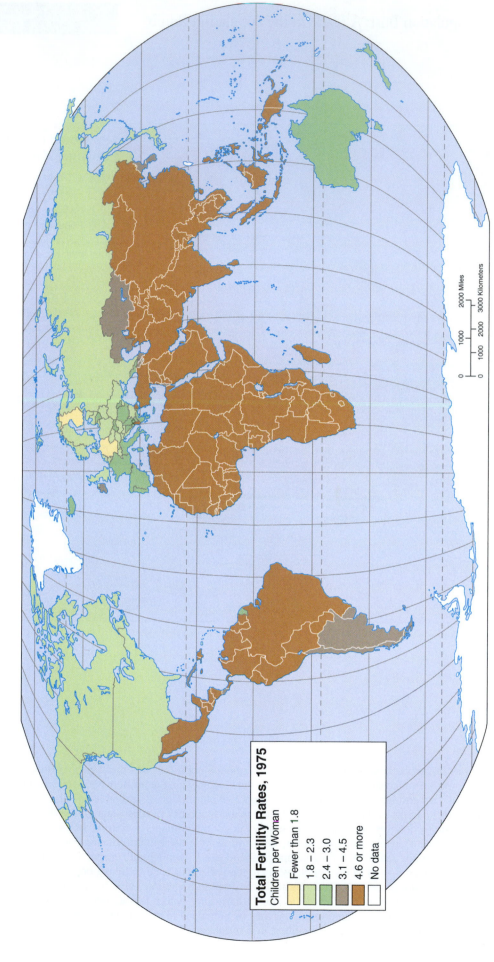

Total Fertility Rates, 1975
Children per Woman

- Fewer than 1.8
- 1.8 – 2.3
- 2.4 – 3.0
- 3.1 – 4.5
- 4.6 or more
- No data

Fertility rates measure the number of children that might be expected to be given birth by a single woman during her child-bearing years. It is often a better predictor of potential population growth than such measures as birth rate. More important, it is a major indicator of the "demographic transition" or shift from a population characterized by large families and short lives to one characterized by small families and longer lives. For the last century, the first type of population profile (large families and short lives) has been evident in the lesser developed parts of the world while the developed nations had populations with the smaller family and longer life span profile. What these comparative maps show us is that over the last quarter century, the same kind of demographic transition begun in Europe and America in the nineteenth century is now beginning in Middle and South America, in Africa, and in Asia. There are still areas of concern—African fertility rates are, for example, still far too high. But particularly in South America, Southeast Asia, and East Asia, fertility rates have dropped dramatically, indicating the potential for much slower population growth in the next generation.

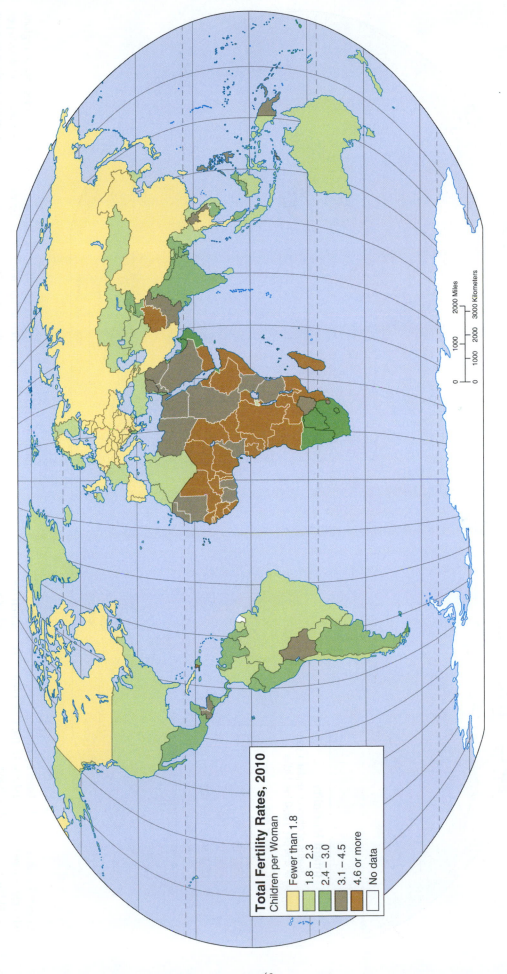

Map 30b Total Fertility Rates, 2010

Total Fertility Rates, 2010
Children per Woman

- Fewer than 1.8
- 1.8 – 2.3
- 2.4 – 3.0
- 3.1 – 4.5
- 4.6 or more
- No data

0 1000 2000 Miles
0 1000 2000 3000 Kilometers

Map 31 Infant Mortality Rate

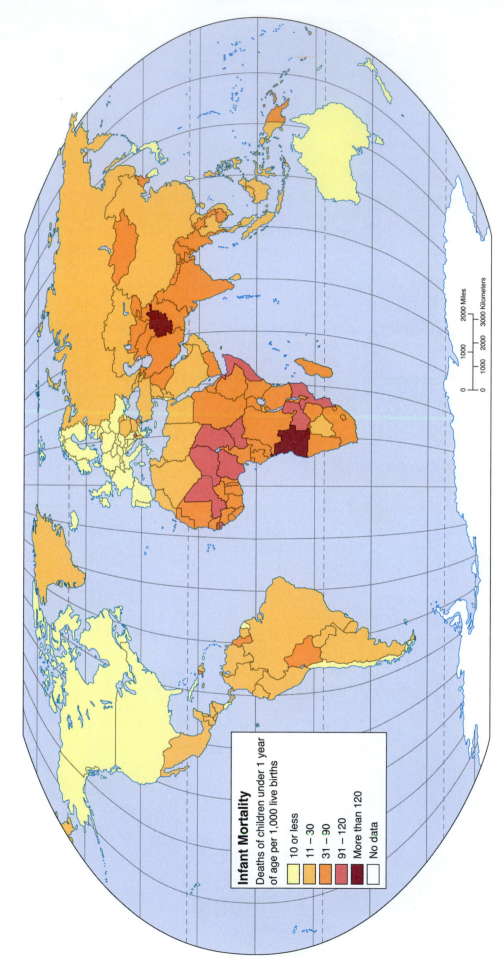

Infant Mortality

Deaths of children under 1 year
of age per 1,000 live births

- 10 or less
- 11 – 30
- 31 – 90
- 91 – 120
- More than 120
- No data

Infant mortality rates are calculated by dividing the number of children born in a given year who die before their first birthday by the total number of children born that year and then multiplying by 1,000; this shows how many infants have died for every 1,000 births. Infant mortality rates are prime indicators of economic development. In highly developed economies, with advanced medical technologies, sufficient diets, and adequate public sanitation, infant mortality rates tend to be quite low. By contrast, in less developed countries, with the disadvantages of poor diet, limited access to medical technology, and the other problems of poverty, infant mortality rates tend to be high. Although worldwide infant mortality has decreased significantly during the last two decades, many regions of the world still experience infant mortality above the 10 percent level (100 deaths per 1,000 live births). Such infant mortality rates not only represent human tragedy at its most basic level, but also are powerful inhibiting factors for the future of human development. Comparing infant mortality rates in the midlatitudes and the tropics shows that children in most African countries are more than 10 times as likely to die within a year of birth as children in European countries.

Map 32 Child Mortality Rate

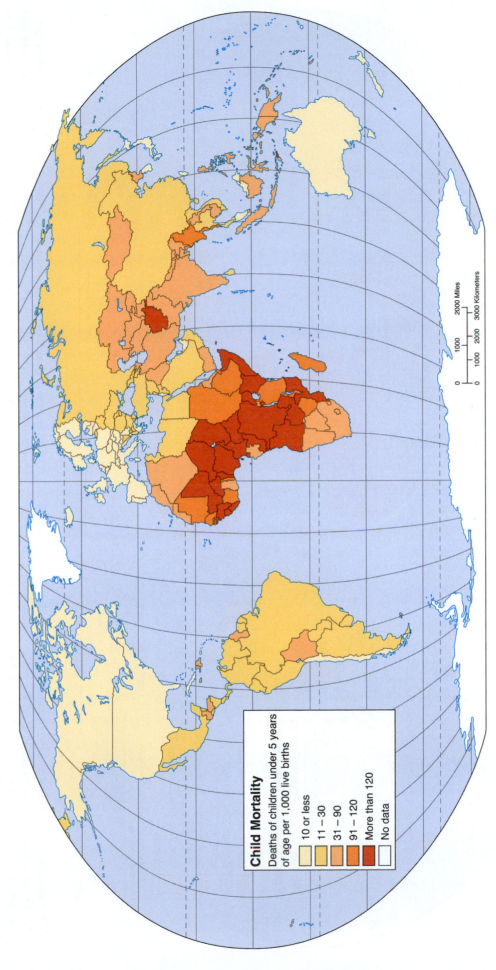

Child Mortality
Deaths of children under 5 years
of age per 1,000 live births

10 or less
11 – 30
31 – 90
91 – 120
More than 120
No data

Child mortality rates are calculated by determining the probability that a child born in a specified year will die before reaching age 5, using current age-specific mortality rates for a population. The major sources of mortality rates are vital registration systems and estimates made from surveys and/or census reports. Along with infant mortality and average life-expectancy rates, child mortality rates, according to the World Bank, "are probably the best general indicators of a community's current health status and are often cited as overall measures of a population's welfare or quality of life." Where infant mortality often reflects health care conditions, child mortality is usually a reflection of the inadequacy of nutrition, leading to early deaths from nutritionally related diseases. In some less developed countries in Africa and Asia, child mortality is also an indicator of the widespread presence of infectious diseases such as malaria, tuberculosis, and HIV/AIDS.

Map 33   Population by Age Group

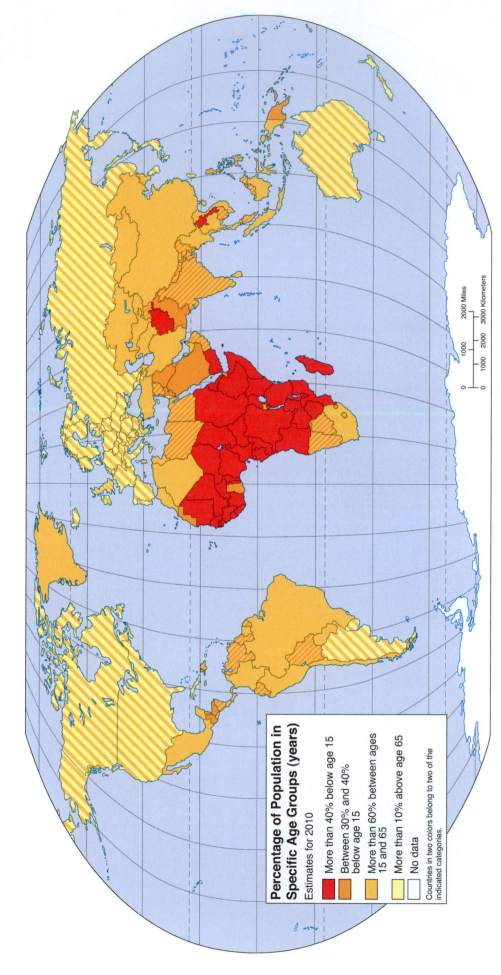

Percentage of Population in Specific Age Groups (years)

Estimates for 2010

- More than 40% below age 15
- Between 30% and 40% below age 15
- More than 60% between ages 15 and 65
- More than 10% above age 65
- No data

Countries in two colors belong to two of the indicated categories.

0	1000	2000 Miles
0	1000	2000	3000 Kilometers

Of all the measurements that illustrate the dynamics of a population, age distribution may be the most significant, particularly when viewed in combination with average growth rates. The particular relevance of age distribution is that it tells us what to expect from a population in terms of growth over the next generation. If, for example, approximately 40–50 percent of a population is below the age of 15, that suggests that in the next generation about one-quarter of the total population will be women of childbearing age. When age distribution is combined with fertility rates (the average number of children born per woman in a population), an especially valid measurement

of future growth potential may be derived. A simple example: Nigeria, with a 2002 population of 130 million, has 43.6 percent of its population below the age of 15 and a fertility rate of 5.5; the United States, with a 2002 population of 280 million, has 21 percent of its population below the age of 15 and a fertility rate of 2.07. During the period in which those women presently under the age of 15 are in their childbearing years, Nigeria can be expected to add a total of approximately 155 million persons to its total population. Over the same period, the United States can be expected to add only 61 million.

Map 34 Average Life Expectancy at Birth

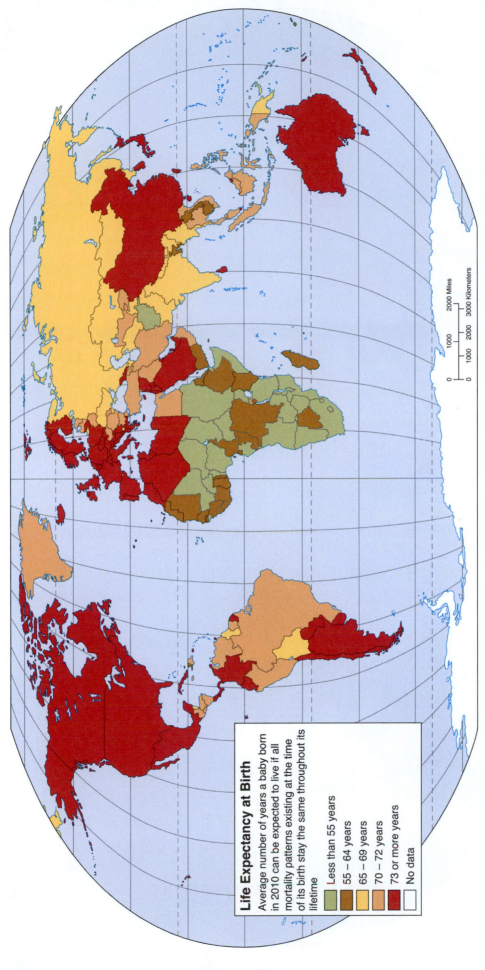

Life Expectancy at Birth

Average number of years a baby born in 2010 can be expected to live if all mortality patterns existing at the time of its birth stay the same throughout its lifetime

- Less than 55 years
- 55 – 64 years
- 65 – 69 years
- 70 – 72 years
- 73 or more years
- No data

Average life expectancy at birth is a measure of the average longevity of the population of a country. Like all average measures, it is distorted by extremes. For example, a country with a high mortality rate among children will have a low average life expectancy. Thus, an average life expectancy of 45 years does not mean that everyone can be expected to die at the age of 45. More normally, what the figure means is that a substantial number of children die between birth and 5 years of age, thus reducing the average life expectancy

for the entire population. In spite of the dangers inherent in misinterpreting the data, average life expectancy (along with infant mortality and several other measures) is a valid way of judging the relative health of a population. It reflects the nature of the health care system, public sanitation and disease control, nutrition, and a number of other key human need indicators. As such, it is a measure of well-being that is significant in indicating economic development and predicting political stability.

Map 35 World Daily Per Capita Food Supply (Kilocalories)

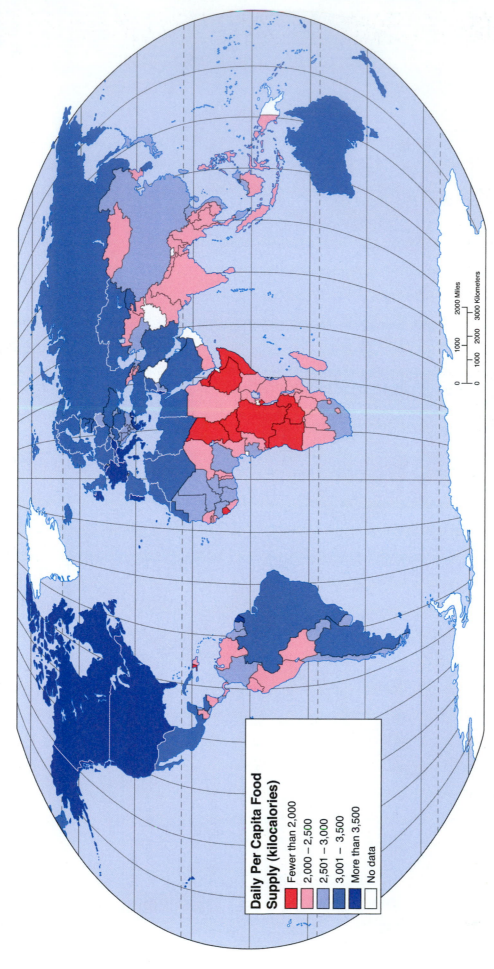

Daily Per Capita Food Supply (kilocalories)

- Fewer than 2,000
- 2,000 – 2,500
- 2,501 – 3,000
- 3,001 – 3,500
- More than 3,500
- No data

The data shown on this map, which indicate the presence or absence of critical food shortages, do not necessarily indicate the presence of starvation or famine. But they certainly do indicate potential problem areas for the next decade. The measurements are in calories from *all* food sources: domestic production, international trade, draw-down on stocks or food reserves, and direct foreign contributions or aid. The quantity of calories available is that amount, estimated by the UN's Food and Agriculture Organization (FAO), that reaches consumers. The calories actually consumed may be lower than the figures shown, depending on how much is lost in a variety of ways: in home storage (to pests such as rats and mice), in preparation and cooking, through consumption by pets and domestic animals, and as discarded foods, for example. The estimate of need is not a global uniform value but is calculated for each country on the basis of the age and sex distribution of the population and the estimated level of activity of the population. Compare this map with Map 60 for a good measure of potential problem areas for food shortages within the next decade.

Map 36   The Earth's Hungry Millions

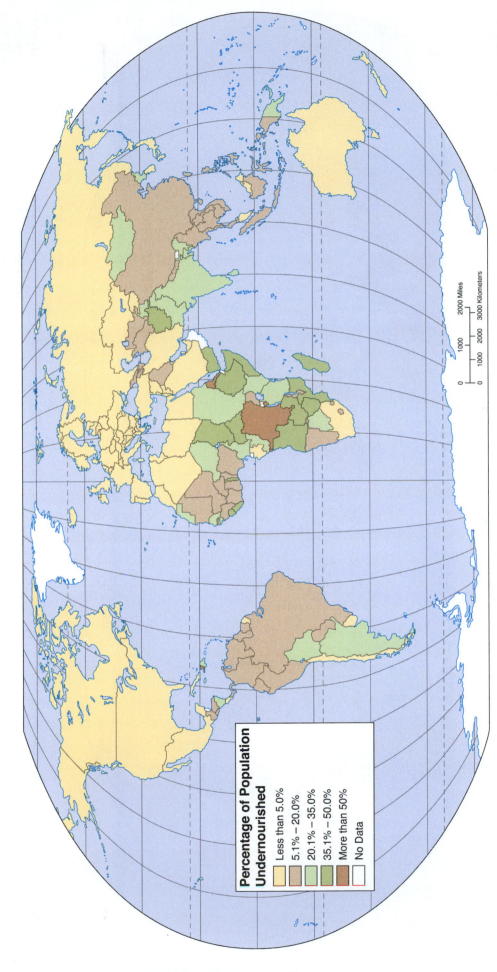

Percentage of Population Undernourished

- Less than 5.0%
- 5.1% – 20.0%
- 20.1% – 35.0%
- 35.1% – 50.0%
- More than 50%
- No Data

2000 Miles

3000 Kilometers

0   1000   2000
0   1000   2000

In order to maintain health and have sufficiently developed immune systems to fight off diseases, humans need to maintain a daily caloric intake of approximately 2,000 calories. Current estimates are that nearly 15% of the world's population have diets that do not constitute the minimum 2,000 calories for basic health maintenance. These countries are particularly susceptible to climate-induced famines which may lower already minimal food intake, or to food emergencies produced by warfare, civil unrest, economic instability, and poor governance. It is little surprise, when looking at the map, that the countries at greatest risk are in Africa and Asia where swelling populations have already approached (or, in some instances, exceeded) carrying capacity or the ability of the environment to sustain a population at a given level. The causes of national malnutrition are as numerous as the countries that experience it but, again, the map is instructive. In Africa, where HIV/AIDS is prevalent among large segments of the working-age population, food production is limited simply by poor health (which, of course, begets low harvests which contributes to poorer health). In South Asia, the problem tends to be magnified by huge populations that have outstripped the ability of the environment to support them. Bangladesh, for example, with an area about the size of West Virginia, is one of the world's ten most populous countries.

Map 37 Child Malnutrition

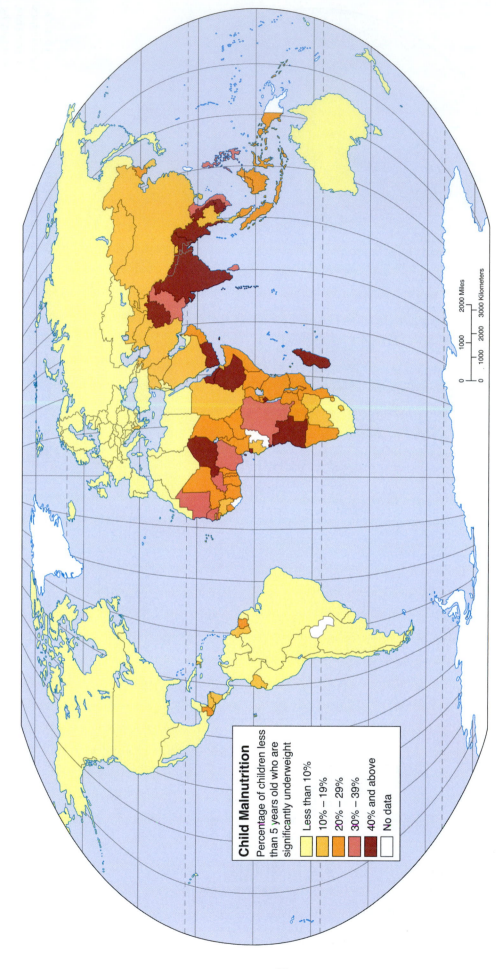

Child Malnutrition

Percentage of children less than 5 years old who are significantly underweight

- Less than 10%
- 10% – 19%
- 20% – 29%
- 30% – 39%
- 40% and above
- No data

The weight of poverty is not evenly spread among the members of a population, falling disproportionately upon the weakest and most disadvantaged members of society. In most societies, these individuals are children, particularly female children. Children simply do not compete as successfully as adults for their (meager) share of the daily food supply. Where food shortages prevail, children tend to have the quality of their future lives severely compromised by poor nutrition, which, in a downward spiral, robs them of the energy necessary to compete more effectively for food. Children who are inadequately fed are less likely to do well in school, are more prone to debilitating disease, and will more often become a drain on scarce societal resources than well-fed children. Recently, health care officials in the more developed world have become concerned over the trend to "overnutrition," leading to obesity and related health problems in the world's economically developed countries. Nevertheless, child malnourishment remains one of the primary distinguishing factors between the "haves" and "have-nots."

-50-

Map 38 Proportion of Daily Food Intake from Animal Products

Meat and Dairy Diets
Percentage of Daily Caloric Intake
from Animal Products

- Less than 5%
- 5% – 10%
- 11% – 15%
- 16% – 20%
- 21% – 25%
- 26% – 30%
- 31% – 35%
- 36% – 40%
- 41% – 45%
- No data

2000 Miles

1000   2000   3000 Kilometers

0   1000   2000

0

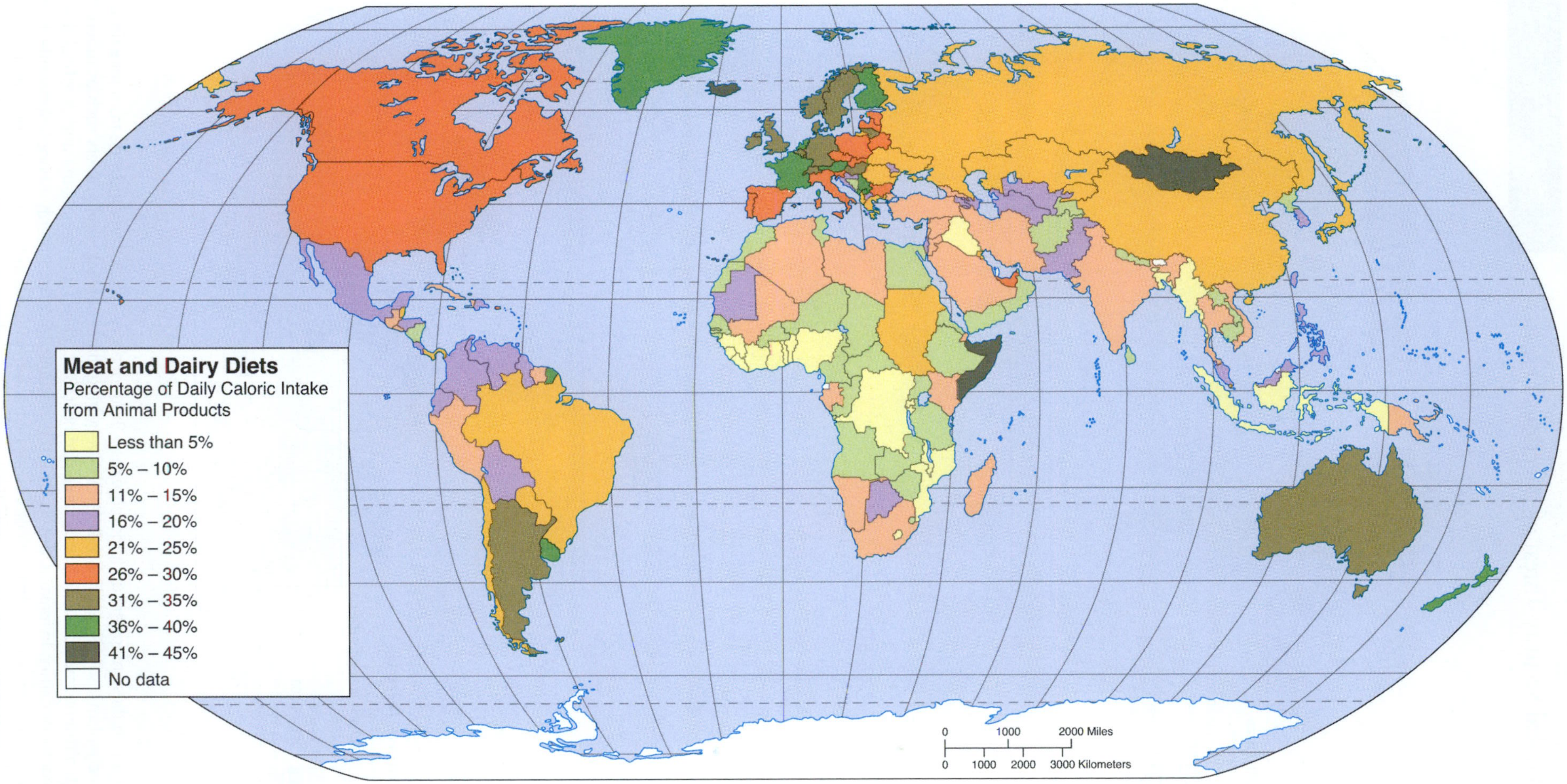

As the world grows proportionally wealthier, the percentage of the diet from meat and fish grows accordingly. While there is precious little scientific evidence to indicate that a vegetarian diet is, by definition, healthier than one richer in animal products, it is indisputable that more reliance on animal products produces a greater strain on environmental systems. Growing plants to feed animals to feed humans simply consumes a great deal more energy (in all forms) than growing plants to feed a human population directly. The long-term implications of altering land-use patterns to fit with greater wealth and, therefore, more meat consumption, are not generally favorable. And we are already beginning to notice the impact on fish population in the world's oceans as the result of an increasing demand for this food resource.

Map 39   The Global Security Threat of Infectious Diseases

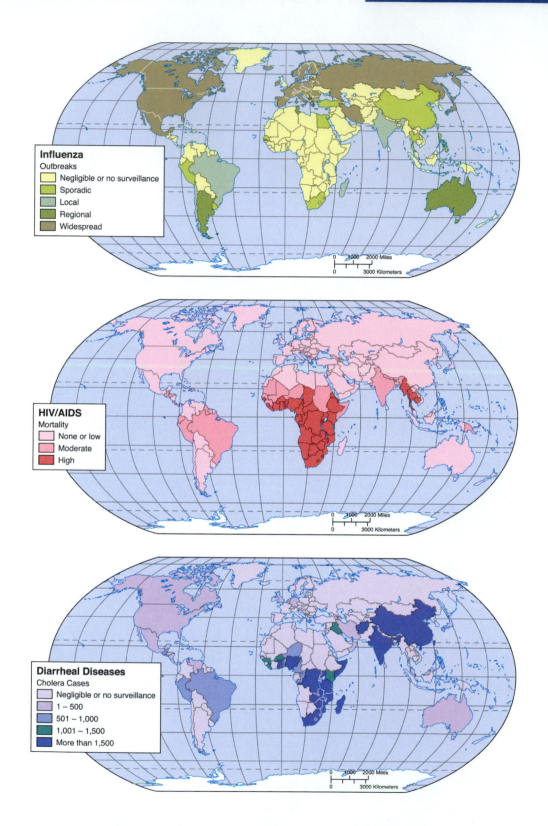

The infectious diseases shown in these maps account for more than 9 out of every 10 deaths worldwide from infectious disease. They are spread through various mechanisms, some—such as HIV/AIDS—requiring direct contact, while others such as tuberculosis or influenza may be contracted through airborne transmissions from an infected host. Still others, such as malaria and diarrheal diseases like cholera, require transmission by a vector or carrier. In malaria, the vector is the mosquito (most commonly *Aenopheles*) while the diarrheal diseases are transmitted by microscopic vectors such as bacteria or viruses that commonly live in the water that people drink. Infectious diseases such as these six do tend to have specific geographic distributions, sometimes as the result of climatic factors (malaria depends upon warm

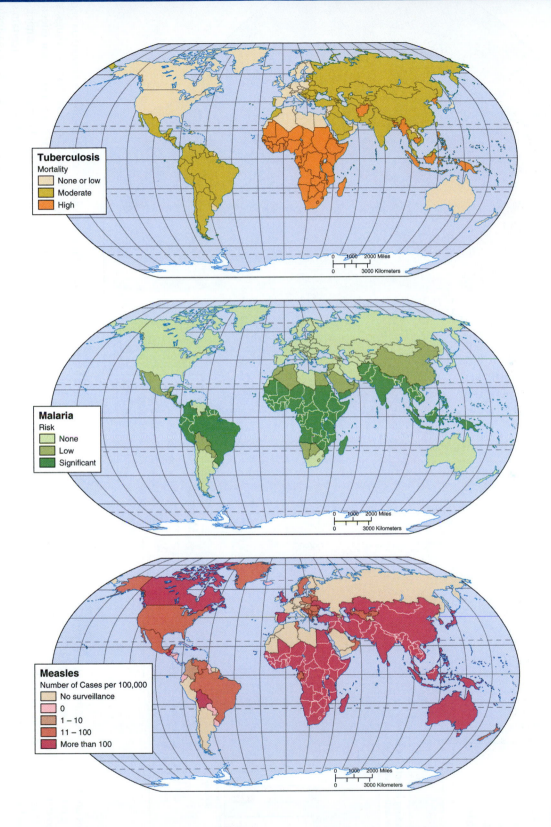

Tuberculosis
Mortality
- None or low
- Moderate
- High

Malaria
Risk
- None
- Low
- Significant

Measles
Number of Cases per 100,000
- No surveillance
- 0
- 1 – 10
- 11 – 100
- More than 100

and moist conditions) and sometimes the result of cultural factors such as rapid and widespread travel. It is the latter that helps to explain why influenza tends to have a greater impact in the developed world, even though the influenza viruses themselves often have their origins in less developed regions of Asia. One good way to avoid influenza is to never travel on a common carrier such as an airplane, train, or bus! None of these diseases has yet been controlled (to the extent that, say, polio myelitis or smallpox have been controlled). But neither have they wrought the havoc caused by the bubonic plague in late medieval Europe. The sites where future concerns about the transmission of infectious diseases ought to be focused are the increasingly large urban concentrations in South America, Africa, and Asia.

Map 40 Global Scourges: Major Infectious Diseases

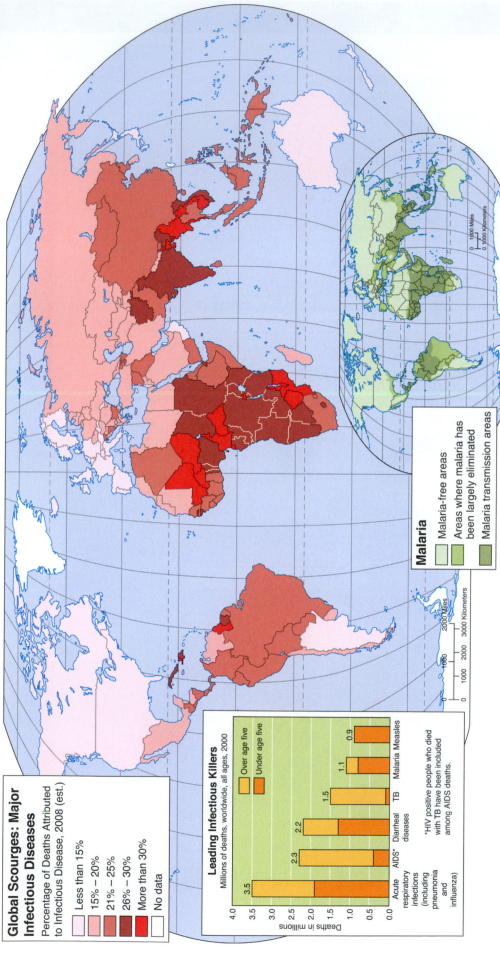

Global Scourges: Major Infectious Diseases

Percentage of Deaths Attributed to Infectious Disease, 2008 (est.)

- Less than 15%
- 15% – 20%
- 21% – 25%
- 26% – 30%
- More than 30%
- No data

Leading Infectious Killers
Millions of deaths, worldwide, all ages, 2000

- Over age five
- Under age five

	Deaths in millions
Acute respiratory infections (including pneumonia and influenza)	3.5
AIDS*	2.3
Diarrheal diseases	2.2
TB	1.5
Malaria	1.1
Measles	0.9

*HIV positive people who died with TB have been included among AIDS deaths.

Malaria

- Malaria-free areas
- Areas where malaria has been largely eliminated
- Malaria transmission areas

Infectious diseases are the world's leading cause of premature death and at least half of the world's population is, at any time, at risk of contracting an infectious disease. Although we often think of infectious diseases as being restricted to the tropical world (malaria, dengue fever), many if not most of them have attained global proportions. A major case in point is HIV/AIDS, which quite probably originated in Africa but has, over the last two decades, spread throughout the entire world. Major diseases of the nineteenth century, such as cholera and tuberculosis, are making a major comeback in many parts of the world, in spite of being preventable or treatable. Part of the problem with infectious diseases is that they tend to be associated with poverty (poor nutrition, poor sanitation, substandard housing, and so on) and, therefore, are seen as a problem

of undeveloped countries, with the consequent lack of funding for prevention and treatment. Infectious diseases are also tending to increase because lifesaving drugs, such as antibiotics and others used in the fight against diseases, are losing their effectiveness as bacteria develop genetic resistance to them. The problem of global warming is also associated with a spread of infectious diseases as many disease vectors (certain species of mosquito, for example) are spreading into higher latitudes with increasingly warm temperatures and are spreading disease into areas where populations have no resistance to them. Infectious diseases have become something greater than simply a health issue of poor countries. They are now major social problems with potentially enormous consequences for the entire world.

-54-

Map 41   Adult Incidence of HIV/AIDS, 2007

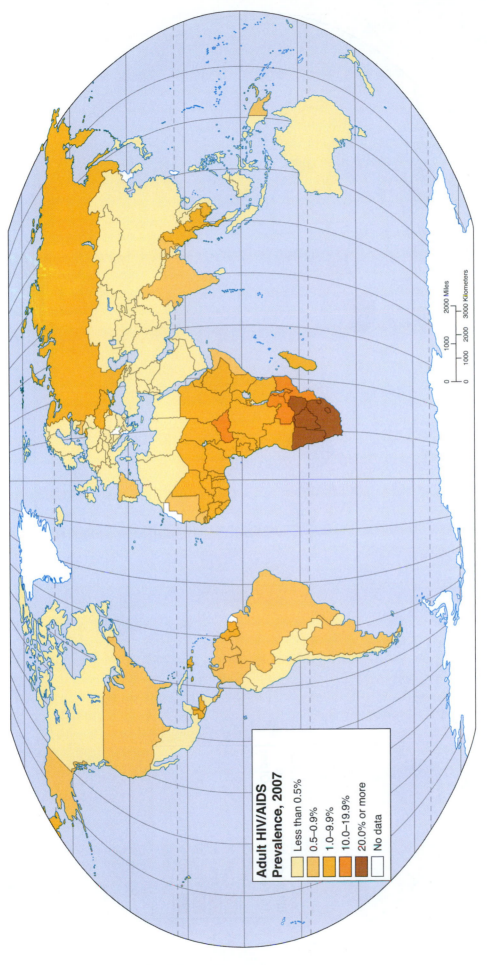

Adult HIV/AIDS Prevalence, 2007

Less than 0.5%
0.5–0.9%
1.0–9.9%
10.0–19.9%
20.0% or more
No data

Of all the infectious diseases, the one that poses the greatest risks to public health worldwide is human immune deficiency which, if untreated, progresses to the deadly autoimmune deficiency, or AIDS. The highest incidence of AIDS among adult populations occurs in sub-Saharan Africa, where public health systems are too poorly funded and developed to treat HIV cases, and the medications needed to preserve health and reasonable longevity are too expensive. Since the rate of adult cases is high, large numbers of children are also infected, having been born HIV-positive. Indeed, among

all the world's children living with HIV, approximately 90 percent live in Sub-Saharan Africa. Other countries in the developing world also are high on the scale of HIV/AIDS incidence and the suspicion is that the official figures could go even higher if accurately reported. China, for example, may have rates that are 3 to 4 times greater than the official figures. Worldwide, HIV/AIDS has a devastating impact on family structures (many children are orphaned by both parents dying of AIDS) and on the economic growth of the developing countries.

Map 42   Illiteracy Rates

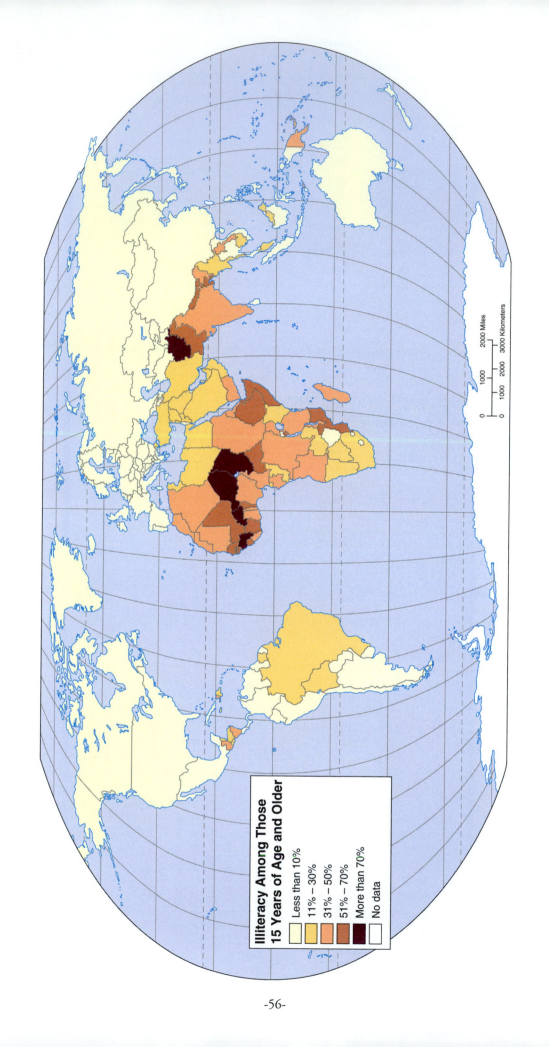

**Illiteracy Among Those
15 Years of Age and Older**

- Less than 10%
- 11% – 30%
- 31% – 50%
- 51% – 70%
- More than 70%
- No data

2000 Miles

3000 Kilometers

0   1000   2000

0   1000   2000

Map 43 Primary School Enrollment

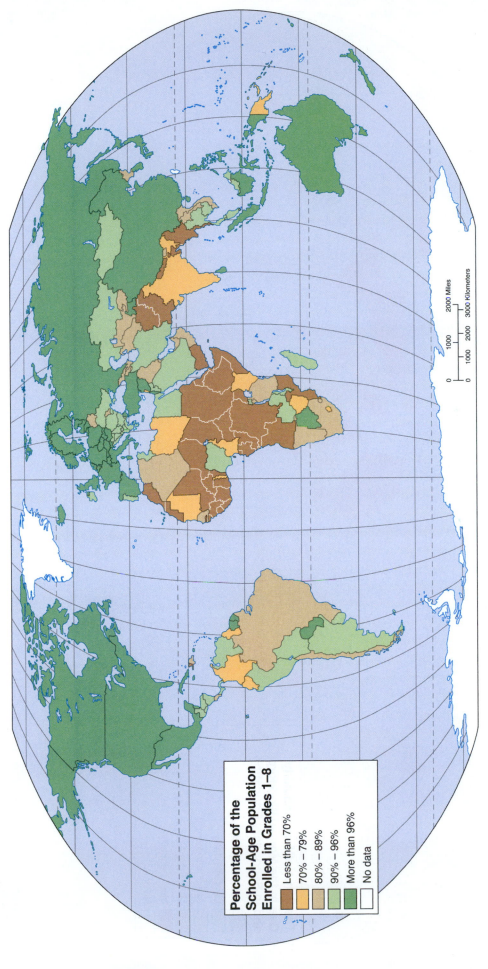

Percentage of the School-Age Population Enrolled in Grades 1–8

- Less than 70%
- 70% – 79%
- 80% – 89%
- 90% – 96%
- More than 96%
- No data

0   1000   2000   2000 Miles

0   1000   2000   3000 Kilometers

Like many of the other measures illustrated in this atlas, primary school enrollment is a clear reflection of the division of the world into "have" and "have-not" countries. It is also a measure that has changed more rapidly over the last decade than demographic and other indicators of development, as countries of even very modest means have made concerted attempts to attain relatively high percentages of primary school enrollment. That they have been able to do so is good evidence of the fact that reasonably respectable levels of human

development are feasible at even modest income levels. High primary school enrollment is also a reflection of the worldwide opinion that a major element in economic development is a well-educated, literate population. The links between human progress, as typified by higher levels of education, and economic growth are not automatic, however, and those countries without programs for maintaining the headway gained by improved education may be on the road to failure in terms of economic development.

Map 44   Female/Male Inequality in Education and Employment

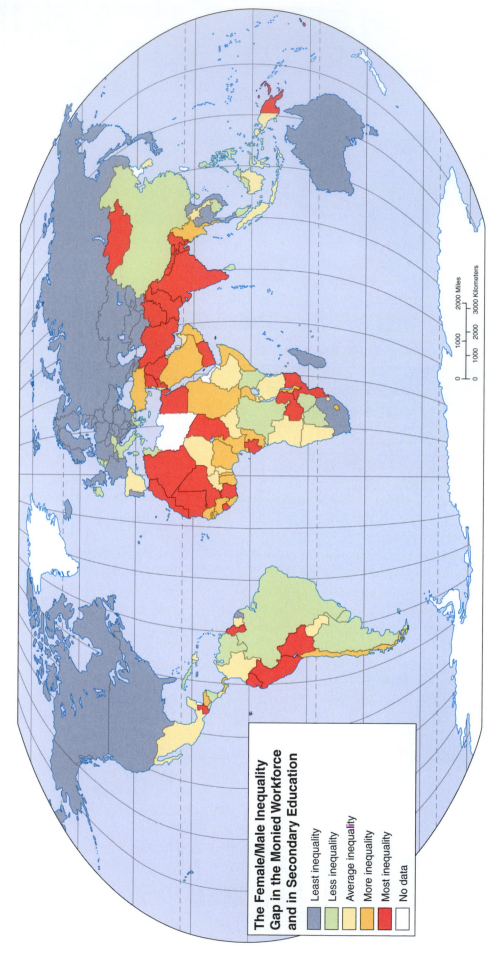

The Female/Male Inequality Gap in the Monied Workforce and in Secondary Education

- Least inequality
- Less inequality
- Average inequality
- More inequality
- Most inequality
- No data

2000 Miles

3000 Kilometers

While women in developed countries, particularly in North America and Europe, have made significant advances in socioeconomic status in recent years, in most of the world they suffer from significant inequality when compared with their male counterparts. Although women have received the right to vote in most of the world's countries, in over 90 percent of these countries that right has only been granted in the last 50 years. In most regions, literacy rates for women still fall far short of those for men; in Africa and Asia, for example, only about half as many women are literate as are men. Women marry considerably younger than men and attend school for shorter periods of time. Inequalities in education and employment are perhaps the most telling indicators of the unequal status of women in most of the world. Lack of secondary education in comparison with men prevents women from entering the workforce with equally high-paying jobs. Even where women are employed in positions similar to those held by men, they still tend to receive less compensation. The gap between rich and poor involves not only a clear geographic differentiation, but a clear gender differentiation as well.

-58-

Map 45 The Index of Human Development

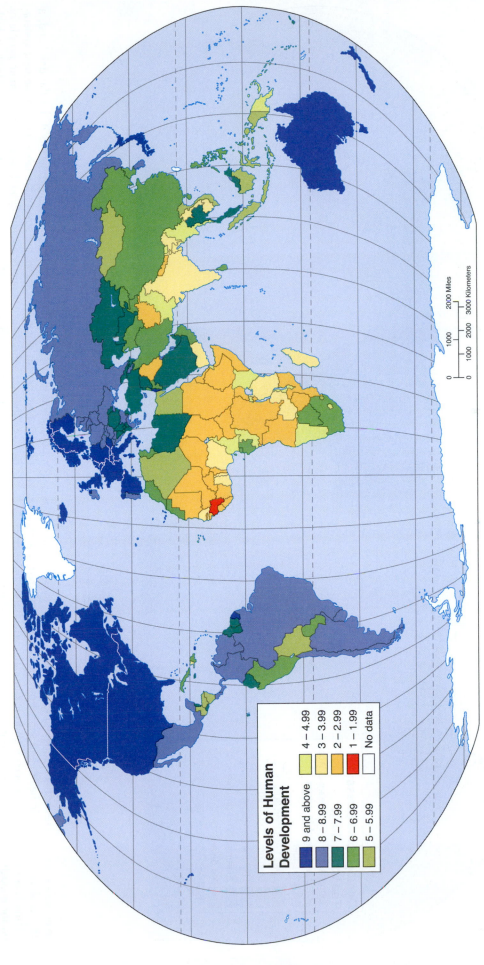

Levels of Human Development

- 9 and above
- 8 – 8.99
- 7 – 7.99
- 6 – 6.99
- 5 – 5.99
- 4 – 4.99
- 3 – 3.99
- 2 – 2.99
- 1 – 1.99
- No data

The development index upon which this map is based takes into account a wide variety of demographic, health, and educational data, including population growth, per capita gross domestic income, longevity, literacy, and years of schooling. The map reveals significant improvement in the quality of life in Middle and South America, although it is questionable whether the gains made in those regions can be maintained in the face of the dramatic population increases expected over the next 30 years. More clearly than anything else, the map illustrates the near-desperate situation in Africa and South Asia. In those regions, the unparalleled growth in population threatens to overwhelm all efforts to improve the quality of life. In Africa, for example, the population is increasing by 20 million persons per year. With nearly 45 percent of the continent's population aged 15 years or younger, this growth rate will accelerate as the women reach child-bearing age. Africa, along with South Asia, faces the very difficult challenge of providing basic access to health care, education, and jobs for a rapidly increasing population. The map also illustrates the striking difference in quality of life between those who inhabit the world's equatorial and tropical regions and those fortunate enough to live in the temperate zones, where the quality of life is significantly higher.

Map 46 Demographic Stress: The Youth Bulge

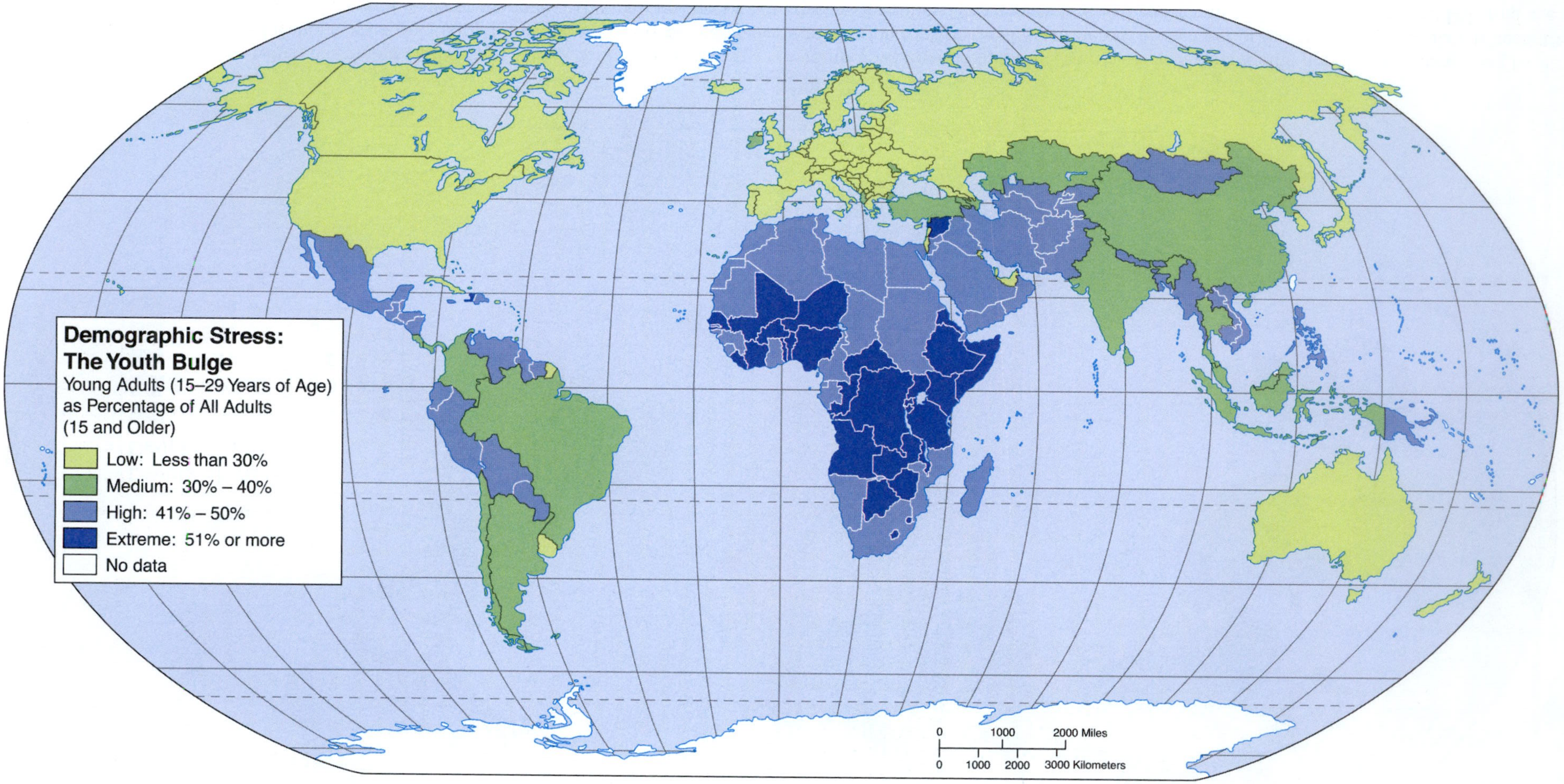
Demographic Stress:
The Youth Bulge
Young Adults (15–29 Years of Age) as Percentage of All Adults (15 and Older)
Low: Less than 30%
Medium: 30% – 40%
High: 41% – 50%
Extreme: 51% or more
No data
0 1000 2000 Miles
0 1000 2000 3000 Kilometers

One of the greatest stresses of the demographic transition is when the death rate drops as the result of better public health and sanitation and the birth rate remains high for the same reasons it has always been high in traditional, agricultural societies: (1) the need for enough children to supply labor, which is one of the few ways to increase agricultural production in a non-mechanized agricultural system, and (2) the need for enough children to offset high infant and child mortality rates so that some children will survive to take care of parents in their old age. With declining death rates and steady birth rates, population growth skyrockets—particularly among the youngest cohorts of a population (between the ages of birth and 15). While this is, on the one hand, a demographic benefit since it increases the size of the labor force in the next generation and thereby helps to accelerate economic growth, it also means more people of child-bearing age in the next generation and, hence, greater numbers of births, which continue to swell the population in the youngest, most vulnerable, and most dependent portion of the population. The literature on population and conflict suggests that the larger the percentage of a population below the age of 25, the greater the chance for political violence and warfare.

Map 47 Demographic Stress: Rapid Urban Growth

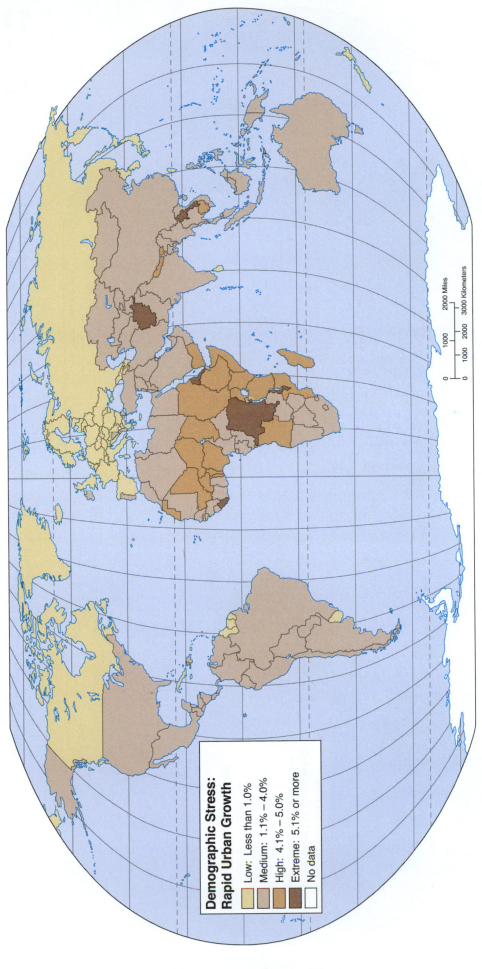

**Demographic Stress:
Rapid Urban Growth**

- Low: Less than 1.0%
- Medium: 1.1% – 4.0%
- High: 4.1% – 5.0%
- Extreme: 5.1% or more
- No data

2000 Miles

3000 Kilometers

0   1000   2000

0   1000   2000

The trend toward urbanization—an increasing percentage of a country's population living in a city—is normally a healthy, modern trend. But in many of the world's developing countries, increasing urbanization is the consequence of high physiologic population densities (too many people for the available agricultural land) and the flight of poorly educated, untrained rural poor to the cities where they hope (often in vain) to find employment. High urbanization exists in many of the world's poorest countries where the urban poor live in conditions that make the worst living conditions in the inner cities of developed countries look positively luxurious. In the urban slums of South America, Africa, and South Asia, millions of people live in temporary housing of cardboard and flattened aluminum cans, with no public services such as water, sewage, or electricity. In Africa, where only 40% of the population is urbanized (in comparison with more than 90% in North America and Europe), the population of urban poor is greater than the total urban population of the United States and the European Union. This represents an increasingly destabilizing element of modern urban societies.

Map 48 Demographic Stress: Competition for Cropland

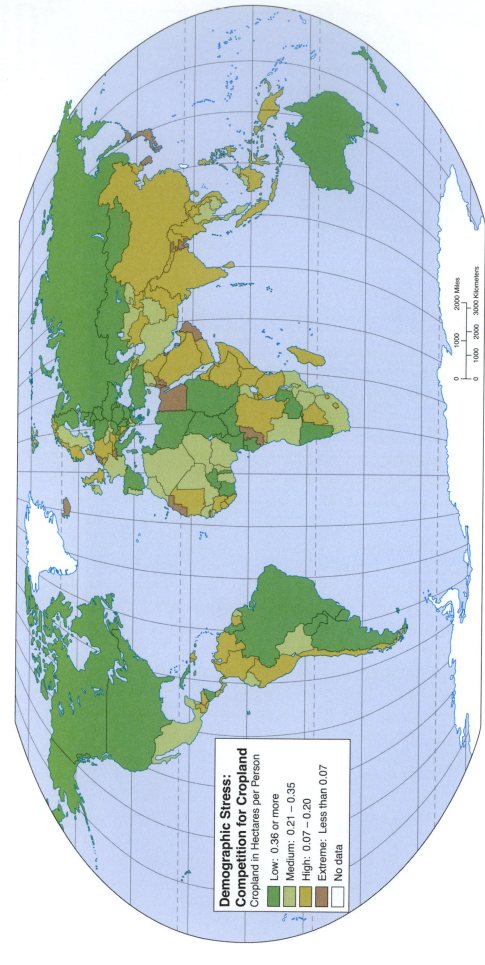

Demographic Stress:
Competition for Cropland
Cropland in Hectares per Person

Low: 0.36 or more

Medium: 0.21 – 0.35

High: 0.07 – 0.20

Extreme: Less than 0.07

No data

0   1000   2000 Miles
0   1000 2000   3000 Kilometers

Many of the world's countries have reached the point where available acres or hectares in cropland are less than the numbers of people wishing to occupy them. For developed countries, who pay for agricultural imports with industrial exports, this is not an alarming trend and therefore countries like Germany or Italy (where the ratio between cropland and farmers is very low) have little to worry about. But in countries in South America, Africa, and South and East Asia, the trend toward more farmers and less available land *is* an alarming trend. In these developing regions, farmers depend on their crops for a relatively meager subsistence diet and—if they are lucky—a few bushels of rice or corn to take to the local market to sell or exchange for the small surpluses of other farmers. Despite the broad global trends toward urbanization and more productive agriculture (and this generally means mechanized agriculture), farm occupation and subsistence cropping remain mainstays of the economy in Africa south of the Sahara, in much of western South America, and in much of South and East Asia. Here, the increasingly small margin between the numbers of farmers and the amount of available farmland can lead to conflict among tribal communities or even among members of a single family.

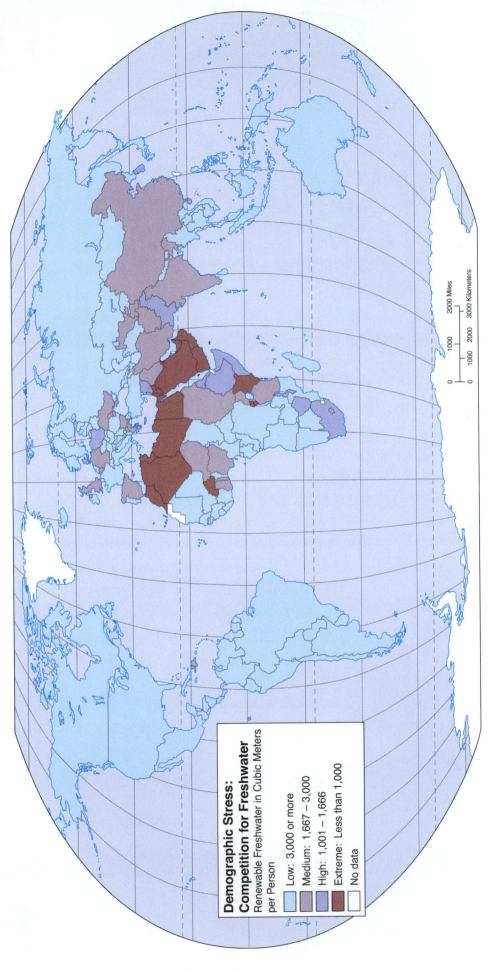

Demographic Stress:
Competition for Freshwater
Renewable Freshwater in Cubic Meters
per Person

Low: 3,000 or more
Medium: 1,667 – 3,000
High: 1,001 – 1,666
Extreme: Less than 1,000
No data

0   1000   2000 Miles
0   1000   2000   3000 Kilometers

As rates of urbanization have increased, the competition between city dwellers and farmers for water has also increased. And as the population of farmers relative to the available surface or ground water supply has accelerated, so rural competition for access to fresh water for irrigation has increased. Obviously, this trend is most obvious in the world's drier regions. On this map, the greatest stress factors relating to water availability are in North and East Africa, the Middle East, and Central and East Asia. The current conflict between Israelis and Palestinians in the West Bank goes far beyond religion or politics: Some is the result of the more affluent Israeli farmers being able to drill wells to tap ground water, which lowers the water table and causes previously accessible surface wells in Palestinian villages to go dry. And part of the horrific Hutu-Tutsi civil war in Rwanda and the taking of European farmlands by the current government of Zimbabwe have more to do with the need for access to fresh water than to racial or ethnic differences. Given that fresh water is, for all practical purposes, a nonrenewable resource, more conflicts over access to water are going to erupt in the future.

-63-

Map 50   Demographic Stress: Death in the Prime of Life

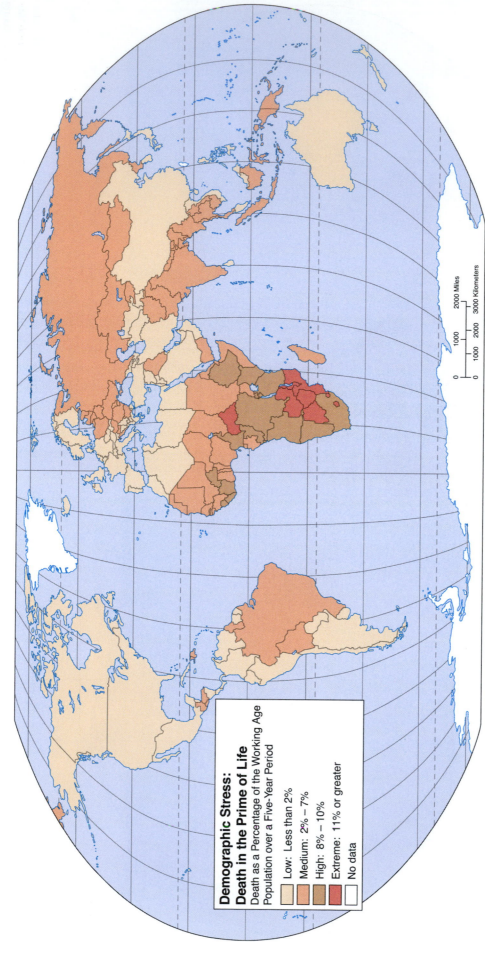

**Demographic Stress:
Death in the Prime of Life**

Death as a Percentage of the Working Age
Population over a Five-Year Period

Low: Less than 2%
Medium: 2% – 7%
High: 8% – 10%
Extreme: 11% or greater
No data

Most infectious diseases lay waste to the oldest and the youngest sectors of a population, leaving the healthier working-age population able to reproduce itself and to continue to grow an economy. Modern infectious diseases—most particularly HIV/AIDS—impacts not the youngest or oldest but those in the prime of life, in their most productive and reproductive years. Indeed, medical data from the United Nations show that 90% of HIV/AIDS-related deaths occur in people of working age. Some of the most severely impacted countries in southern Africa lose between 10% and 20% of their working age population every 5 years. Rates in the developed world are from 0.10% to 0.02% of this level.

A prominent fact of life in much of the developing world—particularly in sub-Saharan Africa—is the long-distance labor migration. In this pattern, a man leaves his wife and family for extended periods of time (10 or more months of the year) to work in mines, on docks, as sailors. These men are highly vulnerable to the acquisition of HIV, which is then transmitted to wives when the husband returns home on his annual visit. Thus both the male and the female portion of the working-age population are impacted by early mortality from AIDS. None of this bodes well for economic development in those affected areas of the world.

2000 Miles

1000   2000   3000 Kilometers

0   1000   2000

Map 51   Demographic Stress: Interactions of Demographic Stress Factors

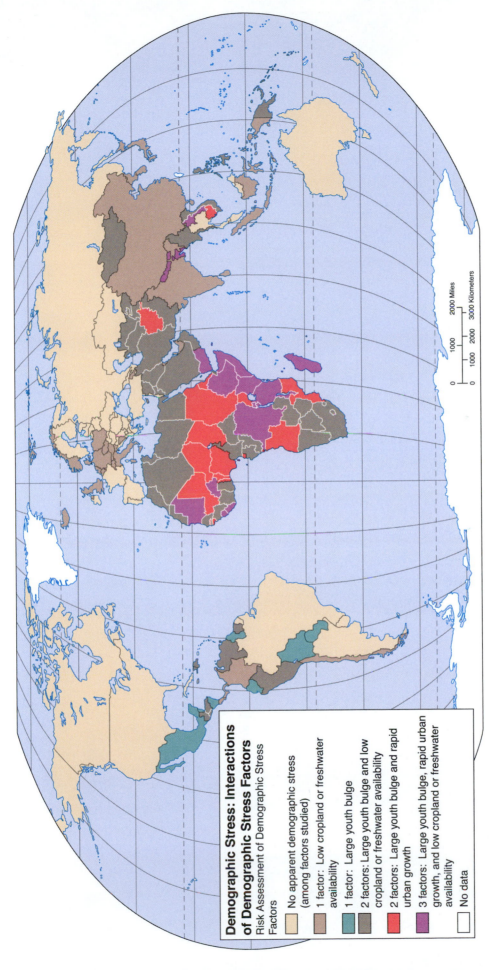

Demographic Stress: Interactions of Demographic Stress Factors
Risk Assessment of Demographic Stress Factors

- No apparent demographic stress (among factors studied)
- 1 factor: Low cropland or freshwater availability
- 1 factor: Large youth bulge
- 2 factors: Large youth bulge and low cropland or freshwater availability
- 2 factors: Large youth bulge and rapid urban growth
- 3 factors: Large youth bulge, rapid urban growth, and low cropland or freshwater availability
- No data

2000 Miles
3000 Kilometers

We have persisted in viewing internal and international conflicts as conflicts produced by religious differences, the acceptance or rejection of "freedom" or "democracy," or even, as argued by a prominent historian, a "Clash of Civilizations." While religious, political, or historical differences cannot be ignored in interpreting the causes of conflict, neither can simple demographic stresses produced by too many new urban dwellers without jobs or hopes of jobs, by too many farmers and too little farmland, by too many deaths among people who should be in the most productive portions of their lives. A massive study by Population Action International, from which these last half-dozen or so maps have been drawn, shows quite clearly that countries suffering from multiple demographic risk factors are, were, and will be much more prone to civil conflict than countries with only one or two demographic risk factors. A look at this map provides a predictor of future civil unrest, violence, and even civil war.

Unit IV

Global Economic Patterns

Map 52   Gross National Income

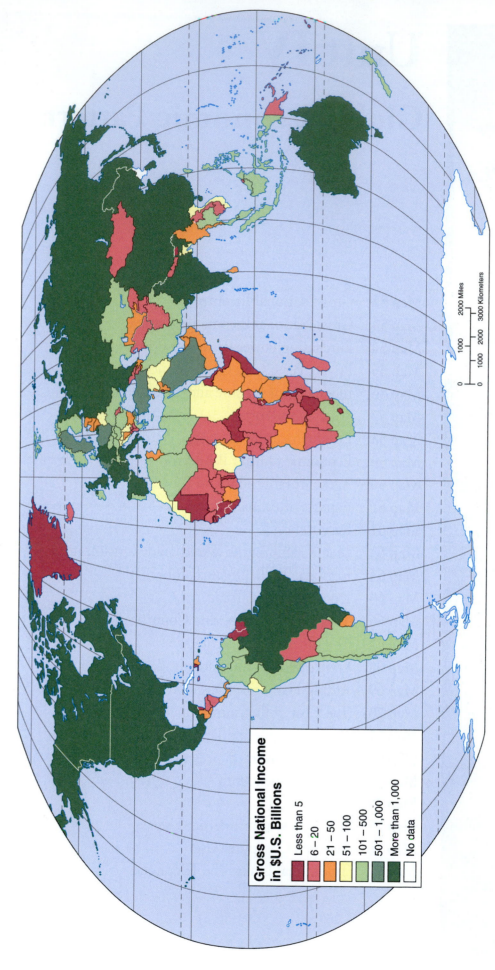

Gross National Income in $U.S. Billions

- Less than 5
- 6 – 20
- 21 – 50
- 51 – 100
- 101 – 500
- 501 – 1,000
- More than 1,000
- No data

Gross National Income (GNI) is the broadest measure of national income and measures the total claims of a country's residents to all income from domestic and foreign products during a year. Although GNI is often misleading and commonly incomplete, it is often used by economists, geographers, political scientists, policy makers, development experts, and others not only as a measure of relative well-being but also as an instrument of assessing the effectiveness of economic and political policies. What is wrong with GNI? First of all, it does not take into account a number of real economic factors, such as environmental deterioration, the accumulation or degradation of human and social capital, or the value of household work. Yet in spite of these deficiencies, GNI is still a reasonable way to assess the relative wealth of nations: the vast differences in wealth that separate the poorest countries from the richest. One of the more striking features of the map is the evidence it presents that such a small number of countries possess so many of the world's riches (keeping in mind that GNI provides no measure of the distribution of wealth within a country).

Map 53 Gross National Income Per Capita

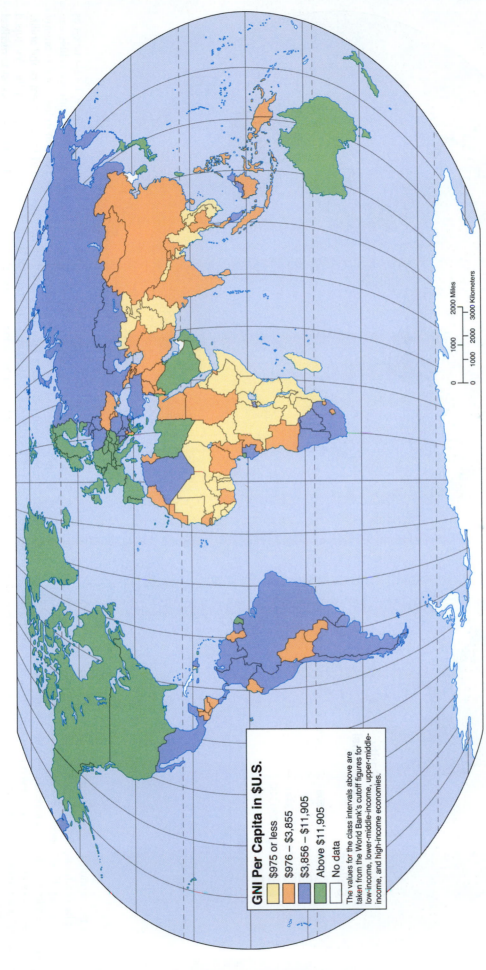

GNI Per Capita in $U.S.

- $975 or less
- $976 – $3,855
- $3,856 – $11,905
- Above $11,905
- No data

The values for the class intervals above are taken from the World Bank's cutoff figures for low-income, lower-middle-income, upper-middle-income, and high-income economies.

0   1000   2000 Miles

0   1000   2000   3000 Kilometers

Gross National Income (GNI) in either absolute or per capita form should be used cautiously as a yardstick of economic strength because it does not measure the distribution of wealth among a population. There are countries (most notably, the oil-rich countries of the Middle East) where per capita GNI is high but where the bulk of the wealth is concentrated in the hands of a few individuals, leaving the remainder in poverty. Even within countries in which wealth is more evenly distributed (such as those in North America or Western Europe), there is a tendency for dollars or pounds sterling or euros to concentrate in the bank accounts of a relatively small percentage of the population.

Yet the maldistribution of wealth tends to be greatest in the less developed countries, where the per capita GNI is far lower than in North America and Western Europe, and poverty is widespread. In fact, a map of GNI per capita offers a reasonably good picture of comparative economic well-being. It should be noted that a low per capita GNI does not automatically condemn a country to low levels of basic human needs and services. There are a few countries, such as Costa Rica and Sri Lanka, that have relatively low per capita GNI figures but rank comparatively high in other measures of human well-being, such as average life expectancy, access to medical care, and literacy.

Map 54   Purchasing Power Parity Per Capita

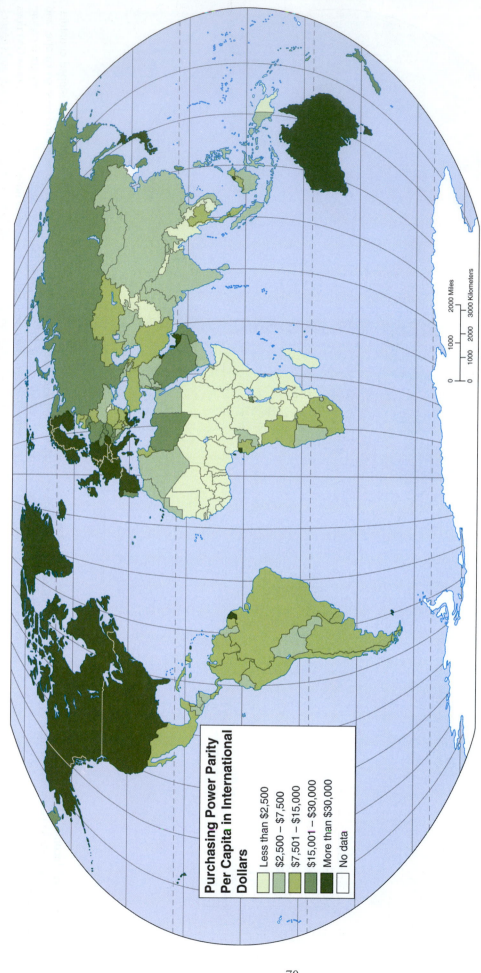

**Purchasing Power Parity
Per Capita in International
Dollars**

Less than $2,500
$2,500 – $7,500
$7,501 – $15,000
$15,001 – $30,000
More than $30,000
No data

Of all the economic measures that separate the "haves" from the "have-nots," perhaps per capita Purchasing Power Parity (PPP) is the most meaningful. While per capita figures can mask significant uneven distributions within a country, they are generally useful for demonstrating important differences between countries. Per capita GNP and GDP (Gross Domestic Product) figures, and even per capita income, have the limitation of seldom reflecting the true purchasing power of a country's currency at home. In order to get around this limitation, international economists seeking to compare national currencies developed the PPP measure, which shows the level of goods and services that holders of a country's money can acquire locally. By converting all currencies to the "international dollar," the World Bank and other organizations using PPP can now show more truly comparative values, since the new currency value shows the number of units of a country's

currency required to buy the same quantity of goods and services in the local market as one U.S. dollar would buy in an average country. The use of PPP currency values can alter the perceptions about a country's true comparative position in the world economy. More than per capita income figures, PPP provides a valid measurement of the ability of a country's population to provide for itself the things that people in the developed world take for granted: adequate food, shelter, clothing, education, and access to medical care. A glance at the map shows a clear-cut demarcation between temperate and tropical zones, with most of the countries with a PPP above $7,500 in the midlatitude zones and most of those with lower PPPs in the tropical and equatorial regions. Where exceptions to this pattern occur, they usually stem from a tremendous maldistribution of wealth among a country's population.

0   1000   2000 Miles
0   1000   2000   3000 Kilometers

Map 55   Private Per Capita Consumption

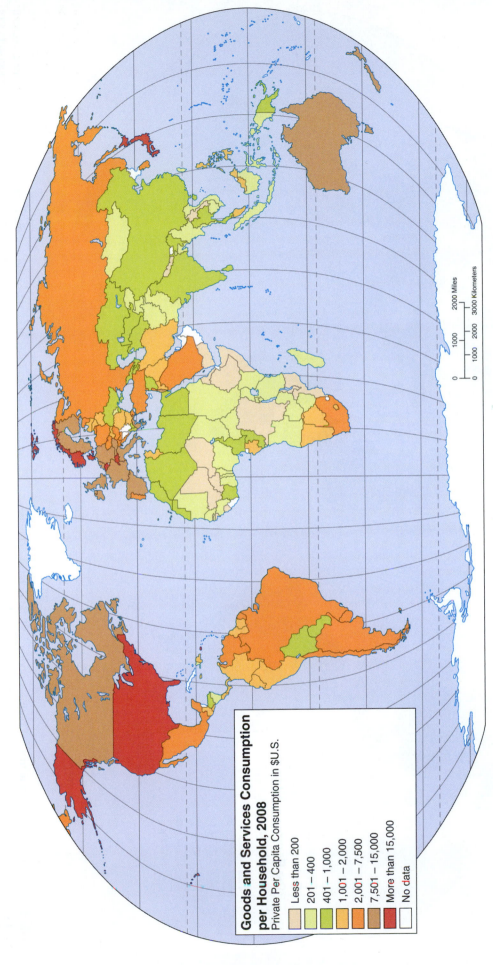

Goods and Services Consumption per Household, 2008

Private Per Capita Consumption in $U.S.

- Less than 200
- 201 – 400
- 401 – 1,000
- 1,001 – 2,000
- 2,001 – 7,500
- 7,501 – 15,000
- More than 15,000
- No data

0   1000   2000   2000 Miles
0   1000   2000   3000 Kilometers

While we often examine production in analyzing global economic patterns, too seldom do we look at patterns of consumption—particularly on a per-person basis. When we do, the dramatic differences between the "rich" and "poor" (of which we are well aware) become even more dramatic. The consumption of key resources—minerals, agricultural products, harvests from the oceans, forestry products, water supplies—is among the most uneven distributions to be found among all the economic patterns of consumption. Those who inhabit the wealthiest nations consume far more than their share of the world. Those who inhabit the wealthiest nations consume far more than their share of these key resources because they can afford to. And they produce far more than

their share of the waste products, including atmospheric and oceanic pollution, of that consumption. Not only do they deplete their own resource bases but those of other countries as well. A trend that can be viewed as alarming by some and encouraging by others is that the number of "rich" or even "moderately well-off" countries is rising, and that number is rising among the world's most populous countries: India and China. What will happen to the global economy and the global environment (and we cannot separate the two) when India and China begin to consume and waste at the same levels as, say, the United States?

Map **56** Inequality of Income and Consumption

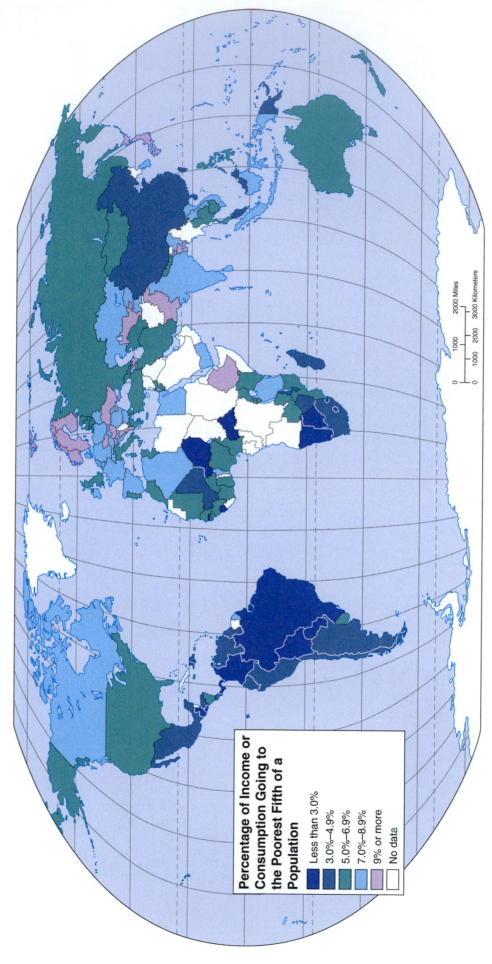

Percentage of Income or Consumption Going to the Poorest Fifth of a Population

- Less than 3.0%
- 3.0%–4.9%
- 5.0%–6.9%
- 7.0%–8.9%
- 9% or more
- No data

0   1000   2000 Miles

0   1000   2000   3000 Kilometers

While it is more than arguably true that the poor get poorer and the rich get richer (and that, by the way, is just as true in some highly developed areas such as the United States and the United Kingdom as it is in the lesser-developed regions), what is often ignored is the breadth of the inequality in distribution of incomes or in the levels of consumption of basic goods. In many of the world's developing countries—despite the rapid increases in economic growth of China and India over the last two decades—the poorest 20% of the population receives less than 7% of the income or consumption share. Although school participation rates have risen worldwide, in countries with the greatest inequalities of income, access to education remains low. If translated into ratios, the inequality ratio of many of the world's countries (including, as noted above, some in the highly developed world) is 8 or higher, meaning that the top 20% of the population spends and consumes at least 8 times as much per person as the bottom 20% of the population.

Map 57   Total Labor Force

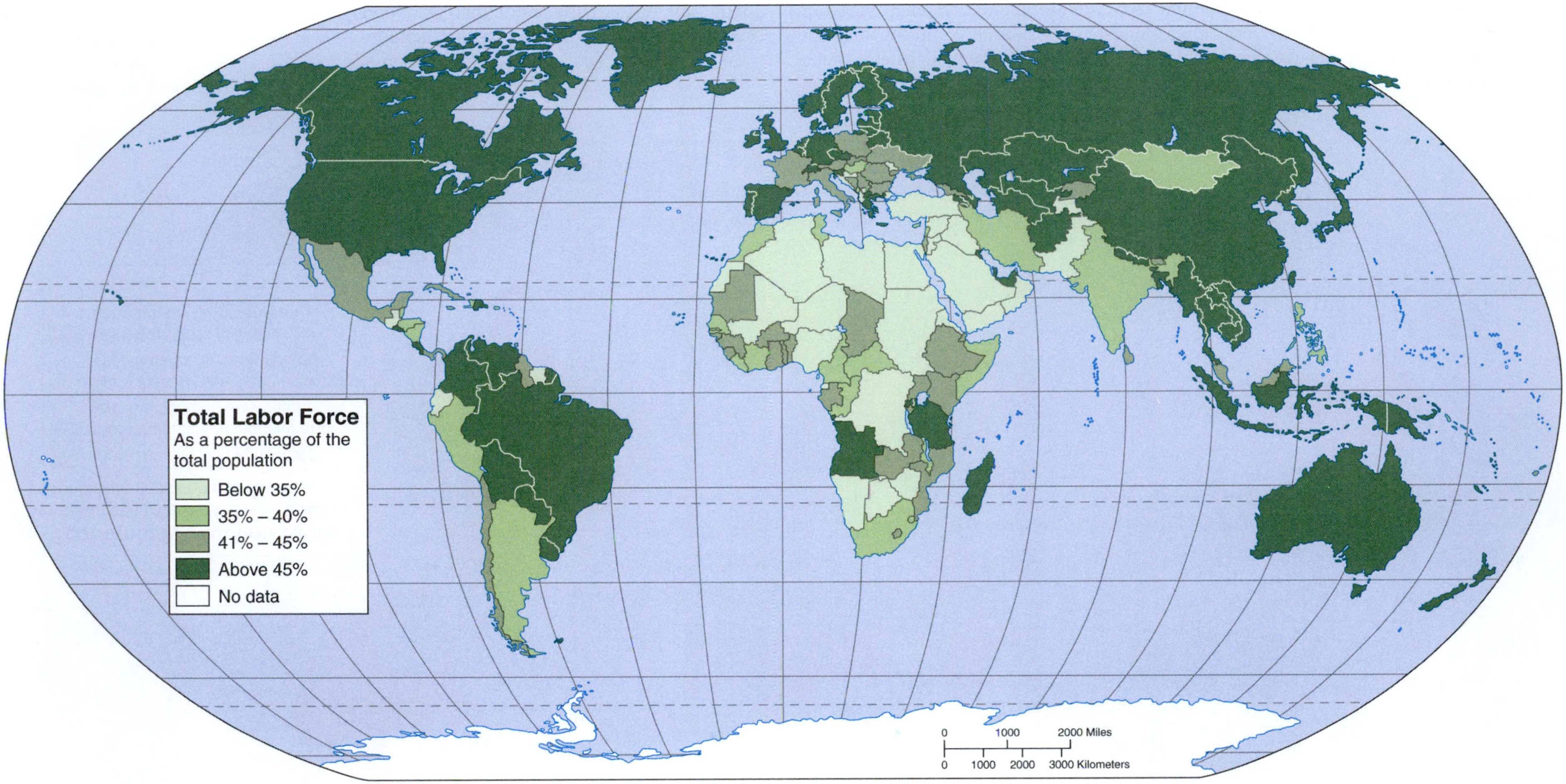

Total Labor Force

As a percentage of the
total population

- Below 35%
- 35% – 40%
- 41% – 45%
- Above 45%
- No data

The term labor force refers to the economically active portion of a population, that is, all people who work or are without work but are available for and are seeking work to produce economic goods and services. The total labor force thus includes both the employed and the unemployed (as long as they are actively seeking employment). Labor force is considered a better indicator of economic potential than employment/ unemployment figures, since unemployment figures will include experienced workers with considerable potential who are temporarily out of work. Unemployment figures will also incorporate persons seeking employment for the first time (many recent college graduates, for example). Generally, countries with higher percentages of total population within the labor force will be countries with higher levels of economic development. This is partly a function of levels of education and training and partly a function of the age distribution of populations. In developing countries, substantial percentages of the total population are too young to be part of the labor force. Also in developing countries a significant percentage of the population consists of women engaged in household activities or subsistence cultivation. These people seldom appear on lists of either employed or unemployed seeking employment and are the world's forgotten workers.

Map **58** Employment by Economic Activity

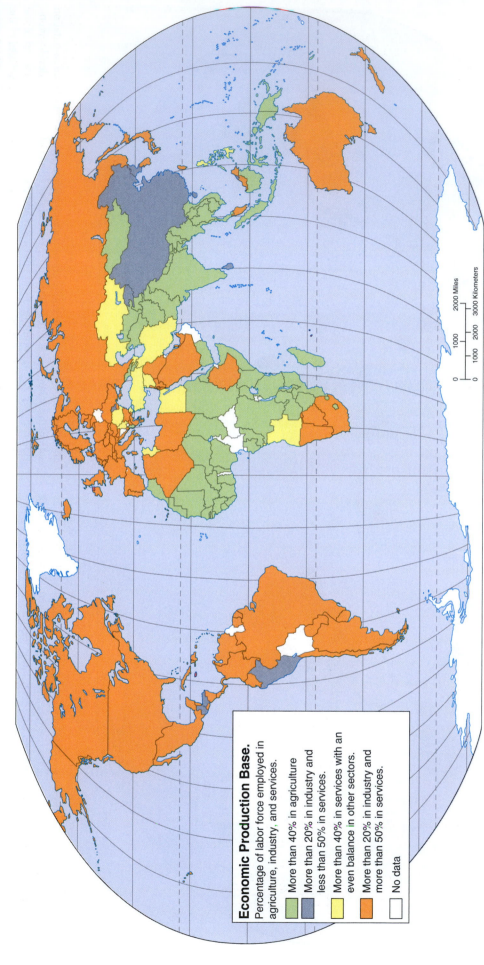

Economic Production Base.
Percentage of labor force employed in agriculture, industry, and services.

- More than 40% in agriculture
- More than 20% in industry and less than 50% in services.
- More than 40% in services with an even balance in other sectors.
- More than 20% in industry and more than 50% in services.
- No data

The employment structure of a country's population is one of the best indicators of the country's position on the scale of economic development. At one end of the scale are those countries with more than 40 percent of their labor force employed in agriculture. These are almost invariably the least developed, with high population growth rates, poor human services, significant environmental problems, and so on. In the middle of the scale are two types of countries: those with more than 20 percent of their labor force employed in industry and those with a fairly even balance among agricultural, industrial, and service employment but with at least 40 percent of their labor force employed in service activities. Generally, these countries have undergone the industrial revolution fairly recently and are still developing an industrial base while building up their service activities. This category also includes countries with a disproportionate share of their economies in service activities primarily related to resource extraction. On the other end of the scale from the agricultural economies are countries with more than 20 percent of their labor force employed in industry and more than 50 percent in service activities. These countries are, for the most part, those with a highly automated industrial base and a highly mechanized agricultural system (the "postindustrial," developed countries). They also include, particularly in Middle and South America and Africa, industrializing countries that are also heavily engaged in resource extraction as a service activity.

2000 Miles
3000 Kilometers

Map 59 Economic Output Per Sector

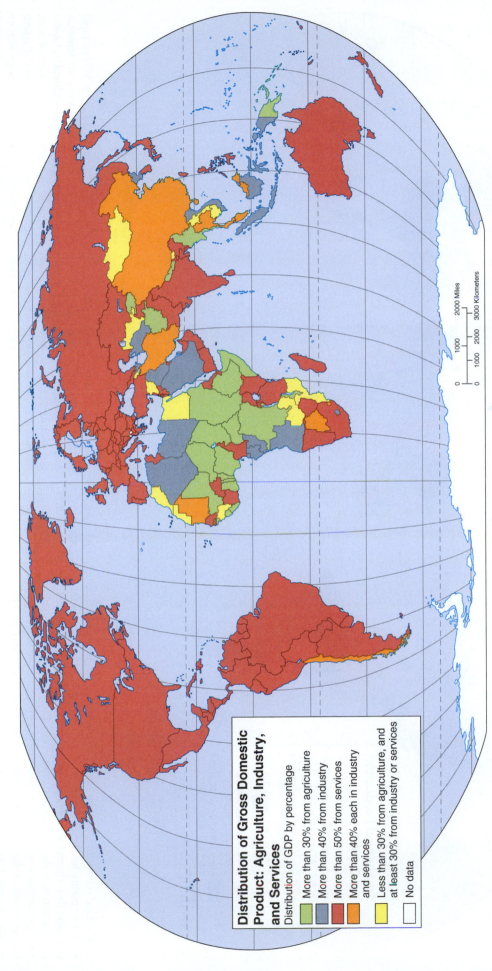

Distribution of Gross Domestic Product: Agriculture, Industry, and Services

Distribution of GDP by percentage

- More than 30% from agriculture
- More than 40% from industry
- More than 50% from services
- More than 40% each in industry and services
- Less than 30% from agriculture, and at least 30% from industry or services
- No data

The percentage of the gross domestic product (the final output of goods and services produced by the domestic economy, including net exports of goods and nonfactor— nonlabor, noncapital—services) that is devoted to agricultural, industrial, and service activities is considered a good measure of the level of economic development. In general, countries with more than 40 percent of their GDP derived from agriculture are still in a *colonial dependency economy*—that is, raising agricultural goods primarily for the export market and dependent upon that market (usually the richer countries). Similarly, countries with more than 40 percent of GDP devoted to both agriculture and services often emphasize resource extractive (primarily mining and forestry) activities. These also tend to be *colonial dependency* countries, providing raw materials for foreign markets. Countries with more than 40 percent of their GDP obtained from industry are normally well along the path to economic development. Countries with more than half of their GDP based on service activities fall into two ends of the development spectrum. On the one hand are countries heavily dependent upon both extractive activities and tourism and other low-level service functions. On the other hand are countries that can properly be termed *postindustrial*: they have already passed through the industrial stage of their economic development and now rely less on the manufacture of products than on finance, research, communications, education, and other service-oriented activities.

Map 60 Agricultural Production Per Capita

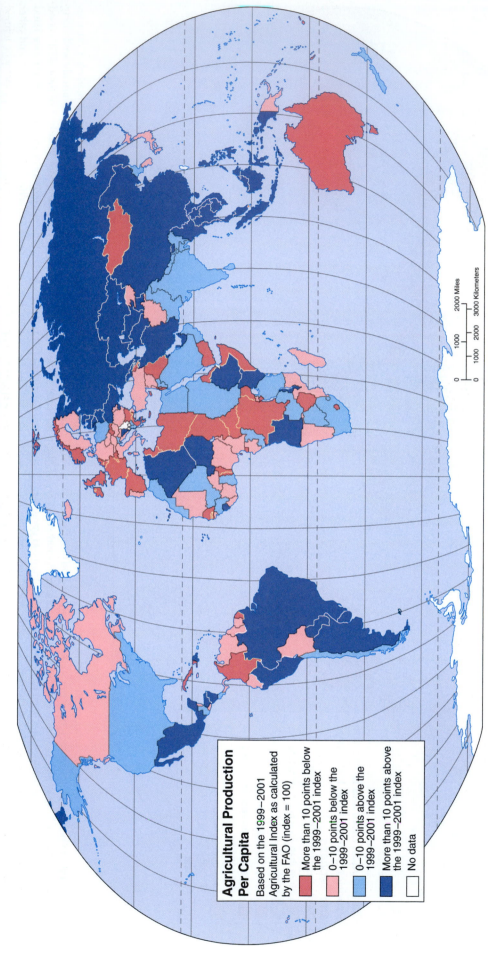

Agricultural Production Per Capita

Based on the 1999–2001 Agricultural Index as calculated by the FAO (index = 100)

- More than 10 points below the 1999–2001 index
- 0–10 points below the 1999–2001 index
- 0–10 points above the 1999–2001 index
- More than 10 points above the 1999–2001 index
- No data

Agricultural production includes the value of all crops and livestock products originating within a country for the base year of 2009. The index value portrays the disposable output (after deductions for livestock feed and seed for planting) of a country's agriculture in comparison with the base period 1999–2001. Thus, the production values show not only the relative ability of countries to produce food but also show whether or not that ability has increased or decreased. In general, global food production has kept up with or very slightly exceeded population growth. However, there are significant regional variations in the trend of food production keeping up with or surpassing population growth. For example, agricultural production in Africa and in Middle America has fallen, while production in South America, Asia, and Europe has risen. In the case of Africa, the drop in production

reflects a population growing more rapidly than agricultural productivity. Where rapid increases in food production per capita exist (as in certain countries in South America, Asia, and Europe), most often the reason is the development of new agricultural technologies that have allowed food production to grow faster than population. In much of Asia, for example, the so-called Green Revolution of new, highly productive strains of wheat and rice made positive index values possible. Also in Asia, the cessation of major warfare allowed some countries (Cambodia, Laos, and Vietnam) to show substantial increases over the 1999–2001 index. In some cases, a drop in production per capita reflects government decisions to limit production in order to maintain higher prices for agricultural products. Several countries in western Europe fall into this category.

-76-

Map 61 Exports of Primary Products

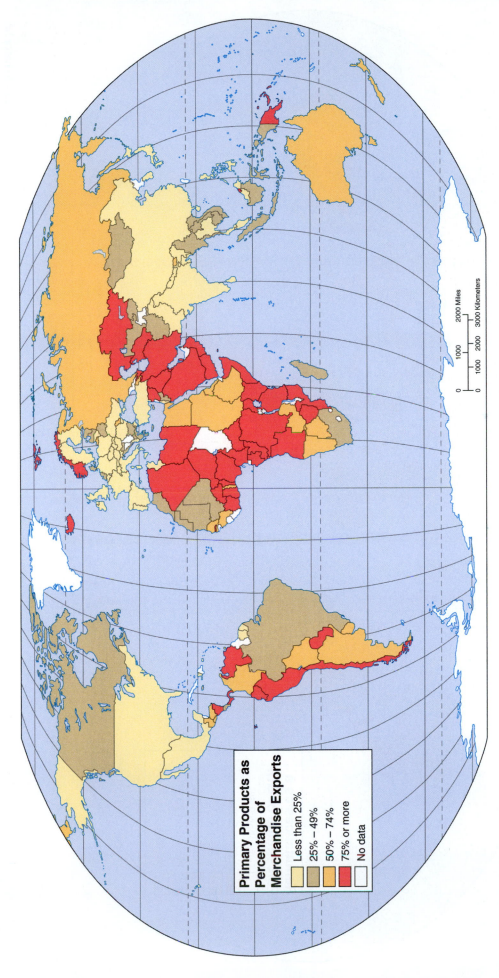

Primary Products as Percentage of Merchandise Exports

- Less than 25%
- 25% – 49%
- 50% – 74%
- 75% or more
- No data

Primary products are those that require additional processing before they enter the consumer market: metallic ores that must be converted into metals and then into metal products such as automobiles or refrigerators; forest products such as timber that must be converted to lumber before they become suitable for construction purposes; and agricultural products that require further processing before being ready for human consumption. It is an axiom in international economics that the more a country relies on primary products for its export commodities, the more vulnerable its economy is to market fluctuations. Those countries with only primary products to export are hampered in their economic growth. A country dependent on only one or two products for export revenues is unprotected from economic shifts, particularly a changing market demand for its products. Imagine what would happen to the thriving economic status of the oil-exporting states of the Persian Gulf, for example, if an alternate source of cheap energy were found. A glance at this map shows that those countries with the lowest levels of economic development tend to be concentrated on primary products and, therefore, have economies that are especially vulnerable to economic instability.

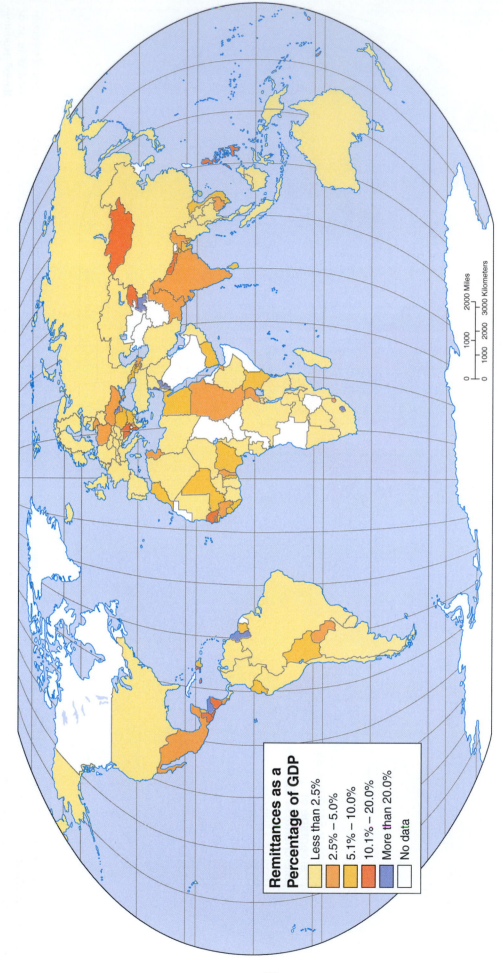

Map 62 Remittances

Remittances as a Percentage of GDP

- Less than 2.5%
- 2.5% – 5.0%
- 5.1% – 10.0%
- 10.1% – 20.0%
- More than 20.0%
- No data

0   1000   2000 Miles
0   1000   2000   3000 Kilometers

As seen in Map 27, the wealthy countries of the world also have sizeable immigrant populations. In countries where economic pull factors are particularly strong, there may also be a large population of foreign workers who ultimately may return to the home country. Whether the move to a new country is temporary or permanent, the ties to the home country oftentimes remain strong. Economic ties can be particularly strong, as demonstrated by remittances. A remittance simply is the transfer of money from an individual to his or her home country. During the mid-2000s remittances exceeded $250 billion. In some smaller countries like Honduras and Tajikistan, remittances comprise more than a quarter of GDP. As would be expected, the bulk of this money is coming from workers employed in North America, Western Europe, and Japan.

Map 63   Membership in the World Trade Organization

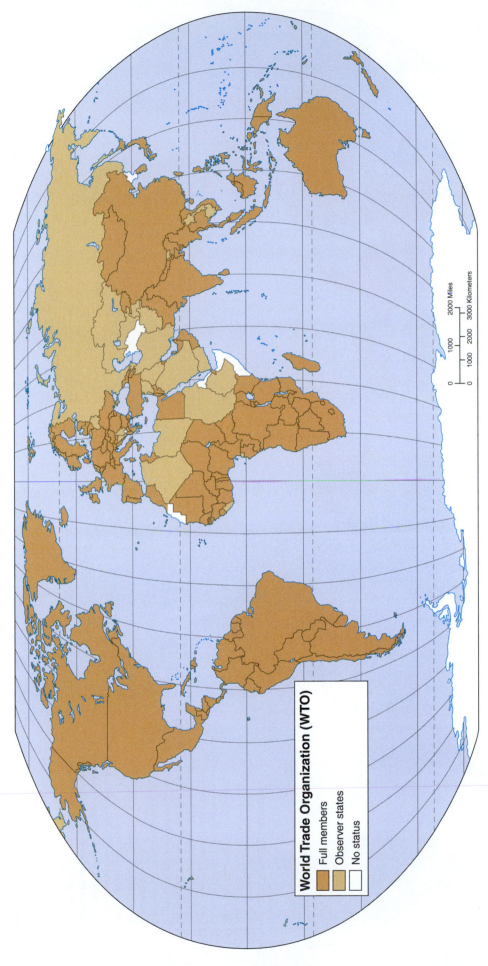

World Trade Organization (WTO)

- Full members
- Observer states
- No status

2000 Miles

1000   3000 Kilometers

1000   2000

0   1000   2000

0

After World War II, the General Agreement on Tariffs and Trade (GATT) sponsored several rounds of negotiations, especially related to lower tariffs but also considering issues such as dumping and other nontariff questions. The last round of negotiations under GATT took place in Uruguay in 1986–1994 and set the stage for the World Trade Organization (WTO), which was formally established in 1995. Today the WTO has 153 members with more than 30 "observer" governments. With the exception of the Holy See (Vatican), observer governments are expected to begin negotiations for full membership within five years of becoming observers. The objective of the WTO is to help international trade flow smoothly and fairly and to assure more stable and secure supplies of goods to consumers. To this end, it administers trade agreements, acts as a forum for trade negotiations, settles trade disputes, reviews national trade policies, assists developing countries through technical assistance and training programs, and cooperates with other international organizations. Increased globalization of the world's economy makes the administrative role of the WTO of increasing importance in the twenty-first century. The headquarters of the WTO is in Geneva, Switzerland.

Map 64 Regional Trade Organizations

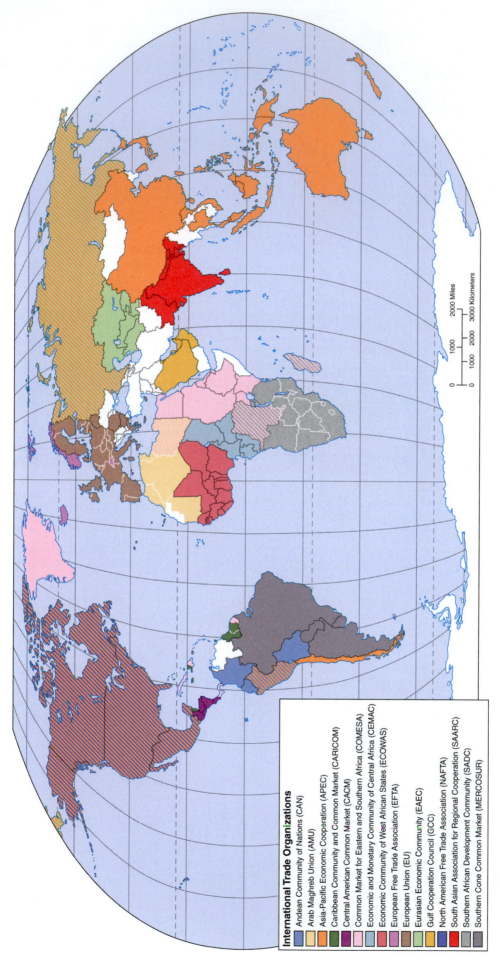

International Trade Organizations

Andean Community of Nations (CAN)
Arab Maghreb Union (AMU)
Asia-Pacific Economic Cooperation (APEC)
Caribbean Community and Common Market (CARICOM)
Central American Common Market (CACM)
Common Market for Eastern and Southern Africa (COMESA)
Economic and Monetary Community of Central Africa (CEMAC)
Economic Community of West African States (ECOWAS)
European Free Trade Association (EFTA)
European Union (EU)
Eurasian Economic Community (EAEC)
Gulf Cooperation Council (GCC)
North American Free Trade Association (NAFTA)
South Asian Association for Regional Cooperation (SAARC)
Southern African Development Community (SADC)
Southern Cone Common Market (MERCOSUR)

0   1000   2000 Miles
0   1000   2000   3000 Kilometers

One of the most pervasive influences in the global economy over the last half century, particularly since the end of the Cold War, has been that of international trade organizations. Pioneered by the European Economic Community, founded in part to assist in rebuilding the European economy after World War II, these organizations have become major players in global movements of goods, services, and labor. Some have integrated to form financial and political unions, such as the European Community, which has grown into the European Union. Others, like the Asia-Pacific Economic Cooperation, and the Southern African

Development Community, incorporate vastly different regions, states, and even economic systems, and are attempts to anticipate the direction of future economic growth. The role of international trade organizations is likely to grow greater in the next fifty years. In terms of purchasing power parity GDP, the North American Free Trade Agreement (NAFTA) is the largest trade bloc in the world and has restructured many segments of the Canadian, American, and Mexican economies (particularly in Mexico).

Map 65 Dependence on Trade

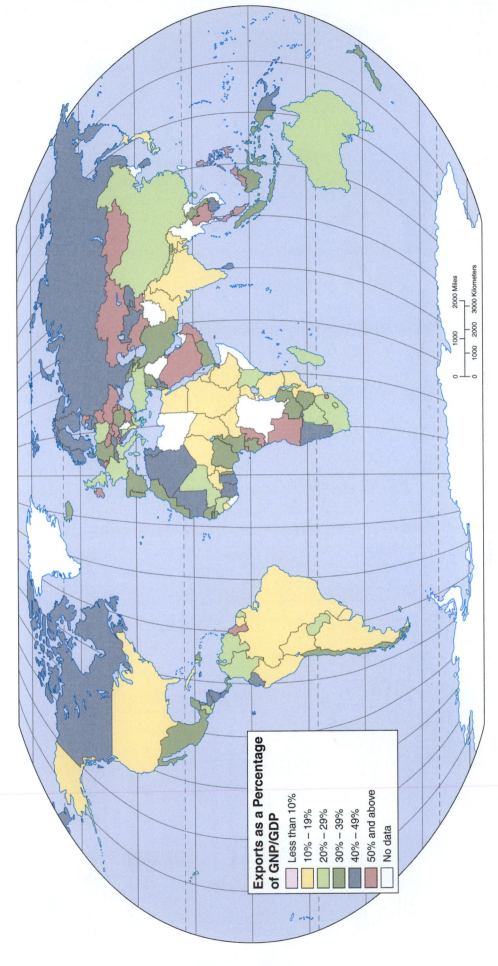

Exports as a Percentage of GNP/GDP

- Less than 10%
- 10% — 19%
- 20% — 29%
- 30% — 39%
- 40% — 49%
- 50% and above
- No data

2000 Miles

1000   2000   3000 Kilometers

0   1000   2000

0

As the global economy becomes more and more a reality, the economic strength of virtually all countries is increasingly dependent upon trade. For many developing nations, with relatively abundant resources and limited industrial capacity, exports provide the primary base upon which their economies rest. Even countries like the United States, Japan, and Germany, with huge and diverse economies, depend on exports to generate a significant percentage of their employment and wealth. Without imports, many products that consumers want would be unavailable or more expensive; without exports, many jobs would be eliminated.

Map 66 Trade with Neighboring Countries

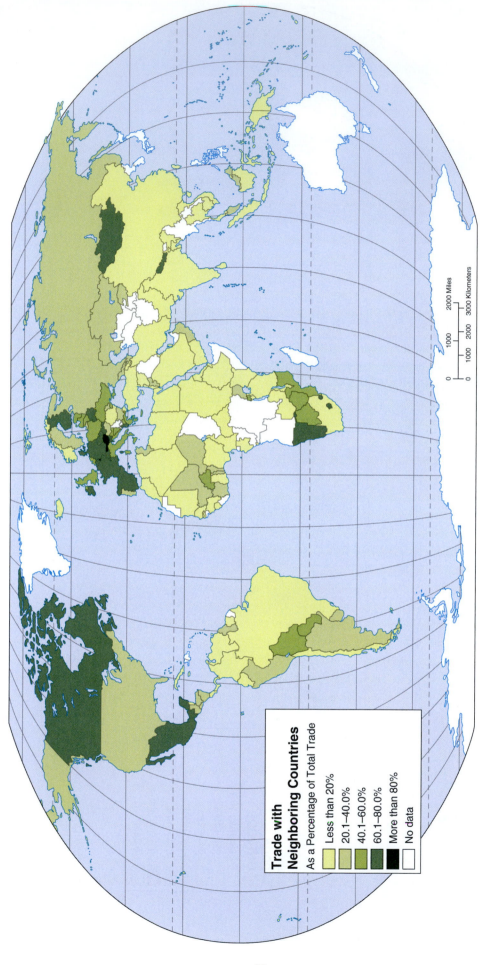

Trade with Neighboring Countries

As a Percentage of Total Trade

- Less than 20%
- 20.1–40.0%
- 40.1–60.0%
- 60.1–80.0%
- More than 80%
- No data

0   1000   2000 Miles

0   1000   2000   3000 Kilometers

Two important factors influencing international trade are proximity and relative wealth of trading partners. Proximity is important because transportation costs can be a large part of an economic activity. Relative wealth is an important influence on the goods and services a country may produce for export. Of course, the percentage of trade with neighboring countries isn't necessarily a strong indicator of a country's wealth or future growth, but the patterns exhibited on this map are telling. The world's countries are increasingly interconnecting through regional trade agreements (see Map 64), and the influence of connectivity among EU and NAFTA states is evident here. Much of Africa, South Asia, and South America trades more with more distant countries such as the United States, Japan, and EU states. As regional trade blocs mature, look for the levels of cross-border trade to increase.

-82-

Map 67   The Indebtedness of States

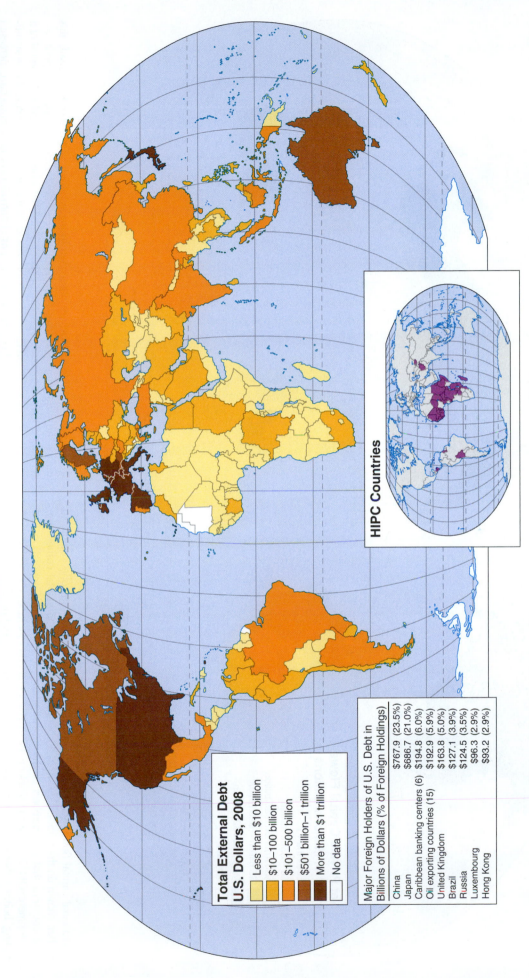

Total External Debt U.S. Dollars, 2008

- Less than $10 billion
- $10–100 billion
- $101–500 billion
- $501 billion–1 trillion
- More than $1 trillion
- No data

Major Foreign Holders of U.S. Debt in Billions of Dollars (% of Foreign Holdings)

China	$767.9 (23.5%)
Japan	$686.7 (21.0%)
Caribbean banking centers (6)	$194.8 (6.0%)
Oil exporting countries (15)	$192.9 (5.9%)
United Kingdom	$163.8 (5.0%)
Brazil	$127.1 (3.9%)
Russia	$124.5 (3.5%)
Luxembourg	$96.3 (2.9%)
Hong Kong	$93.2 (2.9%)

HIPC Countries

External debt is money or credit owed to foreign lenders. It generally is comprised of bonds and treasury bills (in the case of the United States) that are sold to foreign lenders and money owed to banks, governments, and international financial institutions. External debt is highly fluid and many countries of the world have "sustainable debt"—where a country can meet its debt service obligations. Other countries, particularly those in the developing world, have levels of debt, typically to international financial institutions such as the IMF and World Bank, that are beyond the governments' ability to repay. These countries typically are in a constant "catch-up" situation in which they fall increasingly further into debt and cannot meet their obligations without receiving partial or total forgiveness of debt. The IMF and World Bank have jointly put together debt reduction packages for 35 of the 41 countries identified as Heavily Indebted Poor Countries (HIPC). As of 2009 the level of structured debt relief approached $50 billion. Approximately one-third of the U.S. national debt is owed to foreign governments, with China and Japan accounting for over 44 percent of those holdings.

Map 68 Global Flows of Investment Capital

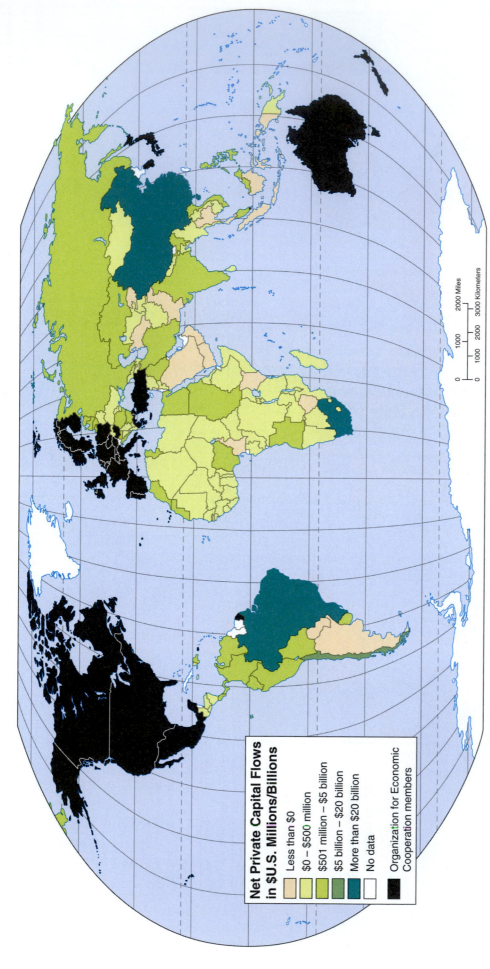

Net Private Capital Flows in $U.S. Millions/Billions

- Less than $0
- $0 – $500 million
- $501 million – $5 billion
- $5 billion – $20 billion
- More than $20 billion
- No data
- Organization for Economic Cooperation members

International capital flows include private debt and nondebt flows from one country to another, shown on the map as flows into a country. Nearly all of the capital comes from those countries that are members of the Organization for Economic Cooperation and Development (OECD), shown in black on the map. Capital flows include commercial bank lending, bonds, other private credits, foreign direct investment, and portfolio investment. Most of these flows are indicators of the increasing influence developed countries exert over the developing economies. Foreign direct investment or FDI, for example, is a measure of the net inflow of investment monies used to acquire long-term management interest in businesses located somewhere other than in the economy of the investor. Usually this means the acquisition of at least 10 percent of the stock of a company by a foreign investor and is, then, a measure of what might be termed "economic colonialism": control of a region's economy by foreign investors that could, in the world of the future, be as significant as colonial political control was in the past. International capital flows have increased greatly in the last decade as the result of the increasing liberalization of developing countries, the strong economic growth exhibited by many developing countries, and the falling costs and increased efficiency of communication and transportation services.

Map 69 Aiding Economic Development

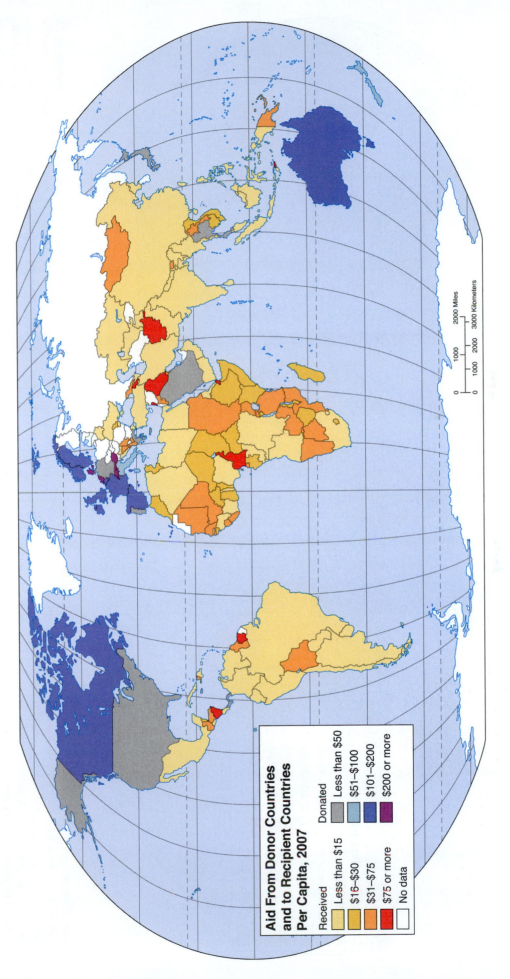

**Aid From Donor Countries
and to Recipient Countries
Per Capita, 2007**

Received	Donated
Less than $15	Less than $50
$16–$30	$51–$100
$31–$75	$101–$200
$75 or more	$200 or more
No data	

Over the last few years, official development assistance to developing countries from the member countries of the Organisation for Economic Co-operation and Development has risen dramatically, with the United States as the world's number one donor country, giving over 25 percent of the total of development assistance. Development assistance—or "foreign aid," as it is sometimes called—is widely recognized as benefiting both the donor and the recipient. Developing countries that increase their levels of per capita income through economic development have more money to spend on products from more highly developed countries. Increased development increases the capacity to foster not just economic but political change. In some parts of the world, such as Sub-Saharan Africa, foreign aid is the largest single source of external finance, far exceeding foreign investments.

Map 70 The Cost of Consumption

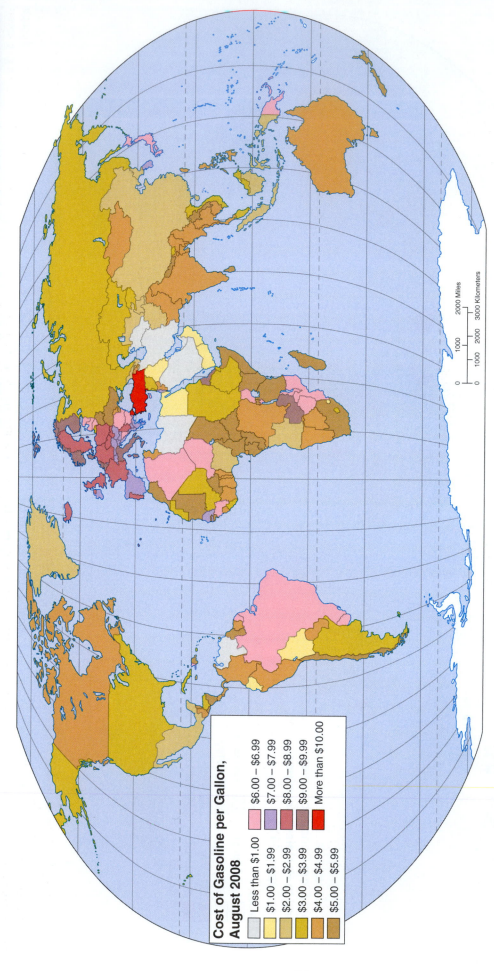

Cost of Gasoline per Gallon, August 2008

Less than $1.00	$6.00 – $6.99
$1.00 – $1.99	$7.00 – $7.99
$2.00 – $2.99	$8.00 – $8.99
$3.00 – $3.99	$9.00 – $9.99
$4.00 – $4.99	More than $10.00
$5.00 – $5.99	

The year 2008 brought massive increases in the price of gasoline at the pump, creating considerable consternation among the American driving population, in particular. It is one thing to point out that gasoline prices are and historically have been considerably higher in Europe than in North America. But it is another to note that European spatial patterns of places of work and places of residence are significantly different than in North America. The North American metropolitan area evolved its spatial patterns in conjunction with the rise of privately owned automobiles; European cities, on the other hand, had spatial patterns well established centuries or even millennia before the automobile emerged as a common mode of transportation. As a consequence, such things as the journey to work are very different for many Europeans who can walk from where they live to where they work than it is for Americans and Canadians who often live considerable distances from their places of employment. A daily commute of 75 miles one way would not be considered unusual in America. In Europe it is virtually unheard of. There is also the component of the scale of organization of human activities: because of the very large country in which Americans live, their spatial movements are customarily more extensive than those of Europeans who live in countries the size of American states. So, yes, gasoline is much more expensive in Europe than in North America. But in North America, the increase in the cost of gasoline to and above $4.00 per gallon produces significantly more economic impact than proportionally similar increases in Europe. Similarly, the reduction in gasoline prices by the end of 2008 had a greater impact in North America—although its impact was overshadowed by larger forces.

Map 71 Energy Production Per Capita

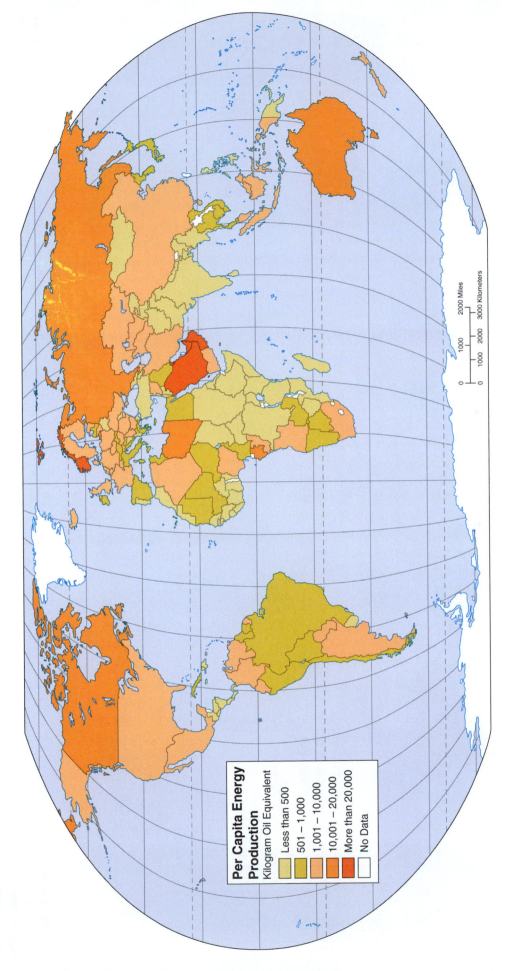

Per Capita Energy Production

Kilogram Oil Equivalent

- Less than 500
- 501 – 1,000
- 1,001 – 10,000
- 10,001 – 20,000
- More than 20,000
- No Data

Energy production per capita is a measure of the availability of mechanical energy to assist people in their work. This map shows the amount of all kinds of energy—solid fuel (primarily coal), liquid fuel (primarily petroleum), natural gas, geothermal, wind, solar, hydroelectric, nuclear, waste recycling, and indigenous heat pumps—produced per person in each country. With some exceptions, wealthier countries produce more energy per capita than poor ones. Countries such as Japan and many European states rank among the world's wealthiest, but are energy-poor and produce relatively little of their own energy.

They have the ability, however, to pay for imports. On the other hand, countries such as those of the Persian Gulf or the oil-producing states of Central and South America may rank relatively low on the scale of economic development but rank high as producers of energy. In many poor countries, especially in Central and South America, Africa, South Asia, and East Asia, large proportions of energy come from traditional fuels such as firewood and animal dung. Indeed for many in the developing world, the real energy crisis is a shortage of wood for cooking and heating.

Map 72 Energy Consumption Per Capita

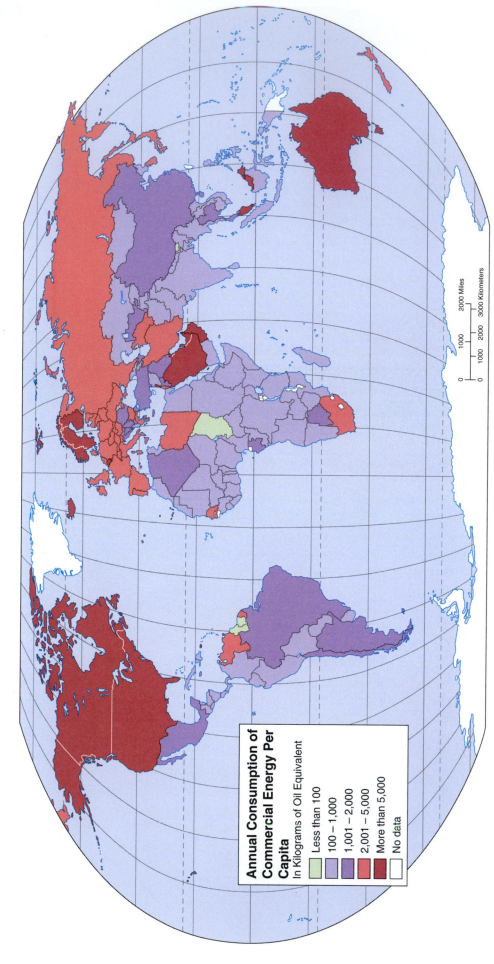

Annual Consumption of Commercial Energy Per Capita
In Kilograms of Oil Equivalent

- Less than 100
- 100 – 1,000
- 1,001 – 2,000
- 2,001 – 5,000
- More than 5,000
- No data

Of all the quantitative measures of economic well-being, energy consumption per capita may be the most expressive. All of the countries defined by the World Bank as having high incomes consume at least 100 gigajoules of commercial energy (the equivalent of about 3.5 metric tons of coal) per person per year, with some, such as the United States and Canada, having consumption rates in the 300 gigajoule range (the equivalent of more than 10 metric tons of coal per person per year). With the exception of the oil-rich Persian Gulf states, where consumption figures include the costly "burning off" of excess energy in the form of natural gas flares at wellheads, most of the highest-consuming countries are in the Northern Hemisphere, concentrated in North America and Western Europe. At the other end of the scale are low-income countries, whose consumption rates are often less than one percent of those of the United States and other high consumers. These figures do not, of course, include the consumption of noncommercial energy—the traditional fuels of firewood, animal dung, and other organic matter—widely used in the less developed parts of the world.

-88-

Map 73 Energy Dependency

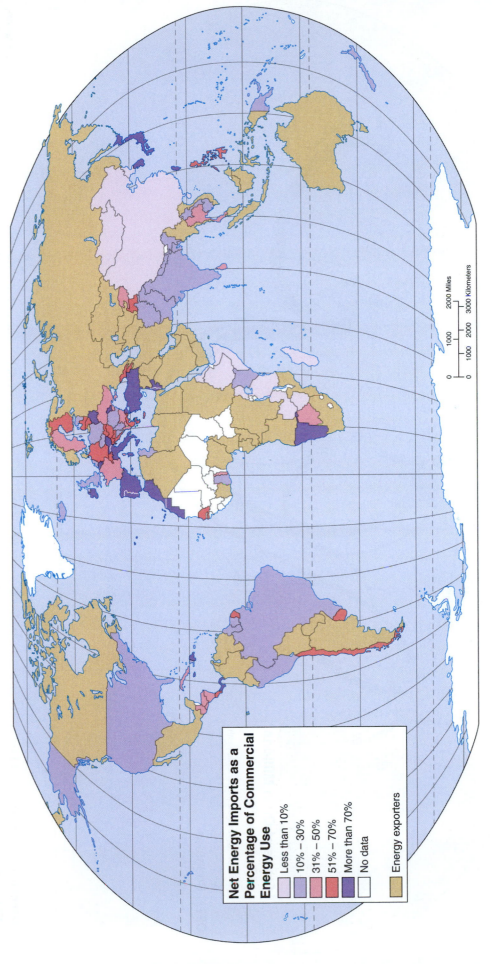

Net Energy Imports as a Percentage of Commercial Energy Use

- Less than 10%
- 10% – 30%
- 31% – 50%
- 51% – 70%
- More than 70%
- No data
- Energy exporters

The patterns on the map show dependence on commercial energy before transformation to other end-use fuels such as electricity or refined petroleum products; energy from traditional sources such as fuelwood or dried animal dung is not included. Energy dependency is the difference between domestic consumption and domestic production of commercial energy and is most often expressed as a net energy import or export. A few of the world's countries are net exporters of energy; most are importers. The growth in global commercial energy use over the last decade indicates growth in the modern sectors of the economy—industry, transportation, and urbanization—particularly in the lesser developed countries. Still, the primary consumers of energy—and those having the greatest dependence on foreign sources of energy—are the more highly developed countries of Europe, North America, and Japan.

Map 74   Flows of Oil

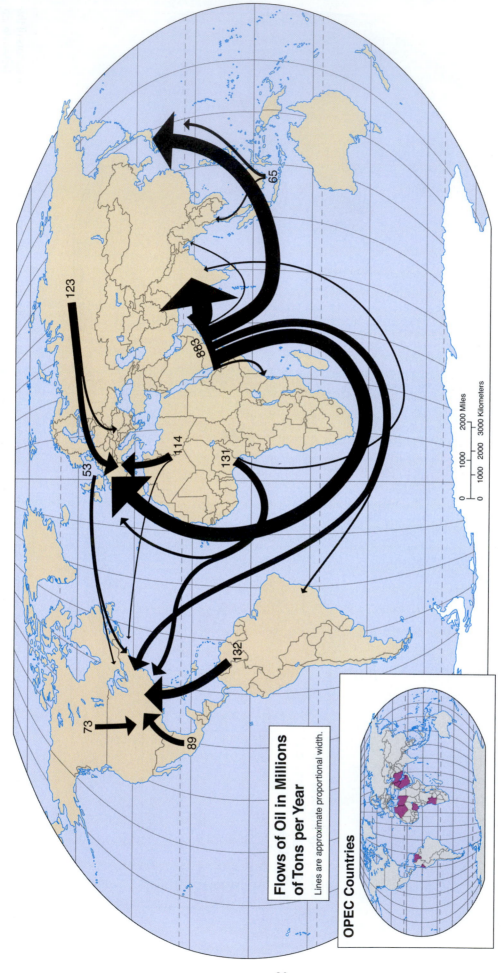

Flows of Oil in Millions of Tons per Year

Lines are approximate proportional width.

OPEC Countries

The pattern of oil movements from producing region to consuming region is one of the dominant facts of contemporary international maritime trade. Supertankers carry a million tons of crude oil and charge rates in excess of $0.10 per ton per mile, making the transportation of oil not only a necessity for the world's energy-hungry countries, but also an enormously profitable proposition. One of the major negatives of these massive oil flows is the damage done to the oceanic ecosystems—not just from the well-publicized and dramatic events like the 2010 *Deepwater Horizon* oil spill but from the incalculable amounts of oil from leakage, scrubbings, purgings, and so on, which are a part of the oil transport technology. As seen above, much of the supply of the world's oil comes from countries belonging to the Organization of the Petroleum Exporting Countries (OPEC). In 2009, OPEC members controlled nearly two-thirds of the world's known oil reserves and over one-third of the world's production. It is clear from the map that the primary recipients of these oil flows are the world's most highly developed economies.

Map 75   A Wired World: Internet Users

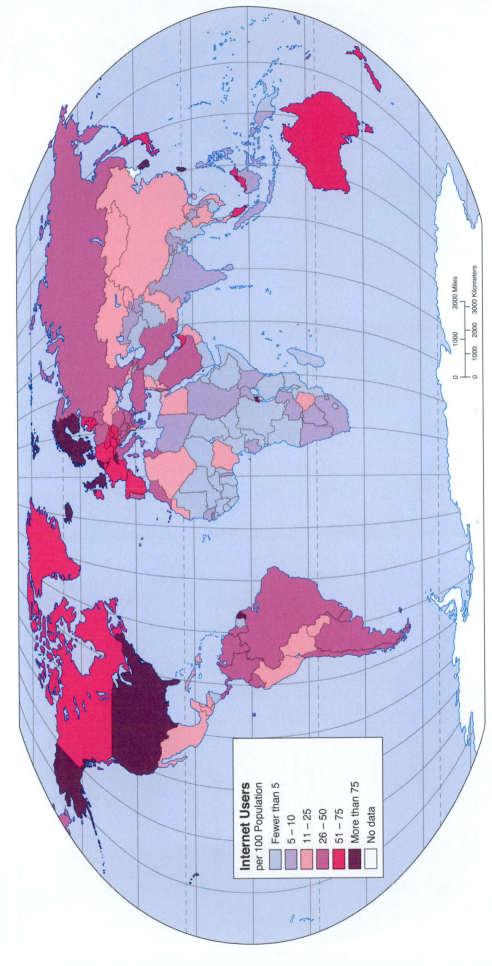

Internet Users
per 100 Population

- Fewer than 5
- 5 – 10
- 11 – 25
- 26 – 50
- 51 – 75
- More than 75
- No data

It is interesting to contemplate that a short quarter of a century ago, such a map could not have been created. The emergence of immediate, long-distance connectivity via the Internet has been one of the most important components of globalization. We now live, as author Thomas Friedman has noted, in a "flat world" where lines of connection are more important than distance and where virtually instantaneous connections have altered—perhaps forever—the way that we do business, exchange information, and transform our cultures. Some of the recent transformations we have seen in the emergence of countries like China and India as major players in the international economy are, in part, the consequence of access to the Internet. Originally conceived as a quick way for academics to exchange information, the Internet has become a cultural and social phenomenon that far exceeds its original purpose. Whether or not this will result in an eventual benefit to human well-being remains to be seen. Does the benefit of quick communication result in a cost to the complexity and richness of human cultures worldwide?

Map 76 Personal Computer Use

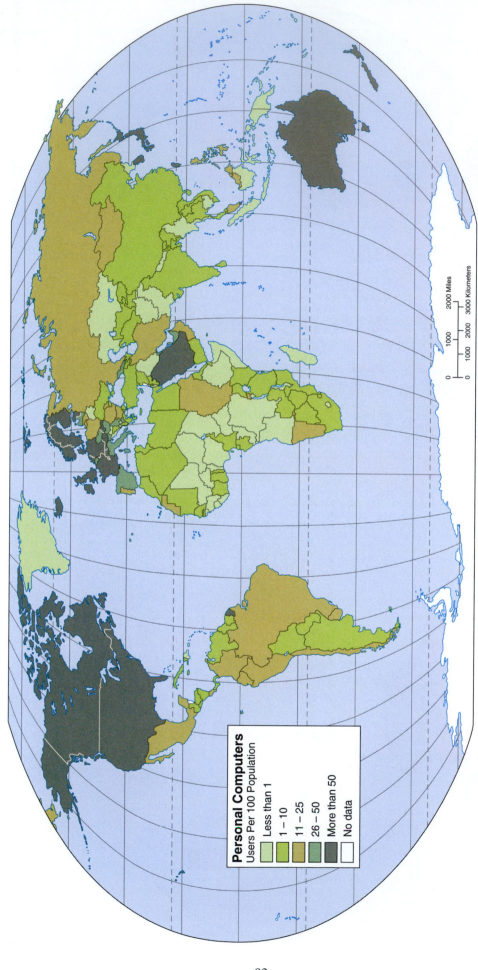

Personal Computers
Users Per 100 Population

- Less than 1
- 1 – 10
- 11 – 25
- 26 – 50
- More than 50
- No data

0   1000   2000 Miles
0 1000 2000 3000 Kilometers

The Internet connections shown in the previous map would not be possible without personal computers—or at least not possible at their present scale. But personal computers do a great deal more than simply act as high-speed transmitters of information. They are incredibly powerful devices for the storage and analysis of data and are becoming, seemingly exponentially, even more powerful. When the use of mainframe computers became common on university campuses in the 1960s, they were used primarily for faculty and graduate student research. Now, an undergraduate can run on his or her laptop computer—in a matter of seconds—a program that would have required hours to run on an institutional computer that occupied spaces similar to those required for medium-sized classrooms. Certainly computers have changed the way that businesses are run. But they have, perhaps, changed the daily lives of personal computer users even more.

Map 77   Traditional Links: The Telephone

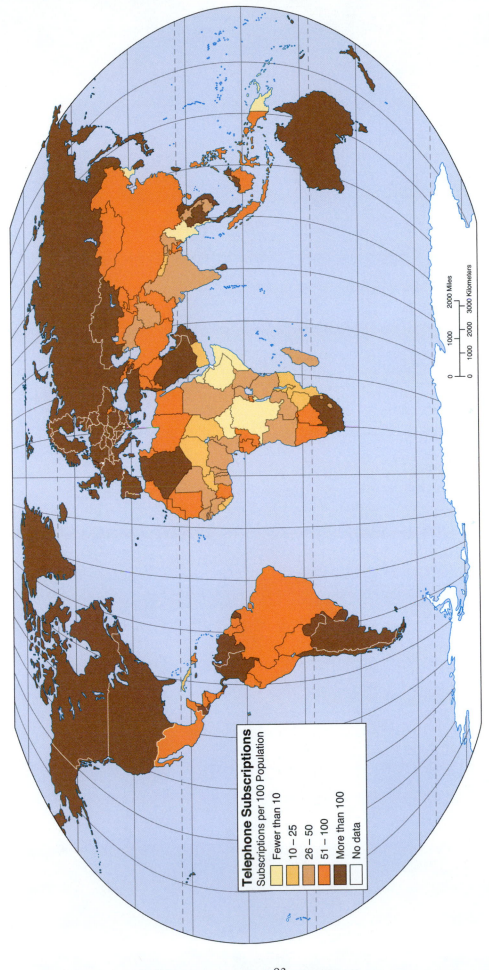

Telephone Subscriptions
Subscriptions per 100 Population

- Fewer than 10
- 10 – 25
- 26 – 50
- 51 – 100
- More than 100
- No data

Not all of the world's communications take place via computers and the Internet. A lot of person-to-person connection is still carried out via the telephone, and access to telephone connections is perhaps as good an indication of economic development or, more important, the potential for economic development as anything else. The map clearly shows the prevalence of telephone connectivity in the developed world. But it also shows an increasingly high degree of access to telephones in major countries in the developing world, such as India and China. If these countries are to continue to develop their economies at the pace of the last decade, then their degree of communication—including of person-to-person connection via the telephone, and access to telephones—will also have to increase. The data shown on this map include users of both land lines and cellular phones. By the end of 2008, for the first time, the number of cellular phone users worldwide exceeded the number of those using the traditional land lines. As cellular phone complexity increases to include e-mail and other computer functions, the gap between cellular users and traditional phone users can be expected to widen.

Unit V

Global Patterns of Environmental Disturbance

Map 78 Global Air Pollution: Sources and Wind Currents

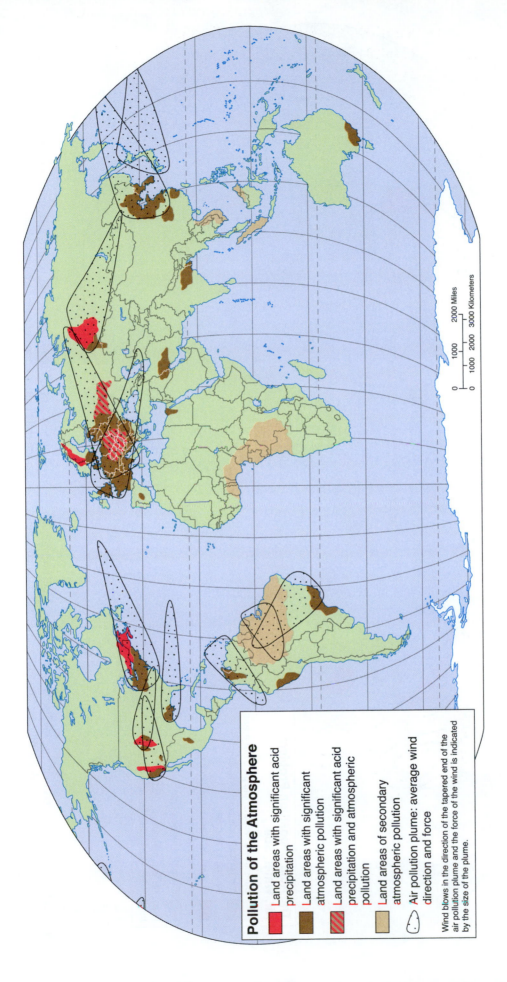

Pollution of the Atmosphere

- ▮ Land areas with significant acid precipitation
- ▮ Land areas with significant atmospheric pollution
- ▨ Land areas with significant acid precipitation and atmospheric pollution
- ▮ Land areas of secondary atmospheric pollution
- ⬭ Air pollution plume: average wind direction and force

Wind blows in the direction of the tapered end of the air pollution plume and the force of the wind is indicated by the size of the plume.

Almost all processes of physical geography begin and end with the flows of energy and matter among land, sea, and air. Because of the primacy of the atmosphere in this exchange system, air pollution is potentially one of the most dangerous human modifications in environmental systems. Pollutants such as various oxides of nitrogen or sulfur cause the development of acid precipitation, which damages soil, vegetation, and wildlife and fish. Air pollution in the form of smog is often dangerous for human health. And most atmospheric scientists believe that the efficiency of the atmosphere in retaining heat—the so-called greenhouse effect—is being enhanced by increased carbon dioxide, methane, and other gases produced by agricultural and industrial activities. The result, they fear, will be a period of global warming that will dramatically alter climates in all parts of the world.

0   1000   2000 Miles
0   1000   2000   3000 Kilometers

Map 79 The Acid Deposition Problem: Air, Water, Soil

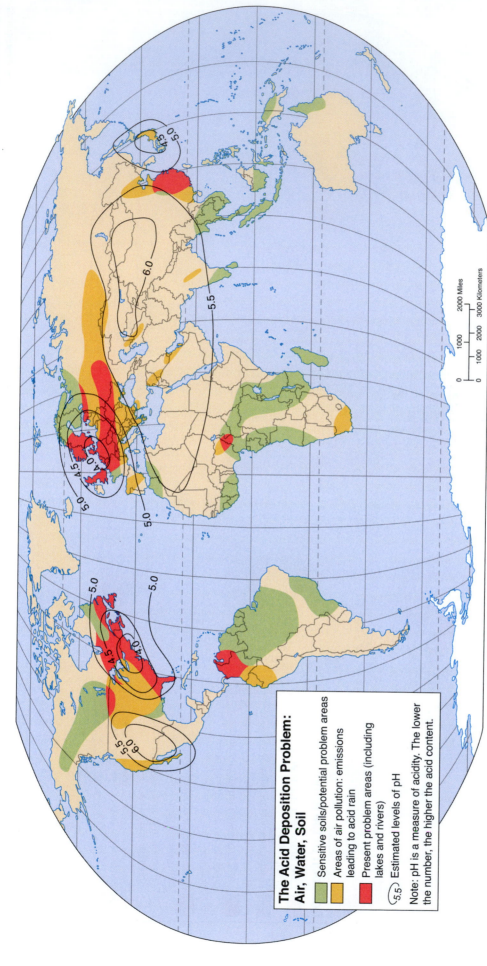

The Acid Deposition Problem: Air, Water, Soil

- Sensitive soils/potential problem areas
- Areas of air pollution: emissions leading to acid rain
- Present problem areas (including lakes and rivers)
- ⌒5.5⌒ Estimated levels of pH

Note: pH is a measure of acidity. The lower the number, the higher the acid content.

The term "acid precipitation" refers to increasing levels of acidity in snowfall and rainfall caused by atmospheric pollution. Oxides of nitrogen and sulfur resulting from incomplete combustion of fossil fuels (coal, oil, and natural gas) combine with water vapor in the atmosphere to produce weak acids that then "precipitate" or fall along with water or ice crystals. Some atmospheric acids formed by this process are known as "dry-acid" precipitates and they too will fall to earth, although not necessarily along with rain or snow. In some areas of the world, the increased acidity of streams and lakes stemming from high levels of acid precipitation or dry acid fallout has damaged or destroyed aquatic life. Acid precipitation and dry acid fallout also harm soil systems and vegetation, producing a characteristic burned appearance in forests that lends the same quality to landscapes that forest fires would. The region most dramatically impacted by acid precipitation is Central Europe where decades of destructive environmental practices, including the burning of high sulfur coal for commercial, industrial, and residential purposes, has produced the destruction of hundreds of thousands of acres of woodlands—a phenomenon described by the German foresters who began their study of the area following the lifting of the Iron Curtain as "Waldsterben": Forest Death.

Map 80 Major Polluters and Common Pollutants

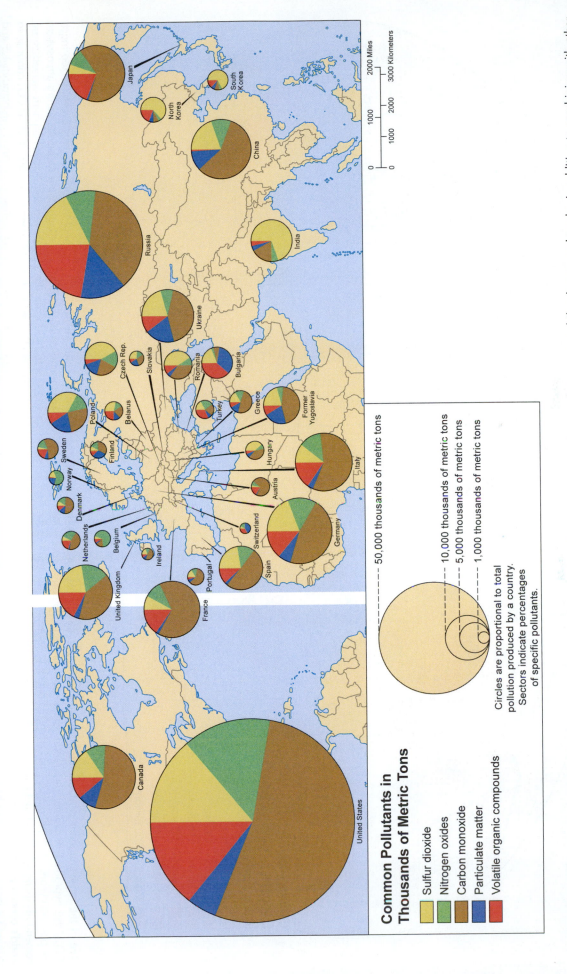

Common Pollutants in Thousands of Metric Tons

- Sulfur dioxide
- Nitrogen oxides
- Carbon monoxide
- Particulate matter
- Volatile organic compounds

50,000 thousands of metric tons

10,000 thousands of metric tons

5,000 thousands of metric tons

1,000 thousands of metric tons

Circles are proportional to total pollution produced by a country. Sectors indicate percentages of specific pollutants.

More than 90 percent of the world's total of anthropogenic (human-generated) air pollutants come from the heavily populated industrial regions of North America, Europe, South Asia (primarily in India), and East Asia (mainly in China, Japan, and the two Koreas). This map shows the origins of the five most common pollutants: sulfur dioxide, nitrogen oxide, carbon monoxide, particulate matter, and volatile organic compounds. These substances are produced both by industry and by the combustion of fossil fuels that generate electricity and power trains, planes, automobiles, buses, and trucks. In addition to combining with other components of the atmosphere and with one another to produce smog, they are the chief ingredients in acid accumulations in the atmosphere, which ultimately result in acid deposition, either as acid precipitation or dry acid fallout. Like other forms of pollutants, these air pollutants do not recognize political boundaries, and regions downwind of major polluters receive large quantities of pollutants from areas over which they often have no control.

Map 81   Global Carbon Dioxide Emissions

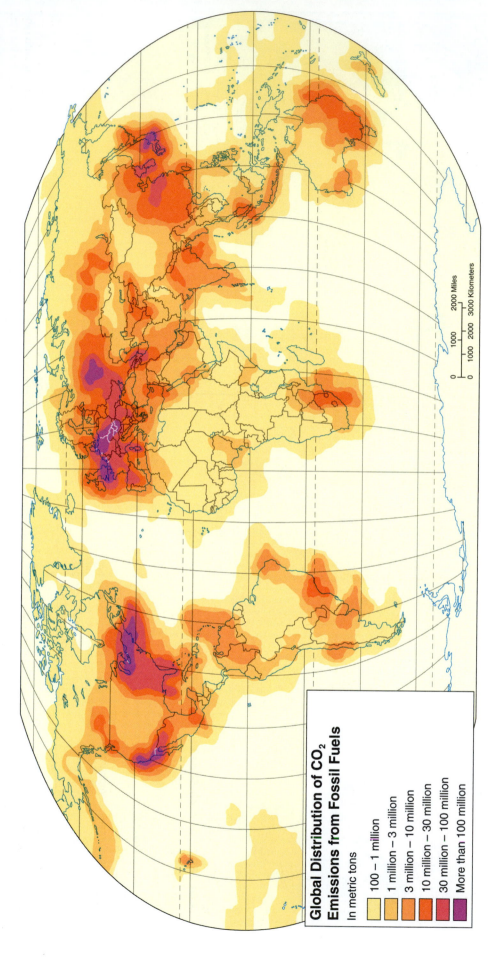

Global Distribution of CO₂ Emissions from Fossil Fuels

In metric tons

- 100 – 1 million
- 1 million – 3 million
- 3 million – 10 million
- 10 million – 30 million
- 30 million – 100 million
- More than 100 million

0   1000   2000 Miles
0   1000   2000   3000 Kilometers

One of the most important components of the atmosphere is the gas carbon dioxide (CO₂), the byproduct of animal respiration, decomposition, and combustion. During the past 200 years, atmospheric CO₂ has risen dramatically, largely as the result of the tremendous increase in fossil fuel combustion brought on by the industrialization of the world's economy and the burning and clearing of forests by the expansion of farming. While CO₂ by itself is relatively harmless, it is an important "greenhouse gas." The gases in the atmosphere act like the panes of glass in a greenhouse roof, allowing light in but preventing heat from escaping. The greenhouse capacity of the atmosphere is crucial for

organic life and is a purely natural component of the global energy cycle. But too much CO₂ and other greenhouse gases such as methane could cause the earth's atmosphere to warm up too much, producing the global warming that atmospheric scientists are concerned about. Researchers estimate that if greenhouse gases such as CO₂ continue to increase at their present rates, the earth's mean temperature could rise between 1.5 and 4.5 degrees Celsius by the middle of the present century. Such a rise in global temperatures would produce massive alterations in the world's climate patterns.

Map 82   The Earth Warms, 1976–2006

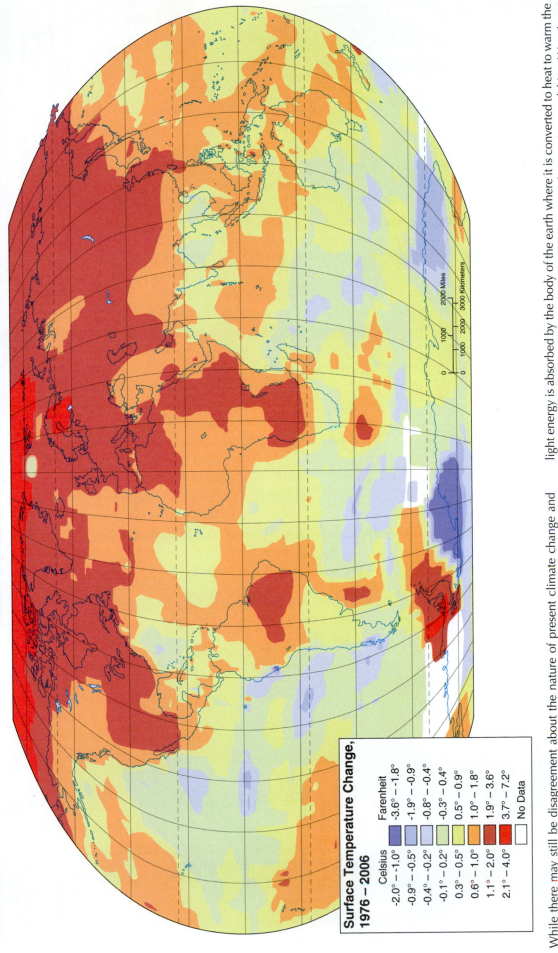

**Surface Temperature Change,
1976 – 2006**

Celsius	Farenheit
-2.0° – -1.0°	-3.6° – -1.8°
-0.9° – -0.5°	-1.9° – -0.9°
-0.4° – -0.2°	-0.8° – -0.4°
-0.1° – -0.2°	-0.3° – -0.4°
0.3° – 0.5°	0.5° – 0.9°
0.6° – 1.0°	1.0° – 1.8°
1.1° – 2.0°	1.9° – 3.6°
2.1° – 4.0°	3.7° – 7.2°
No Data	

While there may still be disagreement about the nature of present climate change and its link to human activities, that disagreement is largely political and economic, rather than scientific. The evidence from a generation of scientific studies of the atmosphere is compelling: the world is warming more rapidly than at any time in the past and the bulk of this warming trend is the consequence of human alterations in the atmospheric proportion of the "greenhouse gases," such as carbon dioxide and methane, that help to trap heat radiated from the earth and release it back into space more slowly. The atmosphere is a massive heat engine: short wave (light) energy from the sun passes through the atmosphere where some of it is absorbed and converted to heat; a larger percentage of the light energy is absorbed by the body of the earth where it is converted to heat to warm the atmosphere. The "greenhouse effect," as this process is known, is a good thing. Were it not for the greenhouse effect, life on Earth as we know it would be impossible. The temperature changes shown on this map are the result of an accelerating greenhouse effect and, in turn, cause changes in precipitation patterns (see next map), storms of greater intensity and frequency, and rising sea levels. The Earth has experienced these changes before. What makes the current changes different is that they are occurring over decades rather than over millennia, making it very difficult (if not impossible) for natural and human systems to adapt to them.

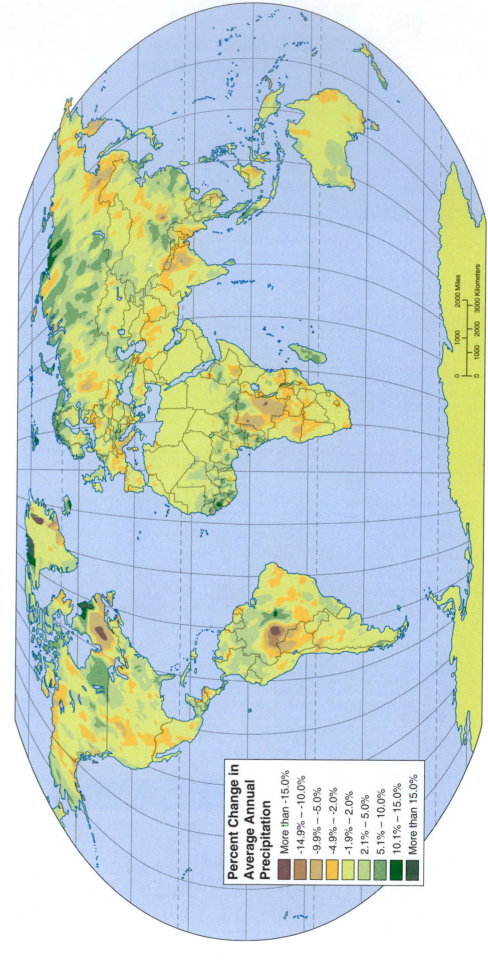

Map 83 Global Precipitation Changes, 1976–2006

Percent Change in Average Annual Precipitation

- More than -15.0%
- -14.9% – -10.0%
- -9.9% – -5.0%
- -4.9% – -2.0%
- -1.9% – 2.0%
- 2.1% – 5.0%
- 5.1% – 10.0%
- 10.1% – 15.0%
- More than 15.0%

The links between global temperature change and alterations in the geographical distribution of precipitation are so complex as to be only marginally understood. What can be said with certainty is that increasing global temperatures will produce altered precipitation patterns. We have already begun to experience some of this in the form of prolonged droughts in the interior regions of North America; increased precipitation in the previously semi-arid region of the Sahel in Africa; and severe and long-term drought in the southern portions of the Amazon Basin, adjacent areas in South America, and in those sections of Africa south of the Sahel. There is an approximate tendency for dry regions to become drier and for wet regions to become wetter with increasing temperatures, although this is not universally the case. The primary problem is that persistent, long-term changes in precipitation patterns will cause significant disruptions in established agricultural patterns such as those in the United States, China, India, and Brazil—four of the world's most productive agricultural countries.

Map 84 Potential Global Temperature Change

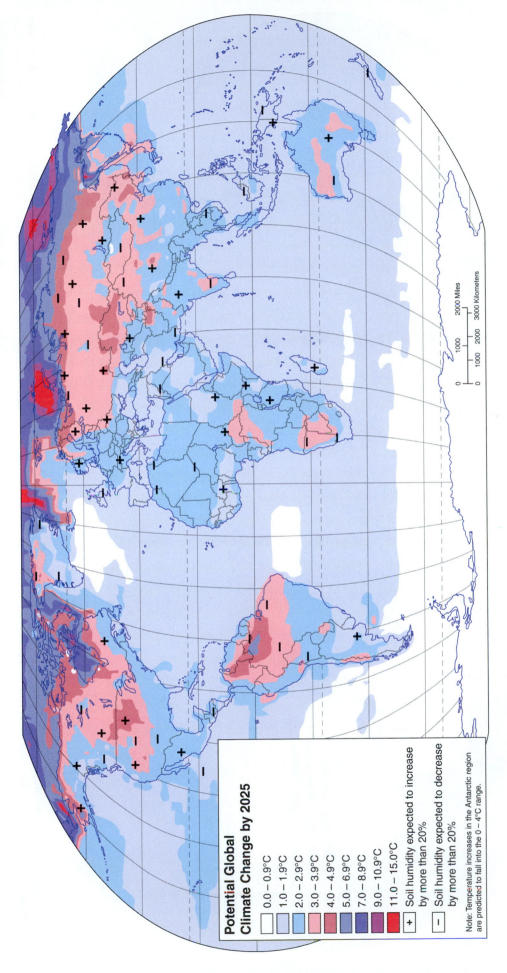

**Potential Global
Climate Change by 2025**

- 0.0 – 0.9°C
- 1.0 – 1.9°C
- 2.0 – 2.9°C
- 3.0 – 3.9°C
- 4.0 – 4.9°C
- 5.0 – 6.9°C
- 7.0 – 8.9°C
- 9.0 – 10.9°C
- 11.0 – 15.0°C

+   Soil humidity expected to increase
by more than 20%

−   Soil humidity expected to decrease
by more than 20%

Note: Temperature increases in the Antarctic region
are predicted to fall into the 0 – 4°C range.

0   1000   2000 Miles

0   1000   2000   3000 Kilometers

By the end of the first quarter of the twenty-first century, the world's population will probably be facing climatic conditions quite different from those of the last quarter of the twentieth century: significantly increased temperatures, particularly in the higher latitudes (closer to the poles), and significantly altered precipitation patterns with some of the world's great agricultural regions experiencing long-term and persistent drought. The increase of temperatures in the higher latitudes means greater melting of the Arctic polar ice sheet (ice floating on water), greater melting of the Greenland ice sheet (ice of incredible thickness now resting on land), and greater melting of the Antarctic ice sheet—similar in structure but larger than that of Greenland. There are some benefits of polar ice melting

(as long as you are not a polar bear), most notably the opening of sea lanes across the Arctic Ocean—the final establishment of the Northwest Passage sought for by European explorers for several centuries. But there are many more hazards in the form of even small increases in sea level—a meter or two would be enough to produce a major increase in damage by even moderate tropical storms, let alone hurricanes. And many low-lying coastal and island communities run the risk of complete inundation by rising sea waters. Is all this inevitable just because a computer model such as the one that generated the above map predicts it? No. But ignoring the evidence gathered and analyzed so carefully by objective scientists would be both foolhardy and arrogant.

-103-

Map 85 Water Resources: Availability of Renewable Water Per Capita

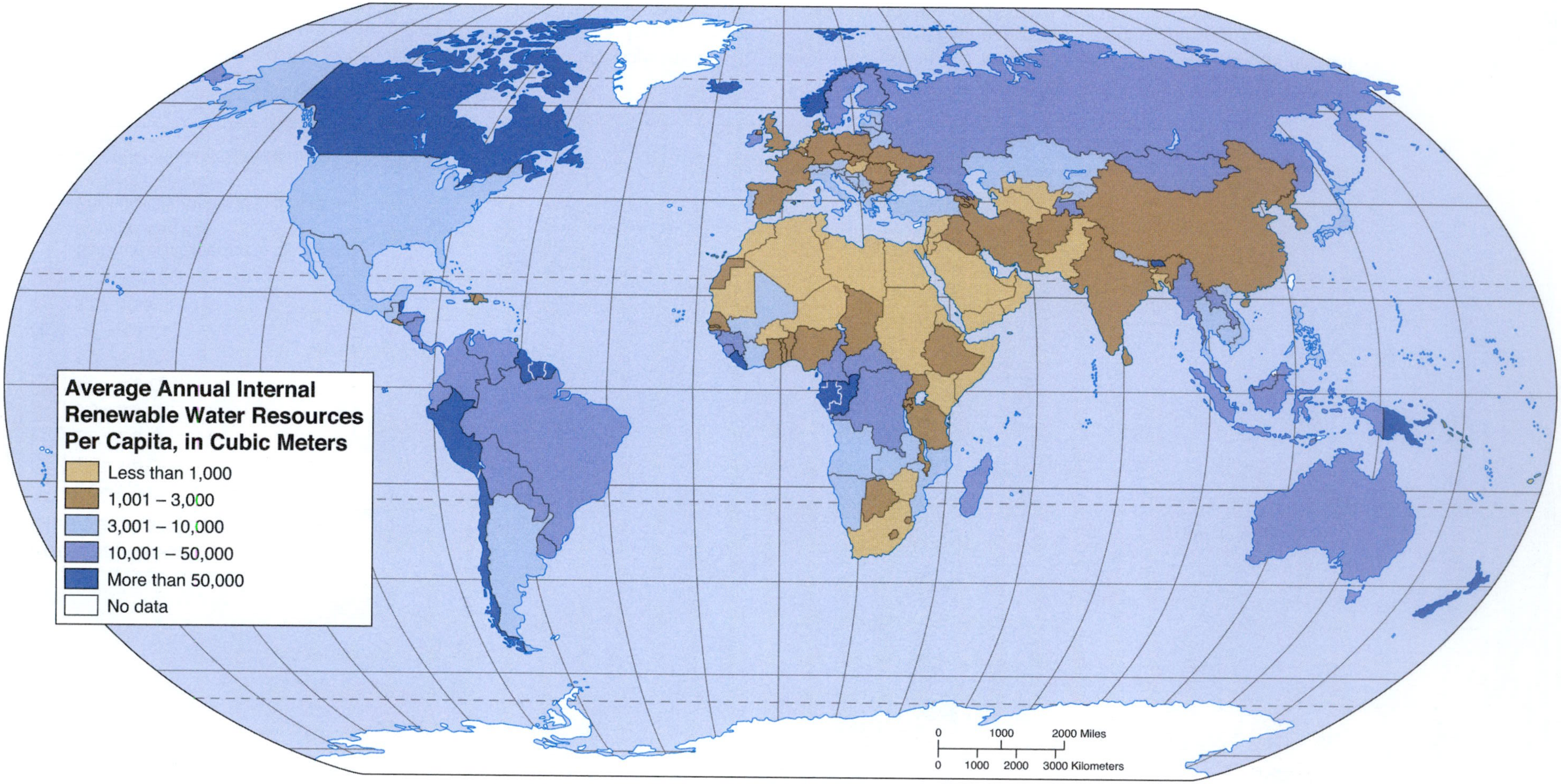
Average Annual Internal Renewable Water Resources Per Capita, in Cubic Meters
Less than 1,000
1,001 – 3,000
3,001 – 10,000
10,001 – 50,000
More than 50,000
No data
0 1000 2000 Miles
0 1000 2000 3000 Kilometers

Renewable water resources are usually defined as the total water available from streams and rivers (including flows from other countries), ponds and lakes, and groundwater storage or aquifers. Not included in the total of renewable water would be water that comes from such nonrenewable sources as desalinization plants or melted icebergs. While the concept of renewable or flow resources is a traditional one in resource management, in fact, few resources, including water, are truly renewable when their use is excessive. The water resources shown here are indications of that principle. A country like the United States possesses truly enormous quantities of water. But the United States also uses enormous quantities of water. The result is that, largely because of excessive use, the availability of renewable water is much less than in many other parts of the world where the total supply of water is significantly less.

Map 86 Water Resources: Annual Withdrawal Per Capita

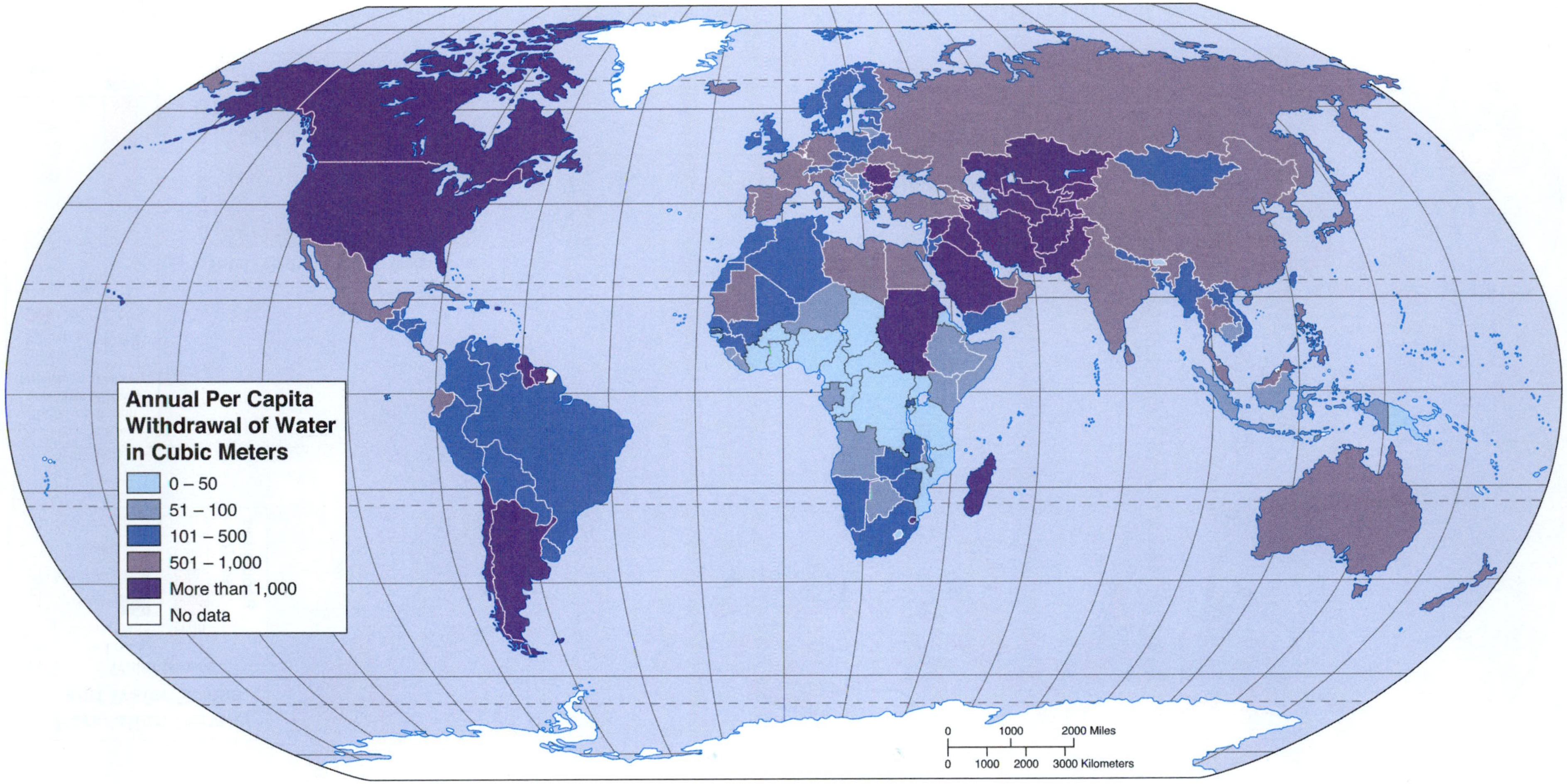
Annual Per Capita Withdrawal of Water in Cubic Meters
0 – 50
51 – 100
101 – 500
501 – 1,000
More than 1,000
No data
0 1000 2000 Miles
0 1000 2000 3000 Kilometers

Water resources must be viewed like a bank account in which deposits and withdrawals are made. As long as the deposits are greater than the withdrawals, a positive balance remains. But when the withdrawals begin to exceed the deposits, sooner or later (depending on the relative sizes of the deposits and withdrawals) the account becomes overdrawn. For many of the world's countries, annual availability of water is insufficient to cover the demand. In these countries, reserves stored in groundwater are being tapped, resulting in depletion of the water supply (think of this as shifting money from a savings account to a checking account). The water supply can maintain its status as a renewable resource only if deposits continue to be greater than withdrawals, and that seldom happens. In general, countries with high levels of economic development and countries that rely on irrigation agriculture are the most spendthrift when it comes to their water supplies.

Map 87 Water Stress: Shortage, Abundance, and Population Density

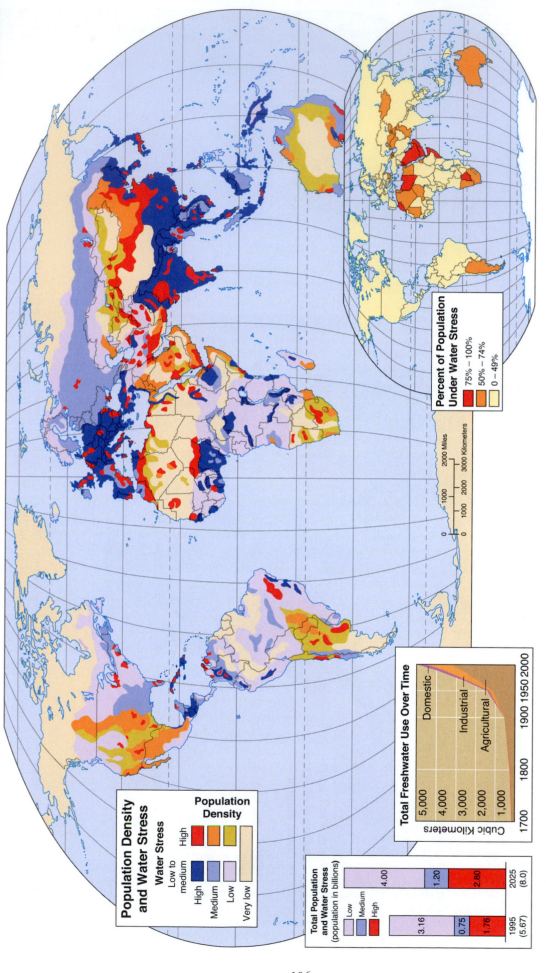

Population Density and Water Stress

Water Stress

Population Density

High	Low to medium
High	
Medium	
Low	
Very low	

Total Population and Water Stress
(population in billions)

Low
Medium
High

	1995 (5.67)	2025 (8.0)
Low	3.16	4.00
Medium	0.75	1.20
High	1.76	2.80

Total Freshwater Use Over Time

Cubic Kilometers

5,000
4,000
3,000
2,000
1,000

Domestic
Industrial
Agricultural

1700 1800 1900 1950 2000

Percent of Population Under Water Stress

75% – 100%
50% – 74%
0 – 49%

0 1000 2000 Miles

0 1000 2000 3000 Kilometers

Maps such as the previous two, based on national-level data for water consumption and availability, should be used only to obtain national-level understanding. Information on water withdrawal and availability are regionally and locally based geographic phenomena and are linked not just with water supplies but with the density of human populations. Even areas (such as New England in the United States) in which water availability is high and withdrawal rates are relatively low show areas of stress in regions of high population density (cities such as Boston). This map, originally produced by scientists at the University of New Hampshire, attempts to show those areas of the world where populations will tend to be at high, medium, and low risk of stress because of water availability. It is important to note that many of the world's prime agricultural regions, such as the Great Plains of the United States or the Argentine Pampas, show the potential for high risk of water stress in the immediate future. Why is this important? Because the greatest single use of water on the planet is for irrigation (nearly 70 percent of the world's water use) and it is the continued expansion of irrigation systems that allows the increase in agricultural production that feeds the Earth's more than 6 billion persons.

-106-

Map 88   Water Stress: The Ogallala Aquifer

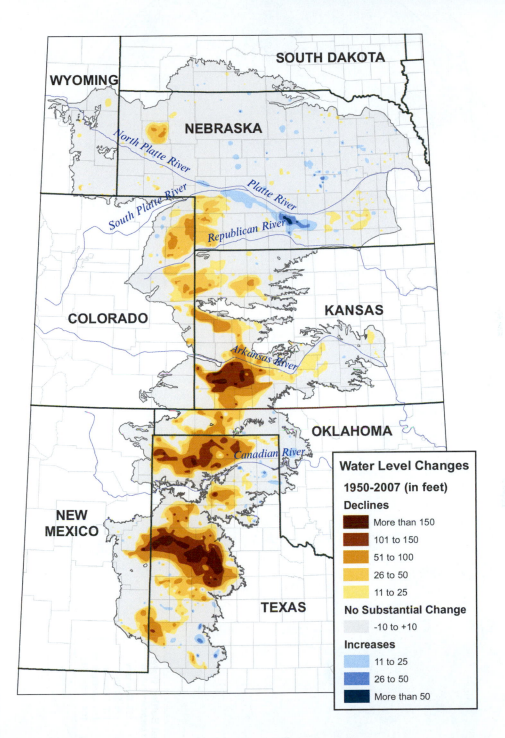

An aquifer is an underground water source, usually consisting of a porous medium of gravel, sand, and bedrock, filled with water trapped by a layer of impermeable stone. Aquifers may feed rivers and lakes through springs. Water is also withdrawn from aquifers for human uses, particularly irrigation. An aquifer is recharged by precipitation and stream run-off. Most aquifers represent thousands of years of accumulation of stored water. The Ogallala aquifer underlies 174,000 square miles of the Great Plains and has total water storage about equal to that of Lake Huron. Although it is a major source of water for agriculture, industry, and human consumption, it is irrigation agriculture that is the dominant use. When aquifers such as the Ogallala are tapped for irrigation, however, the rates of withdrawal far exceed the rates of recharge. Since the beginning of major use of the Ogallala aquifer in the 1950s, water storage has dropped by nearly ten percent and there have been water level declines of more than 150 feet in the Texas panhandle and southwestern Kansas. Although resource managers have often thought of groundwater as a renewable resource, in fact it is just as nonrenewable in the human timeframe as an oil field or iron mine. When the Ogallala aquifer is depleted, it will take tens of thousands of years to restore it.

Map 89 A New Environmental Stress? Consumption of Bottled Water

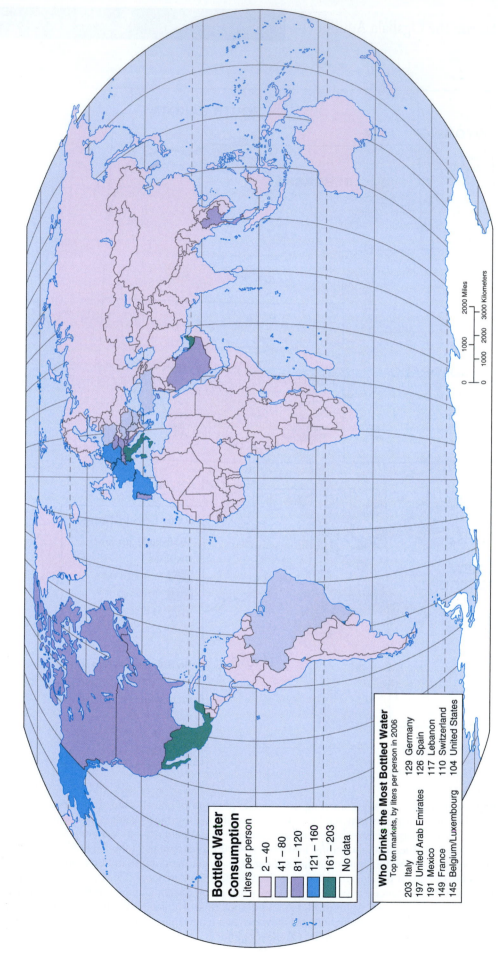

Bottled Water Consumption
Liters per person

- 2 – 40
- 41 – 80
- 81 – 120
- 121 – 160
- 161 – 203
- No data

Who Drinks the Most Bottled Water
Top ten markets, by liters per person in 2006

203	Italy	129	Germany
197	United Arab Emirates	126	Spain
191	Mexico	117	Lebanon
149	France	110	Switzerland
145	Belgium/Luxembourg	104	United States

0   1000   2000 Miles
0 1000 2000 3000 Kilometers

It is probably a truism that most people who commonly use bottled water seldom think of the environmental costs of doing so. Bottled water consumption tends to correlate spatially with higher levels of income (there are, of course, exceptions) and the highest rates of bottled water consumption tend to be in North America and Europe. The consumption of bottled water is one of the most costly environmental practices developed within the last quarter century. Most bottled water comes in plastic bottles that are seldom recycled (it has been estimated that in the United States alone, over 60 million plastic water bottles are simply thrown away each day). Enormous quantities of petroleum are required to produce and ship water bottles and bottled water is more costly, ounce-for-ounce, than gasoline—even when gasoline exceeds $4.00 per gallon. There are much less costly alternatives, including reusable plastic bottles with built-in filters that can be used to hold 200 gallons of water before the filters need to be replaced. In much of the developed world (where the bulk of bottled water is consumed), the easiest solution is to simply drink water from the tap—which is where, advertising notwithstanding, most bottled water comes from originally anyway.

-108-

Map 90   Pollution of the Oceans

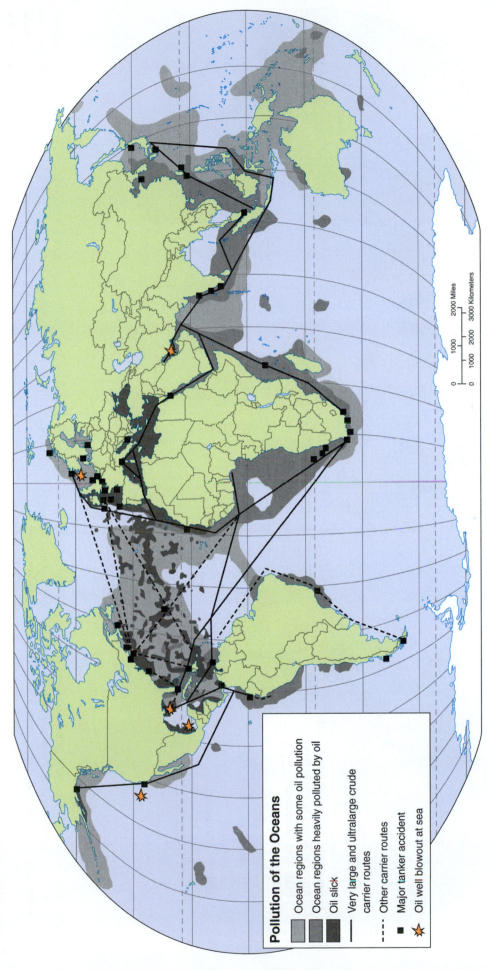

Pollution of the Oceans

- Ocean regions with some oil pollution
- Ocean regions heavily polluted by oil
- Oil slick
- Very large and ultralarge crude carrier routes
- Other carrier routes
- Major tanker accident
- Oil well blowout at sea

The pollution of the world's oceans has long been a matter of concern to physical geographers, oceanographers, and other environmental scientists. The great circulation systems of the ocean are one of the controlling factors of the earth's natural environment, and modifications to those systems have unknown consequences. This map is based on what we can measure: (1) areas of oceans where oil pollution has been proven to have inflicted significant damage to ocean ecosystems and life forms (including phytoplankton, the oceans' primary food producers, equivalent to land-based vegetation) and (2) areas

of oceans where unusually high concentrations of hydrocarbons from oil spills may have inflicted some damage to the oceans' biota. A glance at the map shows that there are few areas of the world's oceans where some form of pollution is not a part of the environmental system. What the map does not show in detail, because of the scale, are the dramatic consequences of large individual pollution events: the devastation produced by the 1991 Gulf War in the Persian Gulf, or the 2010 *Deepwater Horizon* oil spill in the Gulf of Mexico.

Map 91   Food Supply from Marine and Freshwater Systems

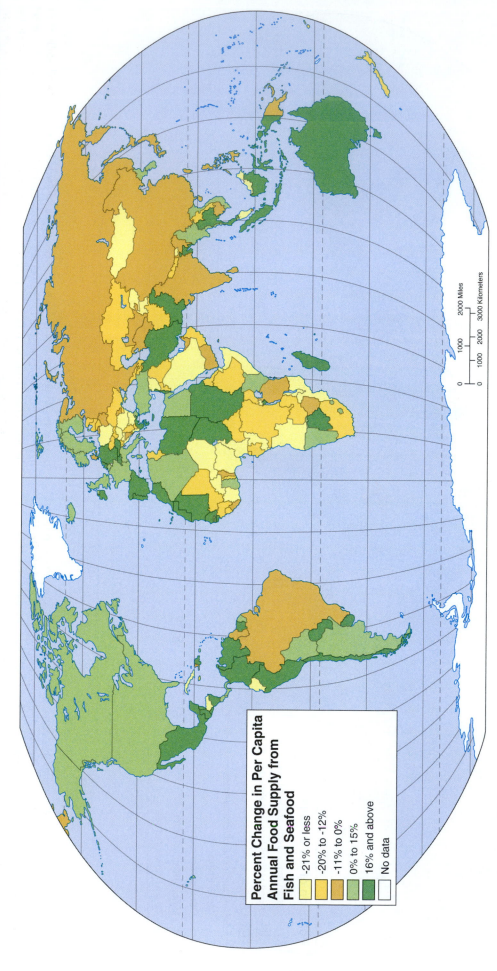

Percent Change in Per Capita Annual Food Supply from Fish and Seafood

- -21% or less
- -20% to -12%
- -11% to 0%
- 0% to 15%
- 16% and above
- No data

2000 Miles
1000   2000   3000 Kilometers
0   1000   2000

Not that many years ago, food supply experts were confidently predicting that the "starving millions" of the world of the future could be fed from the unending bounty of the world's oceans. While the annual catch from the sea helped to keep hunger at bay for a time, by the late 1980s it had become apparent that without serious human intervention in the form of aquaculture, the supply of fish would not be sufficient to offset the population/food imbalance that was beginning to affect so many of the world's regions. The development of factory-fishing with advanced equipment to locate fish and process them before they went to market increased the supply of food from the ocean, but in that increase was sown the seeds of future problems. The factory-fishing system, efficient in terms of economics, was costly in terms of fish populations. In some well-fished areas, the stock of fish that was viewed as near infinite just a few decades ago has dwindled nearly to the point of disappearance. This map shows both increases and decreases in the amount of individual countries' food supplies from the ocean. The increases are often the result of more technologically advanced fishing operations. The decreases are usually the result of the same thing: increased technology has brought increased harvests, which has reduced the supply of fish and shellfish and that, in turn, has increased prices. Most of the countries that have experienced sharp decreases in their supply of food from the world's oceans are simply no longer able to pay for an increasingly scarce commodity.

-110-

Map 92 Cropland Per Capita: Changes, 1996–2007

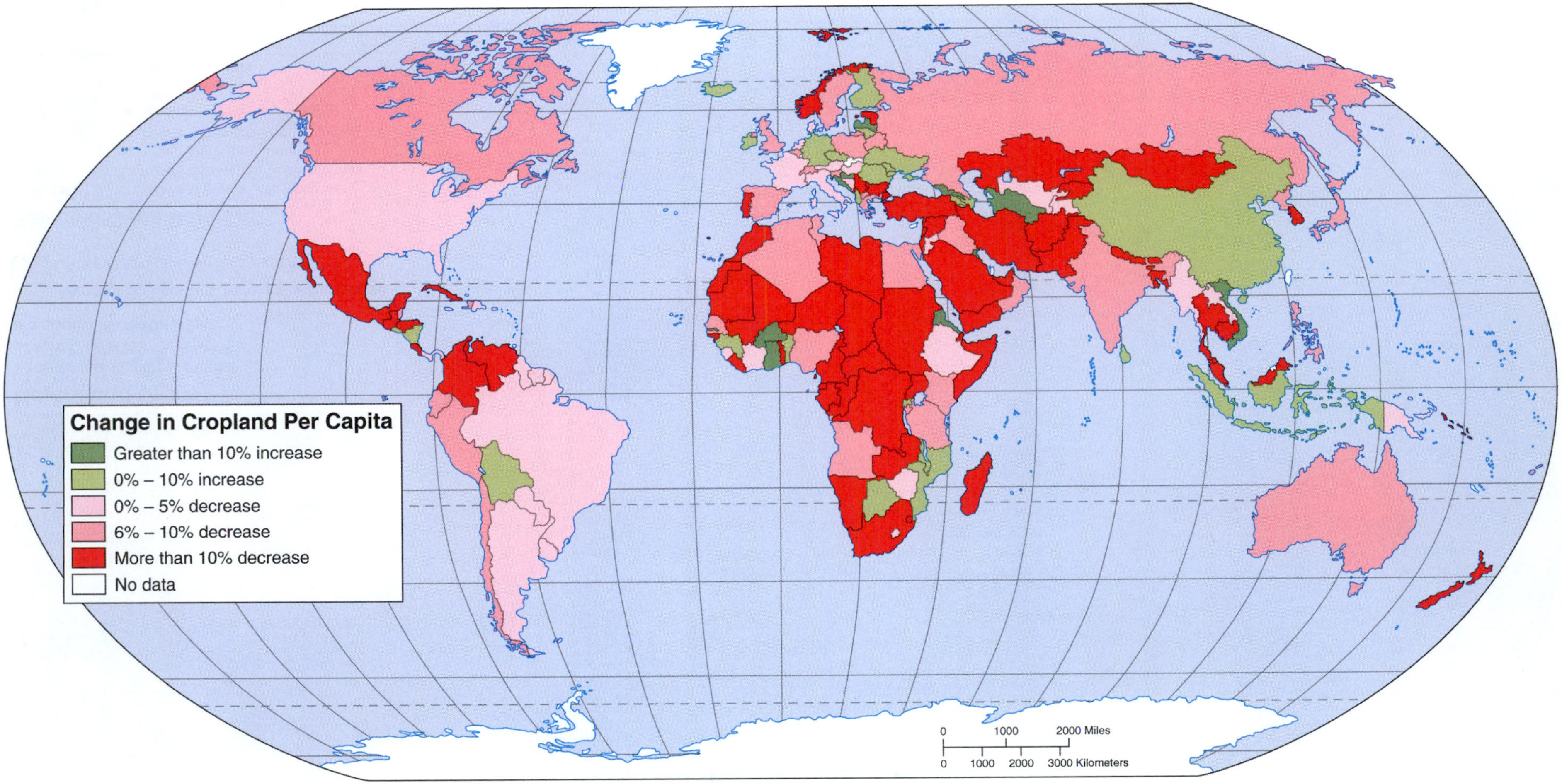

Change in Cropland Per Capita

- Greater than 10% increase
- 0% – 10% increase
- 0% – 5% decrease
- 6% – 10% decrease
- More than 10% decrease
- No data

0   1000   2000 Miles

0   1000   2000   3000 Kilometers

As population has increased rapidly throughout the world, the area of cultivated land has increased at the same time; in fact, the amount of farmland per person has gone up slightly. Unfortunately, the figures that show this also tell us that since most of the best (or even good) agricultural land in 1985 was already under cultivation, most of the agricultural area added since the mid 1990s involves land that would have been viewed as marginal by the fathers and grandfathers of present farmers—marginal in that it was too dry, too wet, too steep to cultivate, too far from a market, and so on. The continued expansion of agricultural area is one reason that serious famine and starvation have struck only a few regions of the globe. But land, more than any other resource we deal with, is finite, and the expansion cannot continue indefinitely. Future gains in agricultural production are most probably going to come through more intensive use of existing cropland, heavier applications of fertilizers and other agricultural chemicals, and genetically engineered crops requiring heavier applications of energy and water, than from an increase in the amount of the world's cropland.

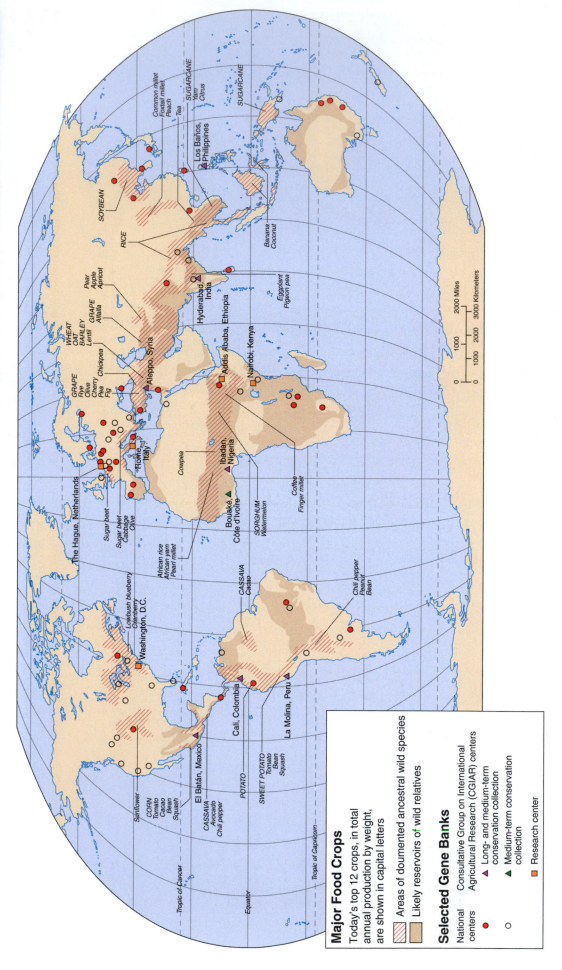

Map 93 Food Staples Under Stress: The Fight Against Genetic Simplicity

Major Food Crops

Today's top 12 crops, in total annual production by weight, are shown in capital letters

- Areas of documented ancestral wild species
- Likely reservoirs of wild relatives

Selected Gene Banks

National centers:
- ● Consultative Group on International Agricultural Research (CGIAR) centers
- ○
- ▲ Long- and medium-term conservation collection
- ▲ Medium-term conservation collection
- ■ Research center

Common millet
Foxtail millet
Peach
Tea
SUGARCANE
Yam
Citrus
SUGARCANE

SOYBEAN

RICE

Los Baños, Philippines

Pear
Apple
Apricot

Banana
Coconut

Hyderabad, India

WHEAT
OAT
BARLEY
Lentil
GRAPE
Alfalfa

Aleppo, Syria

Addis Ababa, Ethiopia

Nairobi, Kenya

Eggplant
Pigeon pea

GRAPE
Rye
Olive
Cherry
Pea
Chickpea
Fig

Rome, Italy

Ibadan, Nigeria

Coffee
Finger millet

The Hague, Netherlands

Sugar beet
Cabbage
Olive

Cowpea

Bouaké, Côte d'Ivoire

SORGHUM
Watermelon

African rice
African yam
Pearl millet

Washington, D.C.

Lowbush blueberry
Cranberry

Chili pepper
Peanut
Bean

CASSAVA
Cacao

Sunflower

CORN
Tomato
Cacao
Bean
Squash

El Batán, Mexico

CASSAVA
Avocado
Chili pepper

POTATO

Cali, Colombia

SWEET POTATO
Tomato
Bean
Squash

La Molina, Peru

0   1000   2000   3000 Kilometers
0   1000   2000 Miles

Tropic of Cancer
Equator
Tropic of Capricorn

The point has been made elsewhere in this atlas that the loss of biological diversity is one of the greatest threats to ecosystem stability. That is certainly true when we speak of the threatened loss of diversity among those plants that provide the Earth's population with the bulk of its food supply. As farming becomes more specialized and certain strains of wheat, rice, corn, etc., achieve dominance because of their increased productivity levels, there is great danger in "putting all our eggs in one basket." Plant geneticists are attempting to protect the world against the loss of genetic diversity and the potential

starvation that could result from crop failure of the favored strains by preserving as much of the genetic material of both domesticated relatives and wild ancestors of the most-used grains and other crops. Protecting the world's genetic reserves of plants is more than just an academic exercise. It is insurance against the biological degradation that future population growth is going to create and, therefore, provides some measure of assurance that future food production will be able to meet future demand.

-112-

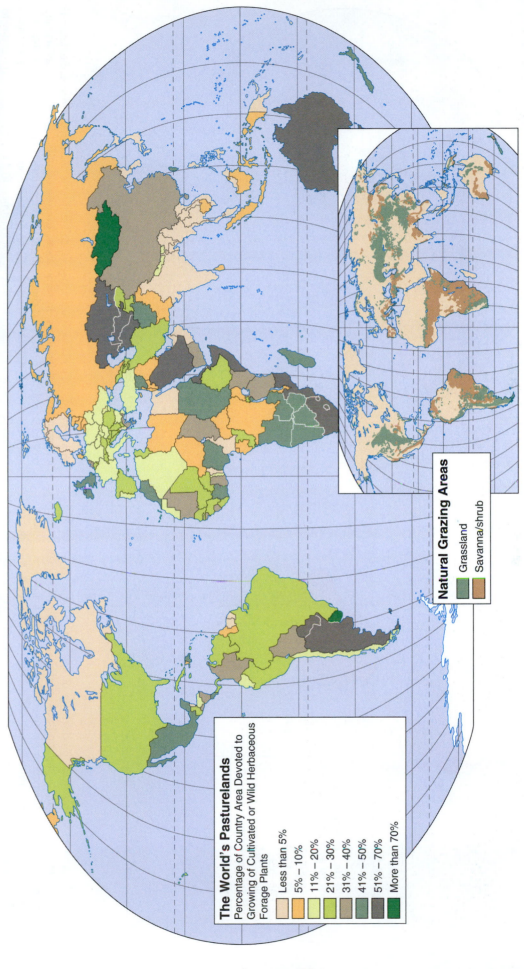

Map 94 World Pastureland, 2005

The World's Pasturelands
Percentage of Country Area Devoted to Growing of Cultivated or Wild Herbaceous Forage Plants

- Less than 5%
- 5% – 10%
- 11% – 20%
- 21% – 30%
- 31% – 40%
- 41% – 50%
- 51% – 70%
- More than 70%

Natural Grazing Areas

- Grassland
- Savanna/shrub

More than 25 percent of the world's surface is considered pastureland—either native wild grasses or human made pastures created by clearing forests and planting grass and other herbaceous forage plants for livestock feed. Two types of pasture predominate: the prairie and steppe grasses of the mid-latitudes and the savanna grasses of the sub-tropics and tropics. These two primary grassland biomes are shown in the inset map. The larger map depicts pastureland as a proportion of land area of individual countries. You will note that many of the countries with high levels of pastureland (Argentina and Australia, for example) are among the world's leading exporters of livestock products. Other countries with high levels of pastureland (Saudi Arabia and the Central Asian states)

consume the bulk of their products domestically. The significance of the distribution and use of pastureland is that—whether it is in the African Sahel, China, Brazil, or the United States—the world's pasturelands are deteriorating rapidly under increasing demands to produce more animal products than even a wealthier world can afford to pay for (see Map 38). Grassland degradation, particularly in areas where pastoral nomads use their animals as their chief source of food and income, creates woody scrublands where the carrying capacity for grazing animals is sharply diminished. In short, most of the world's pasturelands are overgrazed, and the world's supply of animal products is in jeopardy.

Map 95   Fertilizer Use, 2007

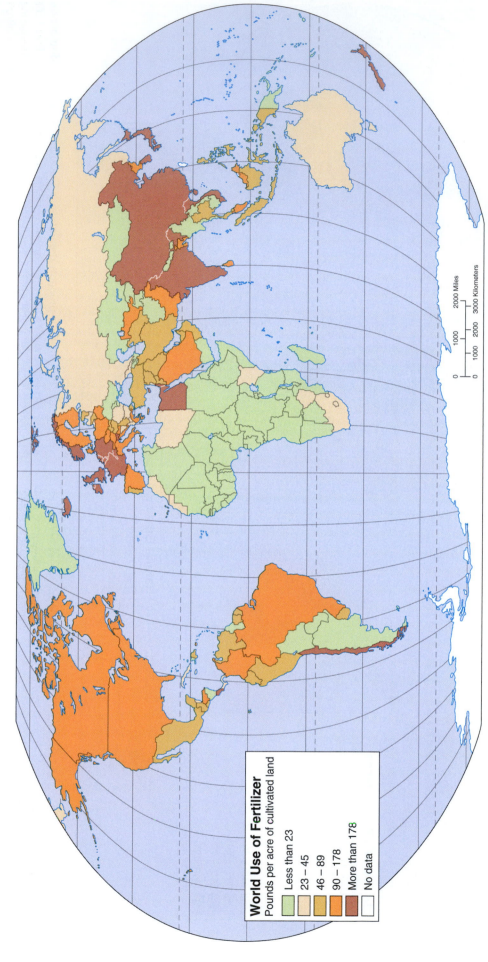

World Use of Fertilizer
Pounds per acre of cultivated land

- Less than 23
- 23 – 45
- 46 – 89
- 90 – 178
- More than 178
- No data

The use of fertilizer to maintain the productivity of agricultural lands is a wonderful agricultural invention—as long as the fertilizers used are natural rather than artificial. In most of the world's developed countries, such as those in Europe and North America, the use of animal manure to fertilize fields has decreased dramatically over the past century, in favor of artificial fertilizers that are cheaper and easier to use and—what is most important—increase crop yields more dramatically. The danger here is that artificial fertilizers normally have high concentrations of nitrates that tend to convert to nitrites in the soil, reducing the ability of soil bacteria to extract "free" nitrogen from the atmosphere.

As more artificial fertilizer is used, natural soil fertility is decreased, creating the demand for more artificial fertilizers. In some areas, overuse of artificial fertilizers has actually created soils that are too "hot" chemically to produce crops. Countries with high fertilizer use in the developing world—northern South America, Southwest, South, and East Asia—still tend to use more natural fertilizers. But as farmers in those areas gain more ability to buy and use artificial fertilizers, their soils will also begin to suffer from overfertilization. Global agriculture needs to come to grips with the need to maintain productivity but to do so in a manner that is sustainable.

Map 96   Annual Deforestation Rates, 2000–2005

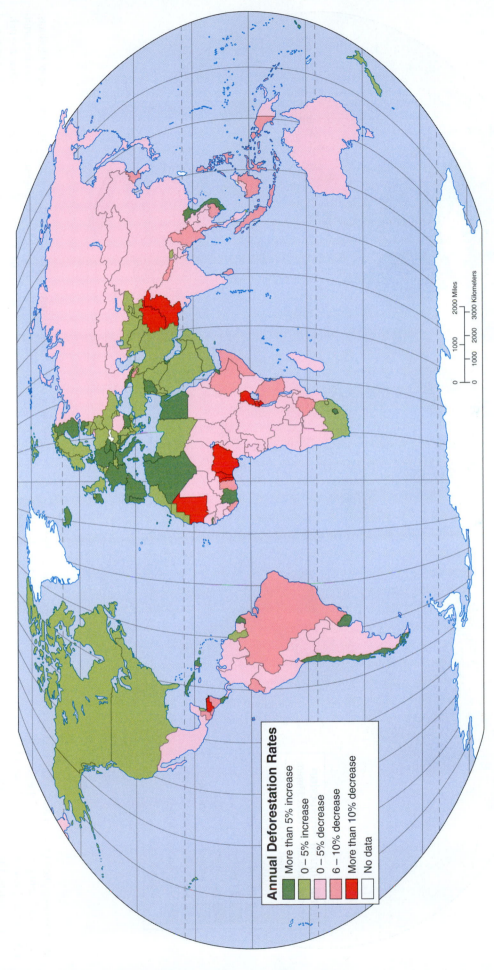

Annual Deforestation Rates

- More than 5% increase
- 0 – 5% increase
- 0 – 5% decrease
- 6 – 10% decrease
- More than 10% decrease
- No data

2000 Miles
1000 2000 3000 Kilometers
0   1000   2000

One of the most discussed environmental problems, is that of deforestation. For most people, deforestation means clearing of tropical rain forests for agricultural purposes. Yet nearly as much forest land per year—much of it in North America, Europe, and Russia—is impacted by commercial lumbering as is cleared by tropical farmers and ranchers. Even in the tropics, much of the forest clearance is undertaken by large corporations producing high-value tropical hardwoods for the global market in furniture, ornaments, and other fine wood products. Still, it is the agriculturally driven clearing of the great rain forests of the Amazon Basin, west and central Africa, Middle America, and Southeast Asia that draws public attention. Although much concern over forest clearance focuses on the relationship between forest clearance and the reduction in the capacity of the world's vegetation system to absorb carbon dioxide (and thus delay global warming), of just as great concern are issues having to do with the loss of biodiversity (large numbers of plants and animals), the near-total destruction of soil systems, and disruptions in water supply that accompany clearing.

Map 97 World Timber Production, 2005

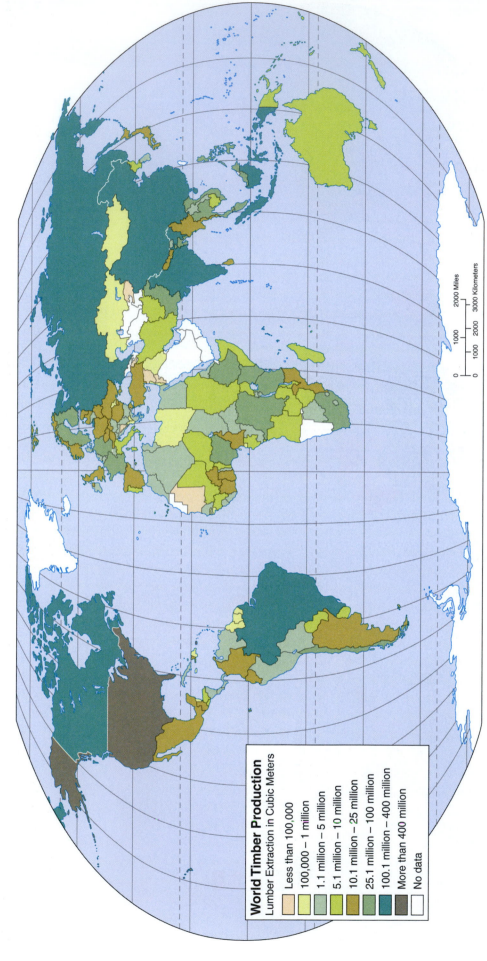

World Timber Production
Lumber Extraction in Cubic Meters

- Less than 100,000
- 100,000 – 1 million
- 1.1 million – 5 million
- 5.1 million – 10 million
- 10.1 million – 25 million
- 25.1 million – 100 million
- 100.1 million – 400 million
- More than 400 million
- No data

With the exception of water and arable land, forests have played a larger part in the transition from hunting-gathering to the emergence of modern industrial economies than any other natural resource. The chief uses of forests are for lumber, fuelwood, and pulp and paper products. The per capita use of wood does not tend to vary greatly between the developed and developing countries. In developed areas such as North America, Europe, and Russia, the bulk of forest products is used for lumber (much of it exported to forest-poor areas) and for pulp and paper products. In the developing countries in South America, Africa, and Asia (excluding Russia), more wood is used for fuel than for any other purpose.

In those parts of the world where forests are used for lumber and pulp-paper products, more-or-less sustainable forest systems have been developed. These systems are often artificially maintained and are much simpler (and, hence, more fragile) ecosystems than the rich, complex natural ecosystems they have replaced. Where forests provide the chief source of fuel, the forest systems tend to remain healthier but are shrinking in size in proportion to the increase in human populations. Of the major forest ecosystems, those in greatest jeopardy are the tropical forests that are being cleared for agricultural land and the boreal or northern forests of Canada and Russia that are being simplified by overcutting.

Map 98 The Loss of Biodiversity: Globally Threatened Animal Species, 2007

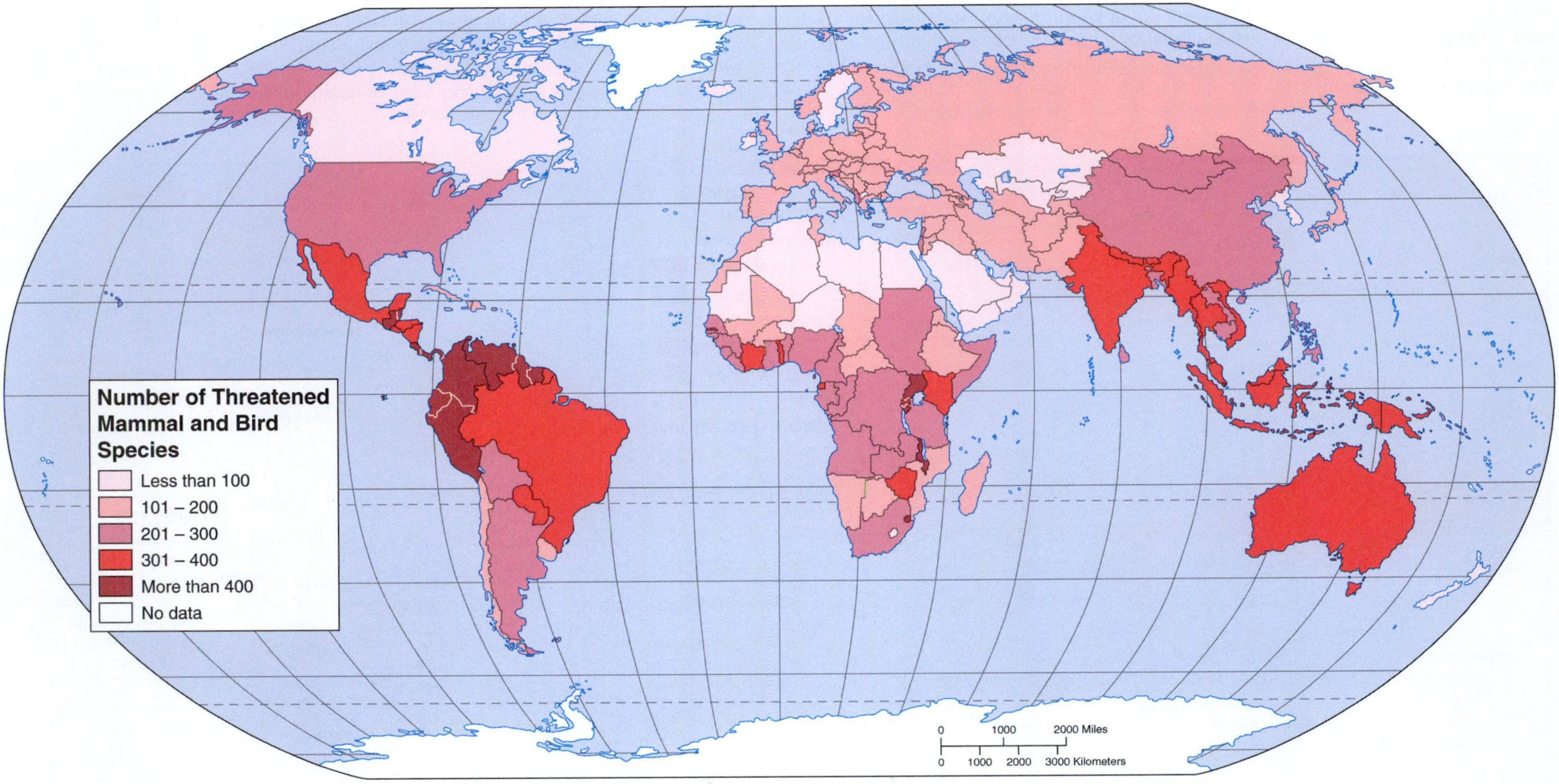

Number of Threatened Mammal and Bird Species

- Less than 100
- 101 – 200
- 201 – 300
- 301 – 400
- More than 400
- No data

Threatened species are those in grave danger of going extinct. Their populations are becoming restricted in range, and the size of the populations required for sustained breeding is nearing a critical minimum. *Endangered species* are in immediate danger of becoming extinct. Their range is already so reduced that the animals may no longer be able to move freely within an ecozone, and their populations are at the level where the species may no longer be able to sustain breeding. Most species become threatened first and then endangered as their range and numbers continue to decrease. When people think of animal extinction, they think of large herbivorous species like the rhinoceros or fierce

carnivores like lions, tigers, or grizzly bears. Certainly these animals make almost any list of endangered or threatened species. But there are literally hundreds of less conspicuous animals that are equally threatened. Extinction is normally nature's way of informing a species that it is inefficient. But conditions in the late twentieth century are controlled more by human activities than by natural evolutionary processes. Species that are endangered or threatened fall into that category because, somehow, they are competing with us or with our domesticated livestock for space and food. And in that competition the animals are always going to lose.

2000 Miles
3000 Kilometers

-117-

Map 99 Disappearing Giants: The Reduction in "Conservative" Species

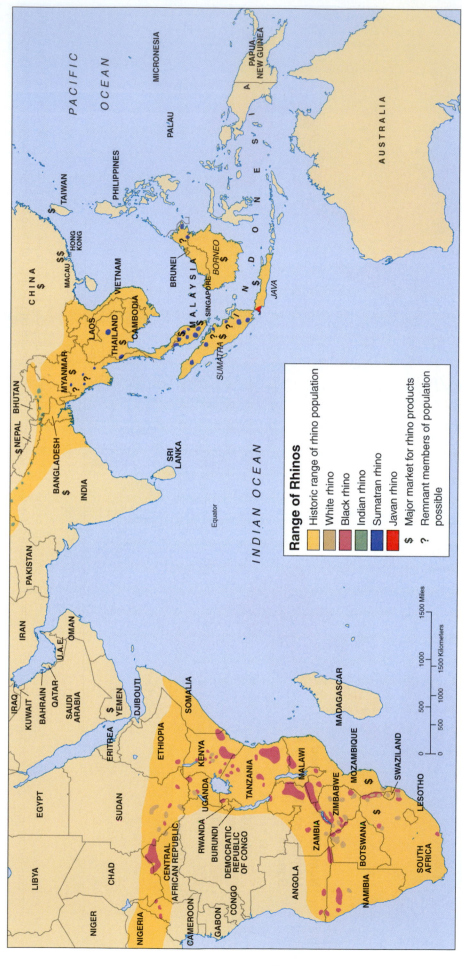

Range of Rhinos

- Historic range of rhino population
- White rhino
- Black rhino
- Indian rhino
- Sumatran rhino
- Javan rhino
- $ Major market for rhino products
- ? Remnant members of population possible

The species singled out for special attention have been given that attention not just because they represent major species in greatest danger of extinction but because they are signal species (indicating the health of ecosystems) as well as conservative or aristocratic species. These species often suffer most from human infringement on traditional land or from overhunting. Rhinos, elephants, and tigers fall into the conservative or aristocratic categories: they are large, they breed slowly, they do not produce large numbers of offspring in a single breeding season, they may have long gestation periods, etc. Conservative species are also often those that are both in great demand for their hides, horns, tusks, or fur (in other words, non-edible products that demand extremely

high prices in the luxury market), and that tend to compete most directly with humans for the same resources—even if that resource is simply space. Part of the rapid loss of such aristocratic species as rhinos, elephants, and tigers is the value of ivory, tiger hides, or rhinoceros horn in the global market; but part of the rapid loss is that these species occupy space that is also needed for cropland or grazing land. In other words, these species compete with humans for the same resources and, in any such competition, it is the humans who are going to win. And a win for humans represents a huge loss for those animals that may have formerly been the ecological dominants in their home ranges.

-118-

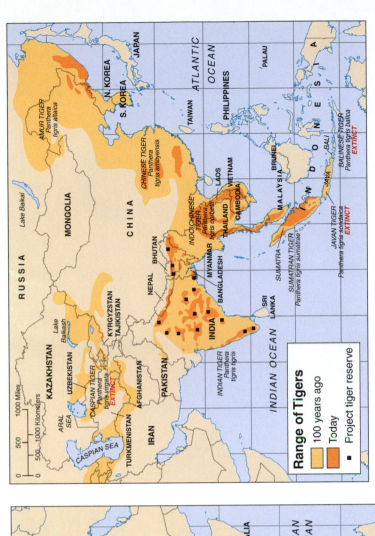

Range of Tigers

	100 years ago
	Today
■	Project tiger reserve

ATLANTIC OCEAN

PALAU

JAPAN

N. KOREA

S. KOREA

TAIWAN

PHILIPPINES

AMUR TIGER
*Panthera
tigris altaica*

CHINESE TIGER
*Panthera
tigris amoyensis*

RUSSIA

Lake Baikal

MONGOLIA

CHINA

BHUTAN

LAOS

VIETNAM

CAMBODIA

THAILAND

MYANMAR

INDOCHINESE
TIGER
*Panthera
tigris corbetti*

MALAYSIA

BRUNEI

I N D O N E S I A

BALI

JAVA

BALINESE TIGER
Panthera tigris balica
EXTINCT

JAVAN TIGER
Panthera tigris sondaica
EXTINCT

SUMATRAN TIGER
Panthera tigris sumatrae

SUMATRA

KAZAKHSTAN

Lake
Balkash

KYRGYZSTAN

TAJIKISTAN

UZBEKISTAN

CASPIAN TIGER
*Panthera
tigris virgata*
EXTINCT

AFGHANISTAN

PAKISTAN

NEPAL

BANGLADESH

INDIA

SRI
LANKA

INDIAN TIGER
*Panthera
tigris tigris*

INDIAN OCEAN

ARAL
SEA

CASPIAN SEA

TURKMENISTAN

IRAN

1000 Miles

500

1000 Kilometers

500

0

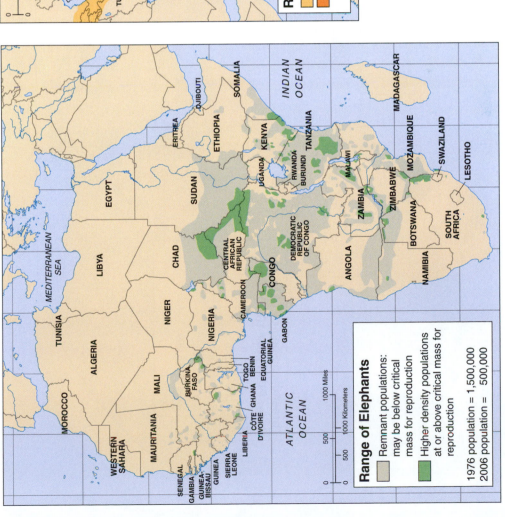

Range of Elephants

| | Remnant populations: may be below critical mass for reproduction |
| | Higher density populations at or above critical mass for reproduction |

1976 population = 1,500,000
2006 population = 500,000

MEDITERRANEAN
SEA

MOROCCO

WESTERN
SAHARA

TUNISIA

ALGERIA

LIBYA

EGYPT

MAURITANIA

MALI

NIGER

CHAD

SUDAN

ERITREA

DJIBOUTI

SOMALIA

ETHIOPIA

INDIAN
OCEAN

MADAGASCAR

SENEGAL

GAMBIA

GUINEA-
BISSAU

GUINEA

SIERRA
LEONE

LIBERIA

CÔTE
D'IVOIRE

GHANA

TOGO

BENIN

BURKINA
FASO

NIGERIA

CAMEROON

EQUATORIAL
GUINEA

GABON

CONGO

CENTRAL
AFRICAN
REPUBLIC

DEMOCRATIC
REPUBLIC
OF CONGO

UGANDA

KENYA

RWANDA

BURUNDI

TANZANIA

MALAWI

ANGOLA

ZAMBIA

ZIMBABWE

MOZAMBIQUE

NAMIBIA

BOTSWANA

SOUTH
AFRICA

SWAZILAND

LESOTHO

ATLANTIC
OCEAN

1000 Miles

1000 Kilometers

500

500

0

Map 100   The Loss of Biodiversity: Globally Threatened Plant Species, 2007

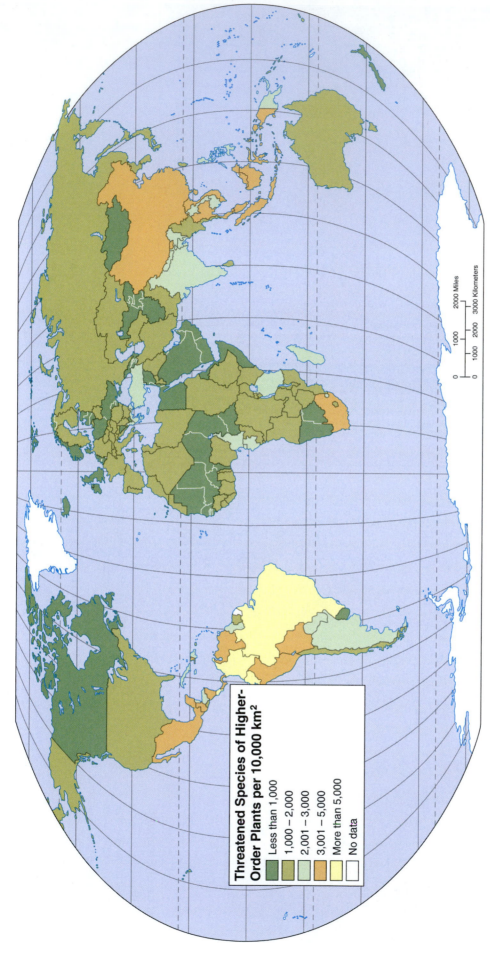

Threatened Species of Higher-Order Plants per 10,000 km²

- Less than 1,000
- 1,000 – 2,000
- 2,001 – 3,000
- 3,001 – 5,000
- More than 5,000
- No data

0   1000   2000 Miles
0   1000   2000   3000 Kilometers

While most people tend to be more concerned about the animals on threatened and endangered species lists, the fact is that many more plants are in jeopardy, and the loss of plant life is, in all ecological regions, a more critical occurrence than the loss of animal populations. Plants are the primary producers in the ecosystem; that is, plants produce the food upon which all other species in the food web, including human beings, depend for sustenance. It is plants from which many of our critical medicines come, and it is plants that maintain the delicate balance between soil and water in most of the world's regions. When environmental scientists speak of a loss of biodiversity, what they are most often describing is a loss of the richness and complexity of plant life that lends stability to ecosystems. Systems with more plant life tend to be more stable than those with less. For these and other reasons, the scientific concern over extinction is greater when applied to plants than to animals. It is difficult for people to become as emotional over a teak tree as they would over an elephant. But as great a tragedy as the loss of the elephant would be, the loss of the teak would be greater.

Map 101 The Areas of Greatest Loss and Hotspots of Biodiversity

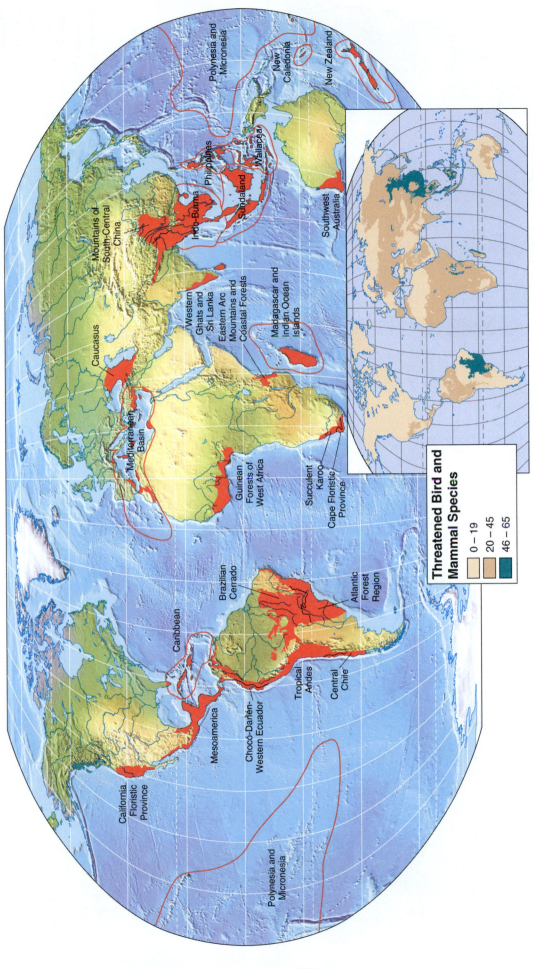

Polynesia and Micronesia

New Caledonia

New Zealand

Mountains of South-Central China

Philippines

Indo-Burma

Sundaland

Wallacea

Southwest Australia

Western Ghats and Sri Lanka

Eastern Arc Mountains and Coastal Forests

Madagascar and Indian Ocean Islands

Caucasus

Mediterranean Basin

Guinean Forests of West Africa

Succulent Karoo

Cape Floristic Province

Brazilian Cerrado

Atlantic Forest Region

Caribbean

Tropical Andes

Central Chile

Mesoamerica

Chocó-Darién-Western Ecuador

California Floristic Province

Polynesia and Micronesia

Threatened Bird and Mammal Species

- 0 – 19
- 20 – 45
- 46 – 65

Where we have normally thought of tropical forest basins such as Amazonia as the world's most biologically diverse ecosystems, recent research has discovered the surprising fact that a number of hotspots of biological diversity exist outside the major tropical forest regions. These hotspot regions contain slightly less than 2 percent of the world's total land area but may contain up to 60 percent of the total world's terrestrial species of plants and animals. Geographically, the hotspot areas are characterized by vertical zonation (that is, they tend to be hilly to mountainous regions), long known to be a factor in biological complexity. They are also in coastal locations or near large bodies of water, locations that stimulate climatic variability and, hence, biological complexity. Although some of the hotspots are sparsely populated, others, such as Sundaland, are occupied by some of the world's densest populations. Protection of the rich biodiversity of these hotspots is, most biologists feel, of crucial importance to the preservation of the world's biological heritage. The greatest actual or potential losses of bird and mammal species tend to occur in those areas of the world where there is the most active competition for living spaces between humans and animals. Thus highly populated areas tend to suffer the greatest loss of native species. China, Brazil, and Indonesia stand out on the inset map; each of these countries ranks in the world's top 10 in total population.

Map 102

Nationally Protected Areas: Parks and Preserves

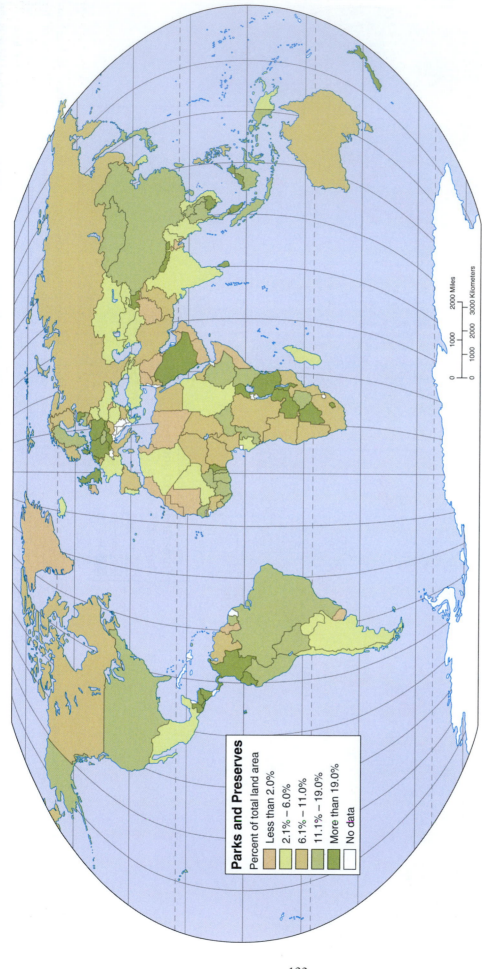

Parks and Preserves

Percent of total land area

- Less than 2.0%
- 2.1% – 6.0%
- 6.1% – 11.0%
- 11.1% – 19.0%
- More than 19.0%
- No data

0   1000   2000 Miles

0   1000   2000   3000 Kilometers

Interpreting the patterns on this map ranges from the obvious to the obscure. Some of the countries with the lowest level of nationally protected land are obvious—Iraq, Afghanistan, Somalia—where internal conflict trumps environmental protection. Other countries with low levels, such as Libya, suggest there is little perceived need for protecting lands in a desert environment (quite the opposite is true). For large countries such as Brazil, China, and the United States (three of the five largest countries in the world), to have such high percentages of nationally protected lands may indicate a high level of conservation awareness—or it may suggest that these countries have recognized the economic benefits of preserved lands in terms of increased tourism, etc. Or it may be a combination of the two. Certainly in the United States (where the first nationally-protected areas

were established) the conservation ethic was the driving mechanism but tourism became increasingly important during the latter half of the twentieth century. Interestingly enough, those countries with the greatest percentages of their land areas in nationally protected spaces may not be countries that many of us would normally predict as being environmentally conservative or preservative. It is heartening that desert countries like Saudi Arabia have recognized the fragility of desert environments by protecting them. It is also heartening that those African countries that serve as the last refuges of the great animals have recognized the importance of attempting to preserve those species by setting aside important national parks and preserves. Countries such as Tanzania and Zambia deserve special credit for setting aside large tracts of land as wildlife refuges and preserves.

Map 103 The Risks of Desertification

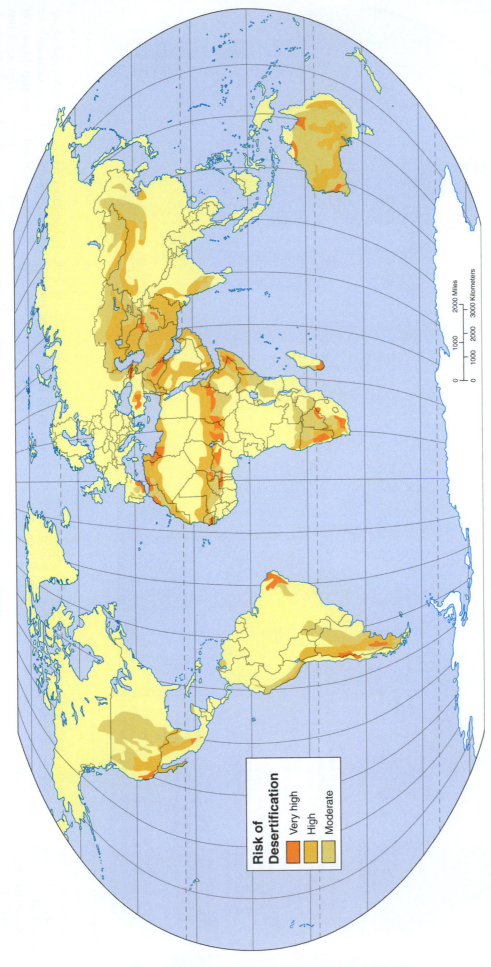

Risk of Desertification

- Very high
- High
- Moderate

The awkward-sounding term *desertification* refers to a reduction in the food-producing capacity of drylands through vegetation, soil, and water changes that culminate in either a drier climate or in soil and plant systems that are less efficient in their use of water. Most of the world's existing drylands—the shortgrass steppes, the tropical savannas, the bunchgrass regions of the desert fringe—are fairly intensively used for agriculture and are, therefore, subject to the kinds of pressures that culminate in desertification. Most desertification is a natural process that occurs near the margins of desert regions. It is caused by dehydration of the soil's surface layers during periods of drought and by high water loss through evaporation in an environment of high temperature and high winds. This natural process is greatly enhanced by human agricultural activities that expose topsoil to wind and water erosion. Among the most important practices that cause desertification are (1) overgrazing of rangelands, resulting from too many livestock on too small an area of land; (2) improper management of soil and water resources in irrigation agriculture, leading to accelerated erosion and to salt buildup in the soil; (3) cultivation of marginal terrain with soils and slopes that are unsuitable for farming; (4) surface disturbances of vegetation (clearing of thorn scrub, mesquite, chaparral, and similar vegetation) without soil protection efforts being made or replanting being done; and (5) soil compaction by agricultural implements, domesticated livestock, and rain falling on an exposed surface.

Map 104 Global Soil Degradation

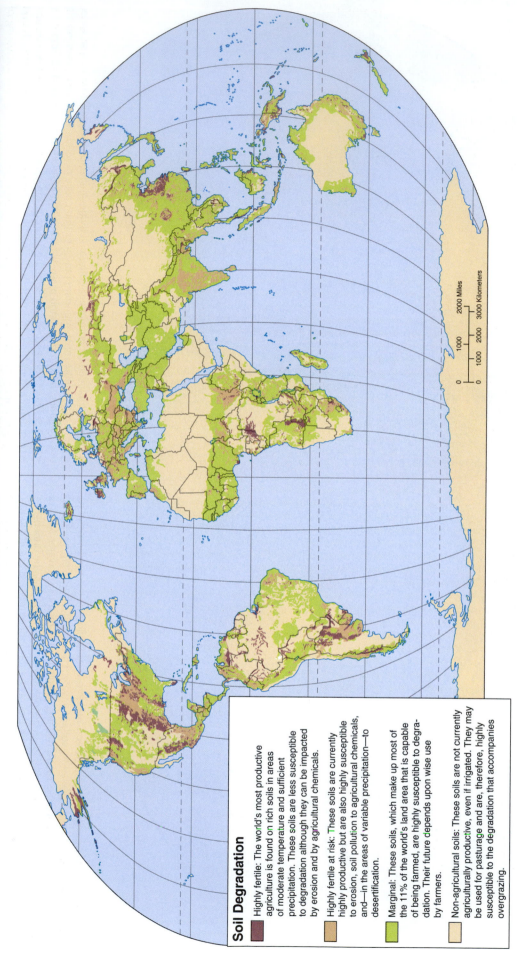

Soil Degradation

Highly fertile: The world's most productive agriculture is found on rich soils in areas of moderate temperature and sufficient precipitation. These soils are less susceptible to degradation although they can be impacted by erosion and by agricultural chemicals.

Highly fertile at risk: These soils are currently highly productive but are also highly susceptible to erosion, soil pollution to agricultural chemicals, and—in the areas of variable precipitation—to desertification.

Marginal: These soils, which make up most of the 11% of the world's land area that is capable of being farmed, are highly susceptible to degradation. Their future depends upon wise use by farmers.

Non-agricultural soils: These soils are not currently agriculturally productive, even if irrigated. They may be used for pasturage and are, therefore, highly susceptible to the degradation that accompanies overgrazing.

0   1000   2000 Miles
0 1000 2000 3000 Kilometers

Recent research has shown that more than 3 billion acres of the world's surface suffer from serious soil degradation, with more than 22 million acres so severely eroded or poisoned with chemicals that they can no longer support productive crop agriculture. Most of this soil damage has been caused by poor farming practices, overgrazing of domestic livestock, and deforestation. These activities strip away the protective cover of natural vegetation forests and grasslands, allowing wind and water erosion to remove the topsoil that contains necessary nutrients and soil microbes for plant growth. But millions of acres of topsoil have been degraded by chemicals as well. In some instances these chemicals are the result of overapplication of fertilizers, herbicides, pesticides, and other agricultural chemicals. In other instances, chemical deposition from industrial and urban wastes and from acid precipitation has poisoned millions of acres of soil. As the map shows, soil erosion and pollution are not problems just in developing countries with high population densities and increasing use of marginal lands. They also afflict the more highly developed regions of mechanized, industrial agriculture. While many methods for preventing or reducing soil degradation exist, they are seldom used because of ignorance, cost, or perceived economic inefficiency.

Map 105 The Degree of Human Disturbance

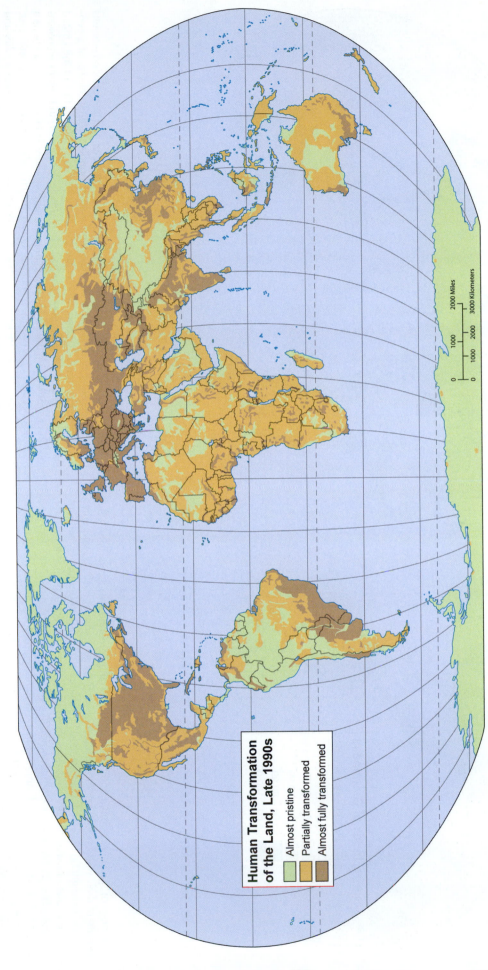

Human Transformation of the Land, Late 1990s

- Almost pristine
- Partially transformed
- Almost fully transformed

0 1000 1000 2000 2000 Miles

0 1000 2000 3000 Kilometers

The data on human disturbance have been gathered from a wide variety of sources, some of them conflicting and not all of them reliable. Nevertheless, at a global scale this map fairly depicts the state of the world in terms of the degree to which humans have modified its surface. The almost pristine areas, covered with natural vegetation, generally have population densities under 10 persons per square mile. These areas are, for the most part, in the most inhospitable parts of the world: too high, too dry, too cold for permanent human habitation in large numbers. The partially transformed areas are normally agricultural areas, either subsistence (such as shifting cultivation) or extensive normally agricultural areas, either subsistence (such as shifting cultivation) or extensive (such as livestock grazing). They often contain areas of secondary vegetation, regrown after removal of original vegetation by humans. They are also sometimes marked by a density of livestock in excess of carrying capacity, leading to overgrazing, which further alters the condition of the vegetation. The almost fully transformed areas are those of permanent and intensive agriculture and urban settlement. The primary vegetation of these regions has been removed, with no evidence of regrowth or with current vegetation that is quite different from natural (potential) vegetation. Soils are in a state of depletion and degradation, and, in drier lands, desertification is a factor of human occupation. The disturbed areas match closely those areas of the world with the densest human populations.

-125-

Map 106   The Green and Not So Green World

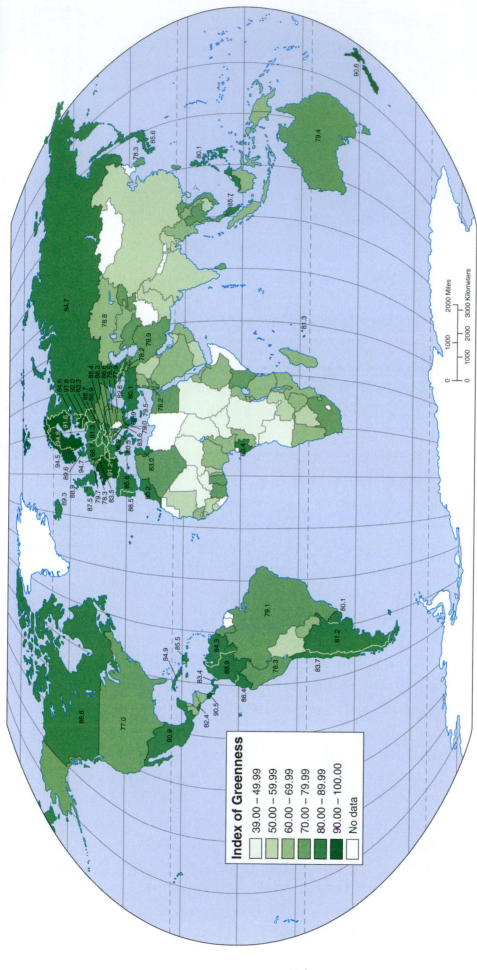

Index of Greenness

- 39.00 – 49.99
- 50.00 – 59.99
- 60.00 – 69.99
- 70.00 – 79.99
- 80.00 – 89.99
- 90.00 – 100.00
- No data

The index of "greenness" ranks countries according to their greenhouse gas emissions, the quality of their water resources, their protection of plant and animal habitats, and other factors. As might be expected, the "greenest" countries also tend to be relatively affluent and in cooler climates. There are obvious exceptions to this generalization (Mexico, for example) but, in general, we find countries like Sweden (no. 1), Switzerland, Norway, Finland, France, New Zealand, and Denmark ranking high on the index while poorer countries in warmer climates (much of South and East Asia, much of Sub-Saharan Africa) generally tend to rank lower on the index. There is a pretty clear correlation between affluence and "greenness"—that is, wealthy countries not only recognize the importance of "being green" but have the ability to pay for it. Still, some of the world's more affluent countries, like the United States and Australia, rank rather lower on the scale. The relatively low ranking (not quite in the top third of the countries in the index) attained by the

United States—despite its excellent record on controlling water and air pollution—results from its persistent reliance on fossil fuels and a correspondingly high release of greenhouse gas emissions. Sweden, whose per capita income is similar to that of the United States, uses renewable resources (particularly hydroelectric power) to keep its emissions low and its ranking on the scale of "greenness" high. Similarly, France (no. 7) maintains a high rank because of a reliance on nuclear power, which curbs greenhouse gas emissions. In Sub-Saharan Africa, the level of carbon dioxide emissions is low but water quality and sanitation are poor. Rapid industrialization in countries like Brazil, India, and China have impacted emissions levels, as well as water and sanitation levels. These countries appear rather lower on the index (China, for example, ranks 134 among 186 countries). The rankings shown on the map are based on 2008 data; it will be interesting to see how the patterns have changed in 20 years.

Unit VI

Global Political Patterns

Map 107   Political Systems

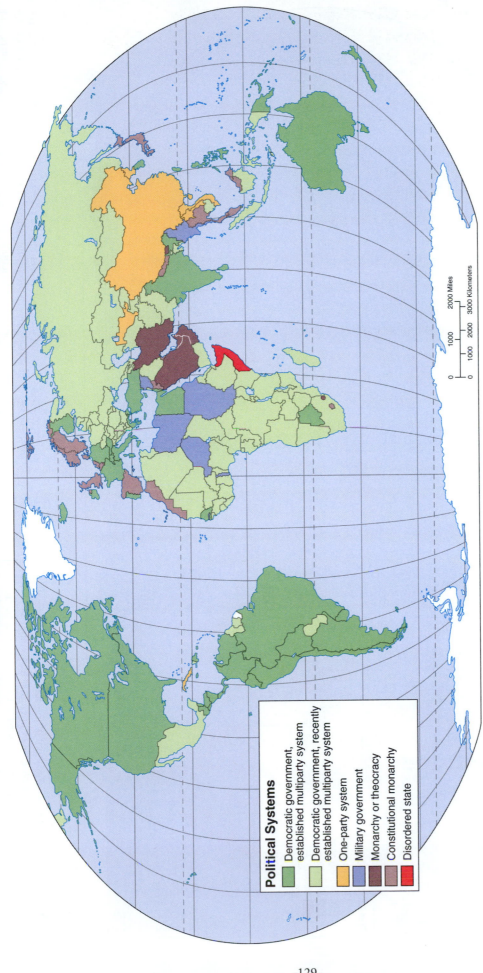

Political Systems

- Democratic government, established multiparty system
- Democratic government, recently established multiparty system
- One-party system
- Military government
- Monarchy or theocracy
- Constitutional monarchy
- Disordered state

2000 Miles

0   1000   2000   3000 Kilometers

0   1000   2000

World political systems have changed dramatically during the last decade and may change even more in the future. The categories of political systems shown on the map are subject to some interpretation: established multiparty democracies are those in which elections by secret ballot with adult suffrage are and have been long-term features of the political landscape; recently established multiparty democracies are those in which the characteristic features of multiparty democracies have only recently emerged. The former Soviet satellites of eastern Europe and the republics that formerly constituted the USSR are in this category; so are states in emerging regions that are beginning to throw off the single-party rule that often followed the violent upheavals of the immediate postcolonial governmental transitions. The other categories are more or less obvious. One-party systems are states where single-party rule is constitutionally guaranteed or where a one-party regime is a fact of political life. Monarchies are countries with heads of state who are members of a royal family. In a constitutional monarchy, such as the U.K. and the Netherlands, the monarchs are titular heads of state only. Theocracies are countries in which rule is within the hands of a priestly or clerical class; today, this means primarily fundamentalist Islamic countries such as Iran. Military governments are frequently organized around a junta that has seized control of the government from civil authority; such states are often technically transitional, that is, the military claims that it will return the reins of government to civil authority when order is restored. Finally, disordered states are countries so beset by civil war or widespread ethnic conflict that no organized government can be said to exist within them.

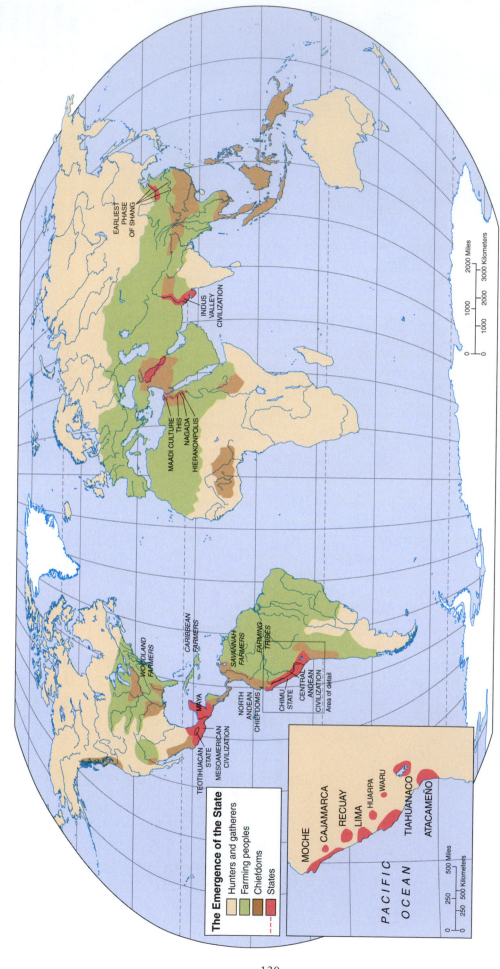

Map 108 The Emergence of the State

The Emergence of the State

- Hunters and gatherers
- Farming peoples
- Chiefdoms
- - - States

EARLIEST
PHASE
OF SHANG

INDUS
VALLEY
CIVILIZATION

MAADI CULTURE
THIS
NAGADA
HIERAKONPOLIS

WOODLAND
FARMERS

CARIBBEAN
FARMERS

SAVANNAH
FARMERS

FARMING
TRIBES

MAYA

TEOTIHUACAN
STATE
MESOAMERICAN
CIVILIZATION

NORTH
ANDEAN
CHIEFDOMS

CHIMU
STATE

CENTRAL
ANDEAN
CIVILIZATION

Area of detail

MOCHE
CAJAMARCA
RECUAY
LIMA
HUARPA
WARU
TIAHUANACO
ATACAMEÑO

PACIFIC
OCEAN

2000 Miles
3000 Kilometers
1000
2000
1000
0
0

500 Miles
500 Kilometers
250
250
0
0

Agriculture is the basis of the development of the state, a form of complex political organization. Social stratification based on wealth and power creates different classes or groups, some of whom no longer work the land. An agricultural surplus supports those who perform other functions for society, such as artisans and craftspeople, soldiers and police, priests and kings. Thus, over time, egalitarian hunters and gatherers shifted to state-level societies in some parts of the world. The first true states are shown in the rust-colored areas of this map.

Agriculture is the basis of the development of the state, a form of complex political organization. Geographers and anthropologists believe that agriculture allowed for larger concentrations of population. Farmers do not need to be as mobile as hunters and gatherers to make a living, and people living sedentarily can have larger families than those constantly on the move. Ideas about access to land and ownership also change as people develop the social and political hierarchies that come with the transition from a hunting-gathering

Map 109 Organized States and Chiefdoms, A.D. 1500

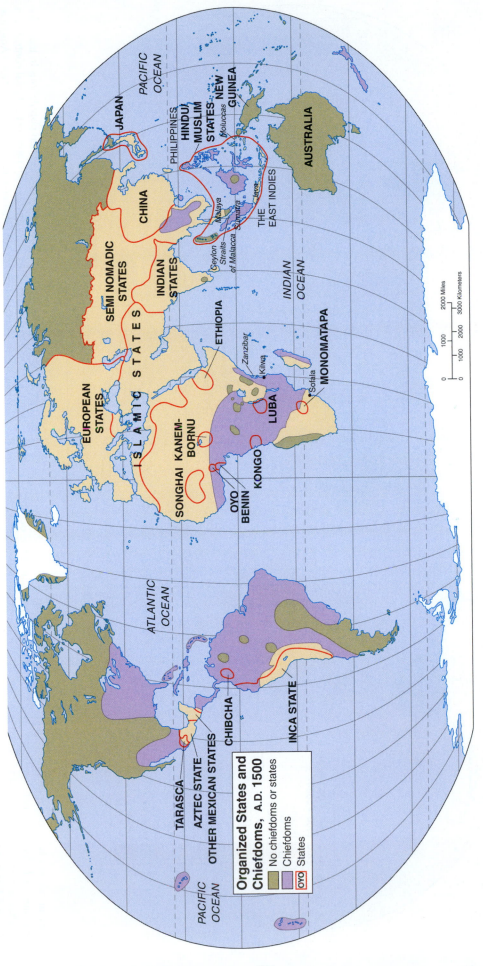

Organized States and Chiefdoms, A.D. 1500

- No chiefdoms or states
- Chiefdoms
- States
- OYO

JAPAN

PACIFIC OCEAN

CHINA

SEMI NOMADIC STATES

INDIAN STATES

PHILIPPINES

HINDU MUSLIM STATES

NEW GUINEA

Moluccas

THE EAST INDIES

Ceylon

Java

Sumatra

Straits of Malacca

Malaya

AUSTRALIA

INDIAN OCEAN

ETHIOPIA

Zanzibar

Kilwa

MONOMATAPA

Sofala

LUBA

KANEM-BORNU

KONGO

BENIN

OYO

SONGHAI

EUROPEAN STATES

ISLAMIC STATES

ATLANTIC OCEAN

2000 Miles

3000 Kilometers

1000

2000

1000

0

1000

TARASCA

AZTEC STATE

OTHER MEXICAN STATES

CHIBCHA

INCA STATE

PACIFIC OCEAN

-131-

As Europeans began to expand outward through exploration and settlement from the late fifteenth to the eighteenth centuries, it was inevitable that they would find the complex political organizations of chiefdoms and states in many different parts of the world. Both chiefdoms and states are large-scale forms of political organization in which some people have privileged access to land, power, wealth, and prestige. Chiefdoms are kin-based societies in which wealth is distributed from upper to lower classes. States are organized on the basis of socioeconomic classes, headed by a centralized government that is led by an elite. States in non-European areas included, just as they did in Europe, full-time bureaucracies and specialized subsystems for such activities as military action, taxation, operation of state religions, and social control. In many of the colonial regions of the world after the fifteenth century, Europeans actually found it easier to gain control of organized states and chiefdoms because those populations were already accustomed to some form of institutionalized central control.

Map 110 European Colonialism, 1500–2000

European powers have controlled many parts of the world during the last 500 years. The period of European expansion began when European explorers sailed the oceans in search of new trading routes and ended after World War II when many colonies in Africa and Asia gained independence. The process of colonization was very complex but normally involved the acquisition, extraction, or production of raw materials (including minerals, forest products, products from the sea, agricultural products, and animal furs/pelts) from the areas being controlled by the European colonial power in exchange for items of European manufacture. The concept of colonial dependency implied an economic structure in which the European country obtained raw materials from the colonial country in exchange for those manufactured items upon which populations in the colonial areas quickly came to depend. The colors on this map represent colonial control at its maximum extent and do not take into account shifting colonial control. In North America, for example, "New France" became British territory and "Louisiana" became Spanish territory after the Seven Years' (French and Indian) War. There are also two non-European countries—the United States and Japan—shown on the map as colonial powers. Both countries followed the classical European-style colonial model, albeit over relatively small areas and for relatively short periods of time.

European Colonialism
A.D. 1500 – 2000

- Belgian
- British
- Danish
- Dutch
- French
- German
- Italian
- Japanese
- Japanese controlled
- Portuguese
- Russian
- Spanish
- United States
- U.S.S.R.
- U.S.S.R. controlled

Scale: 0 1000 2000 3000 Kilometers — 0 1000 2000 Miles

Map labels: UNION OF SOVIET SOCIALIST REPUBLICS, EUROPE/ASIA, EURO-ASIA, EUROPE, UZBEK S.S.R., TURKMEN S.S.R., KIRGHIZ S.S.R., TADZHIK S.S.R., OUTER MONGOLIA, INNER MONGOLIA, MANCHUKUO, CHOSEN (KOREA), JAPAN, CHINESE REPUBLIC, TAIWAN, TIBET, NEPAL, BHUTAN, BURMA, INDIA, AFGHANISTAN, BALUCHISTAN, ANDAMAN IS., CEYLON, THAILAND, FRENCH INDO-CHINA, MALAYA SARAWAK, NETHERLANDS EAST INDIES, PHILIPPINE ISLANDS, AUSTRALIA, NEW CALEDONIA, NEW ZEALAND, FIJI, TURKEY, SYRIA, IRAN (PERSIA), IRAQ, KUWAIT, OMAN, SAUDI ARABIA, YEMEN, LEBANON, PALESTINE, TRANS-JORDAN, EGYPT, BRITISH HINTERLAND OF ADEN, FRENCH SOMALILAND, BRITISH SOMALILAND, ITALIAN SOMALILAND, ERITREA, ANGLO-EGYPTIAN SUDAN, ETHIOPIA, UGANDA, BRITISH EAST AFRICA, ZANZIBAR, GERMAN EAST AFRICA, COMOROS IS., NYASALAND, MADAGASCAR, MOZAMBIQUE, SWAZILAND, BASUTOLAND, SOUTHERN RHODESIA, NORTHERN RHODESIA, ANGOLA, BELGIAN CONGO, FRENCH EQUATORIAL AFRICA, CABINDA, RIO MUNI, KAMERUN, FERNANDO PO, SAO TOME, GOLD COAST, TOGO, NIGERIA, FRENCH WEST AFRICA, LIBERIA, SIERRA LEONE, PORT. GUINEA, GAMBIA, ALGERIA, LIBYA, TUNISIA, IFNI, SPANISH MOROCCO, MADEIRA, CANARY IS., RIO DE ORO, GERMAN SOUTHWEST AFRICA, WALVIS BAY, BECHUANALAND, UNION OF SOUTH AFRICA, TUNISIA, GREENLAND, NEWFOUNDLAND, NOVA SCOTIA, RUPERT'S LAND, NEW FRANCE, THIRTEEN COLONIES, LOUISIANA, NEW SPAIN, ALASKA, SPANISH FLORIDA, BAHAMAS, CUBA, JAMAICA, HAITI, PUERTO RICO, GUADELOUPE, MARTINIQUE, BARBADOS, BRITISH HONDURAS, GALAPAGOS ISLANDS, NEW GRANADA, DUTCH GUIANA, FRENCH GUIANA, BRAZIL, PERU, LA PLATA, PERU

Map 111 — Sovereign States: Duration of Independence

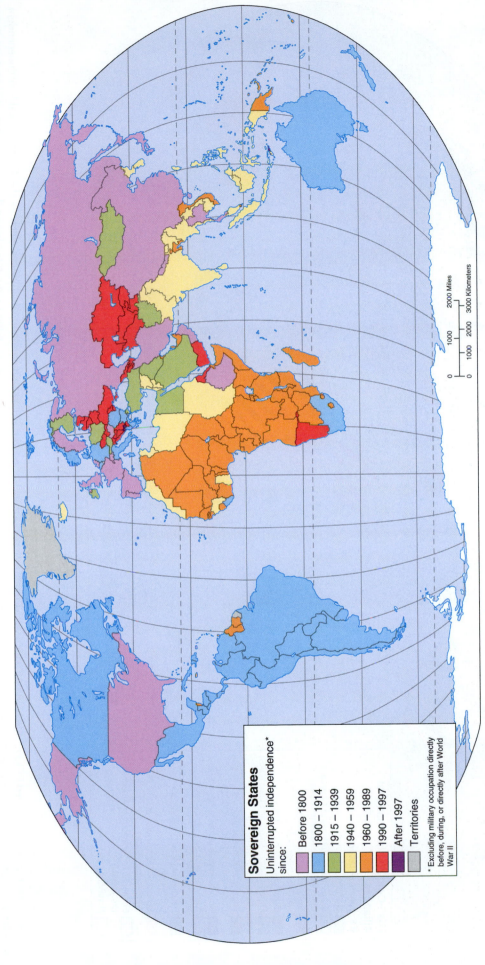

Sovereign States

Uninterrupted independence* since:

- Before 1800
- 1800 – 1914
- 1915 – 1939
- 1940 – 1959
- 1960 – 1989
- 1990 – 1997
- After 1997
- Territories

* Excluding military occupation directly before, during, or directly after World War II

0   1000   2000 Miles
0   1000   2000   3000 Kilometers

Most countries of the modern world, including such major states as Germany and Italy, became independent after the beginning of the nineteenth century. Of the world's current countries, only 27 were independent in 1800. (Ten of the 27 were in Europe; the others were Afghanistan, China, Colombia, Ethiopia, Haiti, Iran, Japan, Mexico, Nepal, Oman, Paraguay, Russia, Taiwan, Thailand, Turkey, the United States, and Venezuela). Following 1800, there have been five great periods of national independence. During the first of these (1800–1914), most of the mainland countries of the Americas achieved independence. During the second period (1915–1939), the countries of Eastern Europe emerged as independent entities. The third period (1940–1959) includes World War II and the years that followed, when independence for African and Asian nations that had been under control of colonial powers first began to occur. During the fourth period (1960–1989), independence came to the remainder of the colonial African and Asian nations, as well as to former colonies in the Caribbean and the South Pacific. More than half of the world's countries came into being as independent political entities during this period. In the last decade of the twentieth century, the breakup of the existing states of the Soviet Union, Yugoslavia, and Czechoslovakia created 22 countries where only 3 had existed before. During the first decade of the twenty-first century, there have been three new additions to the world's states. Timor-Leste became independent in 2002, and the breakup of the former Yugoslavia continued with the separation of Montenegro and Kosovo.

-133-

Map 112   Post–Cold War International Alliances

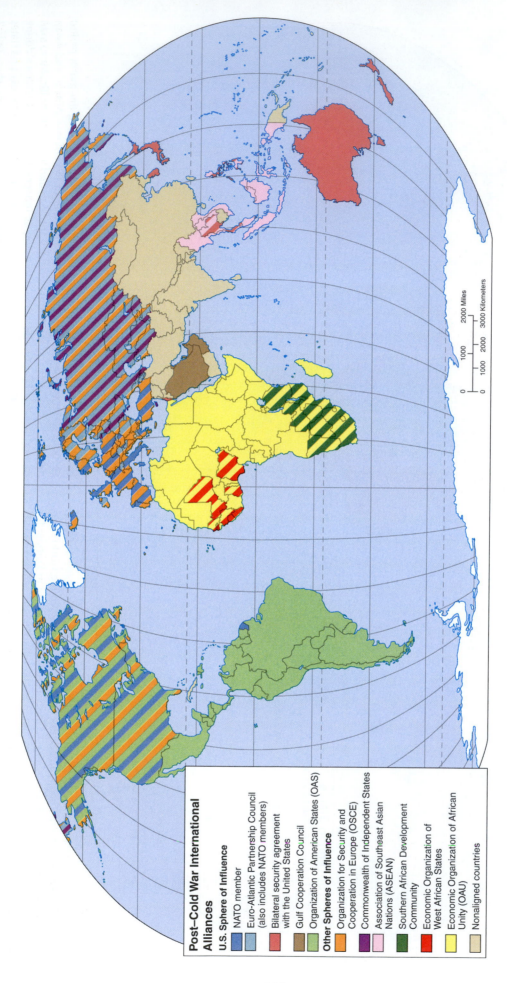

Post–Cold War International Alliances

U.S. Sphere of Influence

- NATO member
- Euro-Atlantic Partnership Council (also includes NATO members)
- Bilateral security agreement with the United States
- Gulf Cooperation Council
- Organization of American States (OAS)

Other Spheres of Influence

- Organization for Security and Cooperation in Europe (OSCE)
- Commonwealth of Independent States
- Association of Southeast Asian Nations (ASEAN)
- Southern African Development Community
- Economic Organization of West African States
- Economic Organization of African Unity (OAU)
- Nonaligned countries

When the Warsaw Pact dissolved in 1992, the North Atlantic Treaty Organization (NATO) was left as the only major military alliance in the world. Some former Warsaw Pact members (Czechia, Hungary, and Poland) have joined NATO, and others are petitioning for entry. The bipolar division of the world into two major military alliances is over, at least temporarily, leaving the United States alone as the world's dominant political and military power. But other international alliances, such as the Commonwealth of Independent States (including most of the former republics of the Soviet Union), will continue to be important. It may well be that during the first few decades of the twenty-first century economic alliances will begin to overshadow military ones in their relevance for the world's peoples.

-134-

Map 113 Political Realms: Regional Changes, 1945–2003

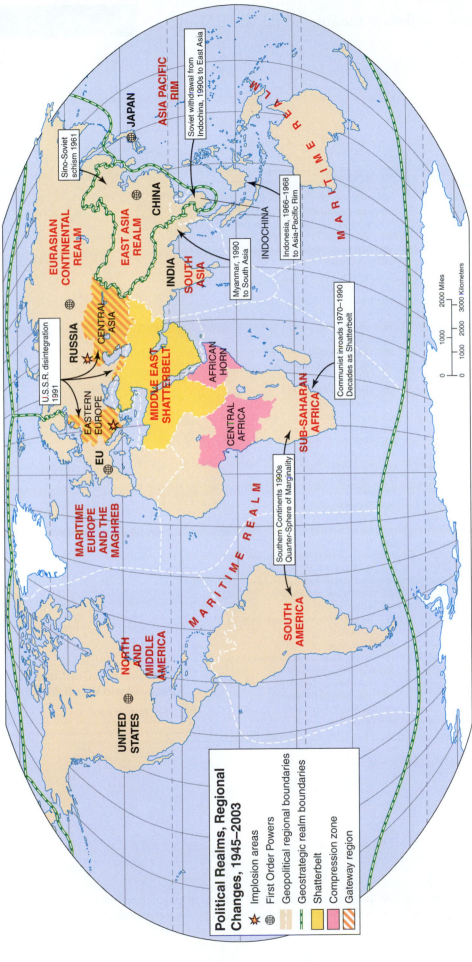

Political Realms, Regional Changes, 1945–2003

- ★ Implosion areas
- ⊕ First Order Powers
- Geopolitical regional boundaries
- Geostrategic realm boundaries
- Shatterbelt
- Compression zone
- Gateway region

EURASIAN CONTINENTAL REALM

EAST ASIA REALM

JAPAN

ASIA PACIFIC RIM

CHINA

Sino-Soviet schism 1961

Soviet withdrawal from Indochina, 1990s to East Asia

RUSSIA

CENTRAL ASIA

INDIA

SOUTH ASIA

U.S.S.R. disintegration 1991

EASTERN EUROPE

EU

MIDDLE EAST SHATTERBELT

AFRICAN HORN

INDOCHINA

Indonesia, 1966–1968 to Asia-Pacific Rim

Myanmar, 1990 to South Asia

MARITIME REALM

MARITIME EUROPE AND THE MAGHREB

CENTRAL AFRICA

SUB-SAHARAN AFRICA

Communist inroads 1970–1990 Decades as Shatterbelt

Southern Continents 1990s Quarter-Sphere of Marginality

MARITIME REALM

NORTH AND MIDDLE AMERICA

SOUTH AMERICA

UNITED STATES

2000 Miles
1000
0 1000 2000 3000 Kilometers

The Cold War following World War II shaped the major outlines of today's geopolitical relations. The Cold War included three phases. In the first, from 1945–1956, the Maritime Realm established a ring around the Continental Eurasian Realm in order to prevent its expansion. This phase included the Korean War (1950–1953), the Berlin Blockade (1948), the Truman doctrine and Marshall Plan (1947), and the founding of NATO (1949) and the Warsaw Pact (1955). Most of the world fell within one of the two realms: the Maritime (dominated by the United States) or the Eurasian Continental Realm (dominated by the Soviet Union). The Soviet Union sought to establish a ring of satellite states to protect it from a repeat of the invasions of World War II. The United States and other Maritime Realm states, in turn, sought to establish a ring of allies around the Continental Realm to prevent its expansion. South Asia was politically independent, but under pressure from both realms. During the second phase (1957–1979), Communist forces from the Continental Eurasian

Realm penetrated deeply into the Maritime Realm. The Berlin Wall went up in 1961, Soviet missiles in Cuba ignited a crisis in 1962, and the United States became increasingly involved in the war in Vietnam (late 1960s). The Soviet Union sought increased political and military presence along important waterways including those in the Middle East, Southeast Asia, and the Caribbean. These regions became especially dangerous shatterbelts. The third phase (1980–1989) saw the retreat of Communist power from the Maritime Realm. China, after ten years of radical Communism and chaos of the Cultural Revolution (1966–1976), broke away from the Continental Eurasian Realm to establish a new East Asian realm. Soviet influence declined in the Middle East, Sub-Saharan Africa, and Latin America. In 1989 the Berlin Wall fell, and Eastern Europe began to establish democratic governments. In 1991 the Soviet Union broke apart as its constituent republics became independent states. Compression zones are areas of conflict, but they are not contested by major powers.

-135-

Map 114 International Conflicts in the Post–World War II World

	Conflict	Start	Belligerents
1	Indonesia	1945	Indonesia, Netherlands, UK
2	Indochina	1946	Viet Minh, France
3	Arab-Israeli	1948	Israel, Egypt, Syria, Jordan, Lebanon, Iraq
4	Korean	1950	North Korea, South Korea, China, USSR, USA, UN Coalition
5	Algerian	1954	Algeria, France
6	Vietnam	1955	North Vietnam, South Vietnam, USA, China
7	Hungarian Revolution	1956	USSR, Hungary
8	Ifni	1957	Spain, France, Morocco
9	Angola	1961	Angola, Portugal
10	Sino-Indian	1962	China, India
11	Guinea-Bissau	1963	Guinea-Bissau, Portugal
12	Indo-Pakistani	1965	India, Pakistan
13	Six-Day	1967	Israel, Egypt, Syria, Jordan
14	Football	1969	El Salvador, Honduras
15	Bangladesh	1971	India, East Pakistan, West Pakistan
16	Indo-Pakistani	1971	India, Pakistan
17	Western Sahara	1973	Spain, Mauritania, Morocco
18	Yom Kippur	1973	Israel, Egypt, Syria
19	East Timor	1975	Indonesia, FRETILIN
20	Ogaden	1977	Ethiopia, Somalia
21	Chad-Libya	1978	Chad, Libya
22	Uganda-Tanzania	1978	Tanzania, Uganda
23	Sino-Vietnamese	1979	China, Vietnam
24	Afghanistan	1979	USSR, Afghanistan
25	Iran-Iraq	1980	Iran, Iraq
26	Falklands	1982	Argentina, UK
27	Lebanon	1982	Israel, PLO, Syria
28	Thai-Laotian	1987	Thailand, Laos
29	Nagorno-Karabakh	1988	Armenia, Azerbaijan
30	Gulf	1990	USA, Iraq, Saudi Arabia, Kuwait, UK, coalition forces
31	Croatia	1991	Croatia, Yugoslavia
32	Somalia	1991	Somalia, Ethiopia, various internal groups
33	Bosnia	1992	Bosnia, Croatian, Yugoslavia, NATO
34	Chechnya	1994	Russia, Chechnya
35	Rwanda	1994	Tutsis and Hutus of Rwanda, Burundi, and Uganda
36	Cenepa	1995	Ecuador, Peru
37	Eritrea-Ethiopia	1998	Ethiopia, Eritrea
38	Dem. Rep. of the Congo	1998	DR Congo, Uganda, Rwanda, 6 other countries
39	Kosovo	1999	Yugoslavia, Albania, NATO forces
40	Afghanistan	2001	USA, NATO, Taliban, etc.
41	Darfur	2003	Sudan, various Darfurian groups
42	Iraq	2003	USA, Iraq, UK, coalition forces
43	North-West Pakistan	2004	Pakistan, al-Qaeda, Taliban
44	Georgia	2008	Georgia, Russia, South Ossetia, Abkhazia

International Conflicts in the Post–World War II World

✳ Area of conflict

The Korean War and the Vietnam War dominated the post–World War II period in terms of international military conflict. But numerous smaller conflicts have taken place, with fewer numbers of belligerents and with fewer battle and related casualties. These smaller international conflicts have been mostly territorial conflicts, reflecting the continual readjustment of political boundaries and loyalties brought about by the end of colonial empires, and the dissolution of the Soviet Union. Many of these conflicts were not wars in the more traditional

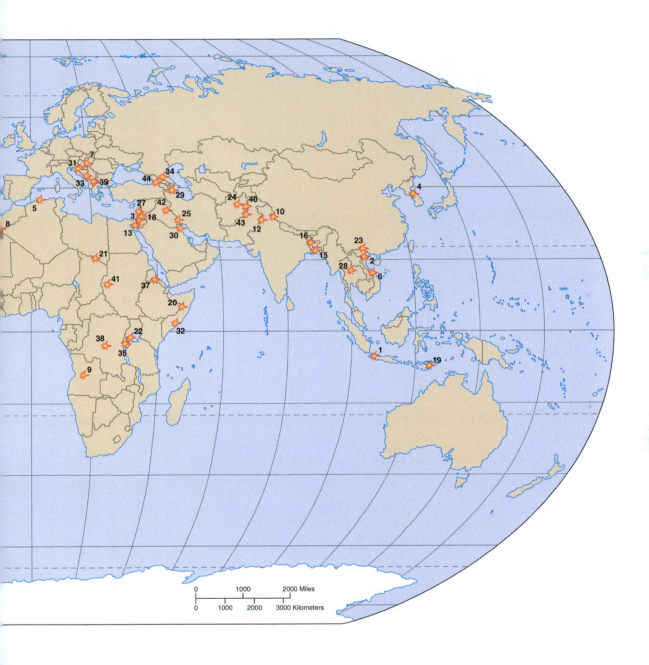

sense, in which two or more countries formally declare war on one another, severing diplomatic ties and devoting their entire national energies to the war effort. Rather, many of these conflicts were and are undeclared wars, sometimes fought between rival groups within the same country with outside support from other countries. The aftermath of the September 11, 2001, terrorist attacks on the United States indicate the dawn of yet another type of international conflict, namely a "war" fought between traditional nation-states and non-state actors.

Map 115 The Geopolitical World in the Twenty-First Century

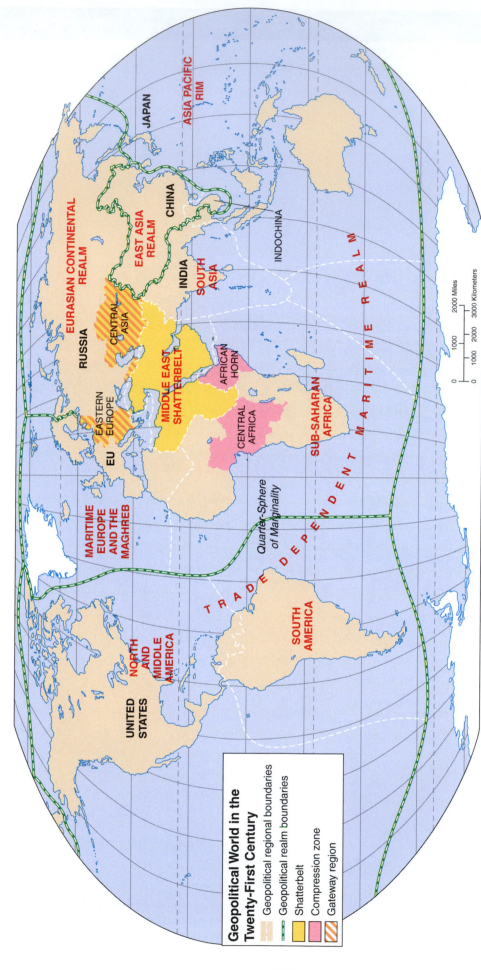

Geopolitical World in the Twenty-First Century

- Geopolitical regional boundaries
- Geopolitical realm boundaries
- Shatterbelt
- Compression zone
- Gateway region

EURASIAN CONTINENTAL REALM

RUSSIA

EAST ASIA REALM

CENTRAL ASIA

EASTERN EUROPE

EU

MARITIME EUROPE AND THE MAGHREB

MIDDLE EAST SHATTERBELT

CENTRAL AFRICA

AFRICAN HORN

SUB-SAHARAN AFRICA

JAPAN

ASIA PACIFIC RIM

CHINA

INDIA

SOUTH ASIA

INDOCHINA

Quarter-Sphere of Marginality

TRADE DEPENDENT MARITIME REALM

NORTH AND MIDDLE AMERICA

UNITED STATES

SOUTH AMERICA

2000 Miles

1000   2000   3000 Kilometers

0   1000   2000   3000 Kilometers

In the geostrategic structure of the world, the largest territorial units are realms. They are shaped by circulation patterns that link people, goods, and ideas. Realms are shaped by maritime and continental influences. Today's Atlantic and Pacific Trade-Dependent Maritime Realm has been shaped by international exchange over the oceans and their interior seas as mercantilism, capitalism, and industrialization gave rise to maritime-oriented states and to economic and political colonialism. The world's leading trading and economic powers are part of this realm. The Eurasian Continental Realm, centered around Russia, is inner-oriented, less influenced by outside economic or cultural forces, and politically closed, even after the fall of Communism. Expansion of NATO in Europe has increased its feeling of being "hemmed in." East Asia has mixed Maritime and Continental influences. China has traditionally been continental, but reforms that began in the late 1970s increased the importance of its maritime-oriented southern coasts. Even so, its trade volume is still low, and it maintains a hold on inland areas like Tibet and Xinjiang. Realms are subdivided into regions, some dependent on others, as South America is on North America. Regions located between powerful realms or regions may be shatterbelts (internally divided and caught up in competition between Great Powers) or gateways (facilitating the flow of ideas, goods, and people between regions).

Map 116 The United Nations

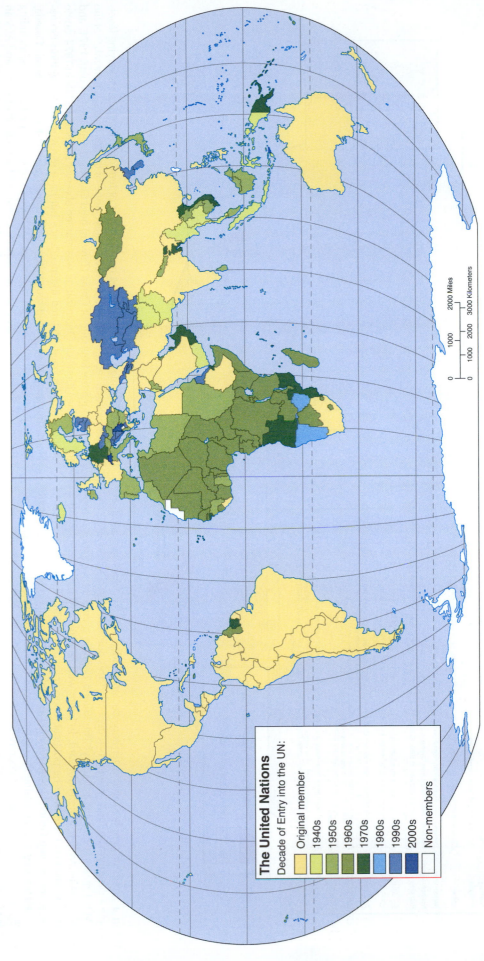

The United Nations

Decade of Entry into the UN:

- Original member
- 1940s
- 1950s
- 1960s
- 1970s
- 1980s
- 1990s
- 2000s
- Non-members

2000 Miles

3000 Kilometers

The United Nations was formed in 1945 after World War II to maintain international peace and security and to promote cooperation involving economic, social, cultural, and humanitarian problems. Originally consisting of 51 member states, the organization has grown to 192 member states in 2009. Most of the African continent and the smaller Caribbean states entered the UN during the 1960s and 1970s following the end of European colonial rule. The 1990s saw the entry of several countries following the dissolution of the Soviet Union and the breakup of Yugoslavia. Montenegro became the most recent country to join the UN in 2006 following its separation from Serbia.

China was represented by the government of the Republic of China at the creation of the United Nations. Following the Communist victory during Chinese Civil War, the government fled to the island of Taiwan. UN representation was maintained by the Republic of China government until 1971 when the government of the People's Republic of China (mainland China) was recognized as the representatives of China to the organization. Western Sahara is not a member of the UN as its sovereignty status is in dispute. Much of the territory of Western Sahara is controlled by Morocco. Kosovo declared its independence in 2008, which was recognized by the United States and 59 other countries. As of 2009, its entry into the UN has been blocked by Russia, which has not recognized its independence from Serbia. The Holy See (Vatican City) and Palestine hold status as observer.

Map 117 Is It a Country?

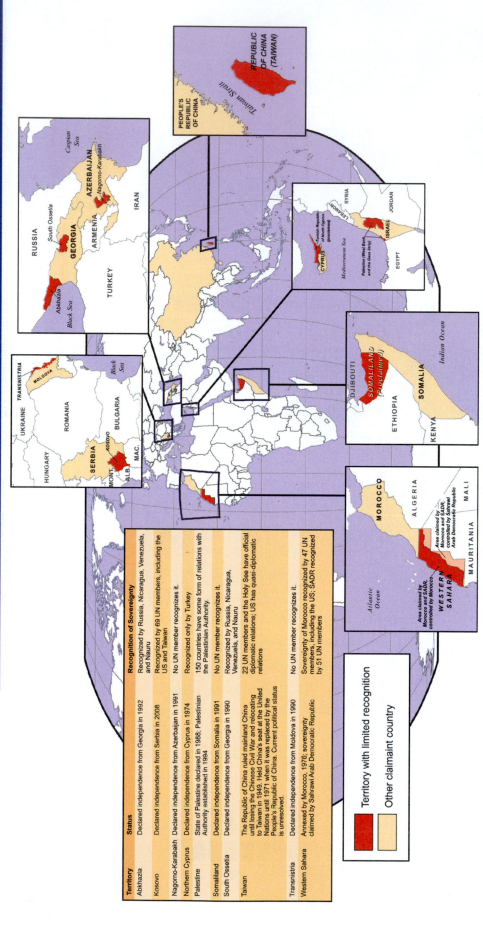

Territory	Status	Recognition of Sovereignty
Abkhazia	Declared independence from Georgia in 1992	Recognized by Russia, Nicaragua, Venezuela, and Nauru
Kosovo	Declared independence from Serbia in 2008	Recognized by 69 UN members, including the US and Taiwan
Nagorno-Karabakh	Declared independence from Azerbaijan in 1991	No UN member recognizes it.
Northern Cyprus	Declared independence from Cyprus in 1974	Recognized only by Turkey
Palestine	State of Palestine declared in 1988; Palestinian Authority established in 1994	150 countries have some form of relations with the Palestinian Authority.
Somaliland	Declared independence from Somalia in 1991	No UN member recognizes it.
South Ossetia	Declared independence from Georgia in 1990	Recognized by Russia, Nicaragua, Venezuela, and Nauru
Taiwan	The Republic of China ruled mainland China until losing the Chinese Civil War and relocating to Taiwan in 1949. Held China's seat at the United Nations until 1971 when it was replaced by the People's Republic of China. Current political status is unresolved.	22 UN members and the Holy See have official diplomatic relations; US has quasi-diplomatic relations
Transnistria	Declared independence from Moldova in 1990	No UN member recognizes it.
Western Sahara	Annexed by Morocco, 1976; sovereignty claimed by Sahrawi Arab Democratic Republic	Sovereignty of Morocco recognized by 47 UN members, including the US; SADR recognized by 51 UN members

Legend:
- Territory with limited recognition
- Other claimant country

When a group of people within a territory declares independence, does that territory become a country? The short answer is "it depends." This question begs a follow-up question: What makes a country a country? The criteria defining a country, or, more appropriately, a *state*, are fairly straightforward. To begin with, a state must have territory and a resident population. There must be political, economic, and social organization. A state possesses *sovereignty*. Generally, sovereignty can be thought of as having complete control over one's territory and possessing the right to defend oneself against external aggression. Finally, a state is *recognized* as a country by other states. Today, that recognition generally is manifested by entry in to the United Nations (Map 116). One cannot overstate the importance of international recognition, because it does not always happen automatically or quickly. For example, international recognition and subsequent entry of Macedonia into the UN was delayed because of objections by Greece to the name of the new country. Even though it is a member of the UN, Israel is not recognized as a state by several UN

members. The map above illustrates ten examples of territories that possess varying levels of international recognition of their sovereignty. None of the territories is a member of the UN, although the Republic of China (Taiwan) held China's seat in the UN until 1971, and the Palestine Liberation Organization was granted observer status to the UN in 1974. Some of the territories have declared independence but have received no recognition (Somaliland, Transnistria) or limited recognition (Abkhazia, Nagorno-Karabakh, South Ossetia, Turkish Republic of North Cyprus). Kosovo has been recognized by 69 countries, including the United States, but not by Serbia, Russia, China, and several other Asian and African states. In the case of Taiwan, it never declared independence from mainland China. Rather, it views itself as the legitimate government of all of China, while the People's Republic of China views it as a breakaway province. A handful of countries have official diplomatic relations with the Republic of China. Much of the rest of the world, including the United States, has quasi-diplomatic relations with the island.

Map 118 United Nations Regions and Sub-Regions

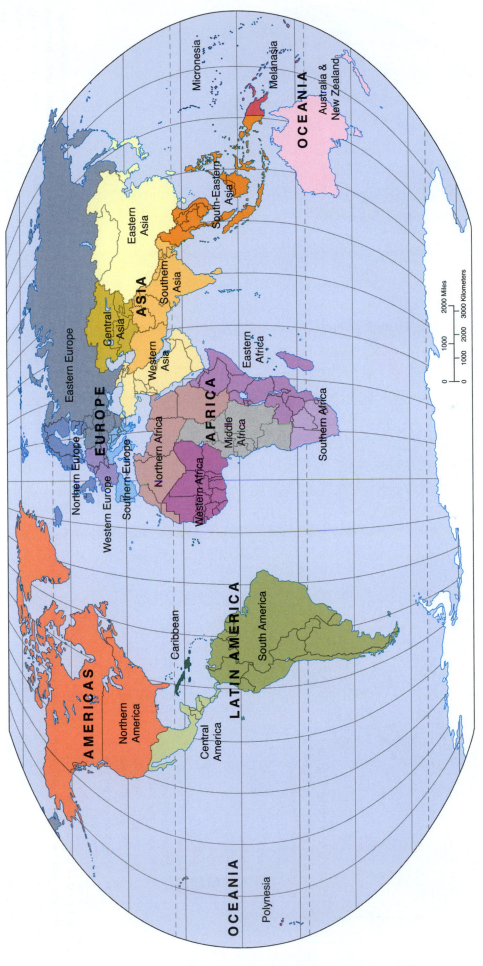

Where in the world is a country? At first, the response to that question seems fairly straightforward: China is in Asia, Angola is in Africa, and so on. The more one learns about the world, however, the more one finds that not everyone groups the world's countries the same way nor do they use standard terminology. For example, is Russia in Europe? Is it in Asia? Is it in both? Some terms commonly used in Western culture tend not to be applicable in a global context. Using "Middle East" to describe the countries of the Arabian Peninsula and the eastern Mediterranean make sense when viewing the world from the United Kingdom, but one would be hard-pressed to characterize these countries as "East" or in the "Middle" of countries to the east if one were in Japan. The map above presents the classification of the world's countries by the United Nations. How well do the names of the regions and sub-regions conform to the names with which you are familiar?

Map 119   Democracy on the Rise

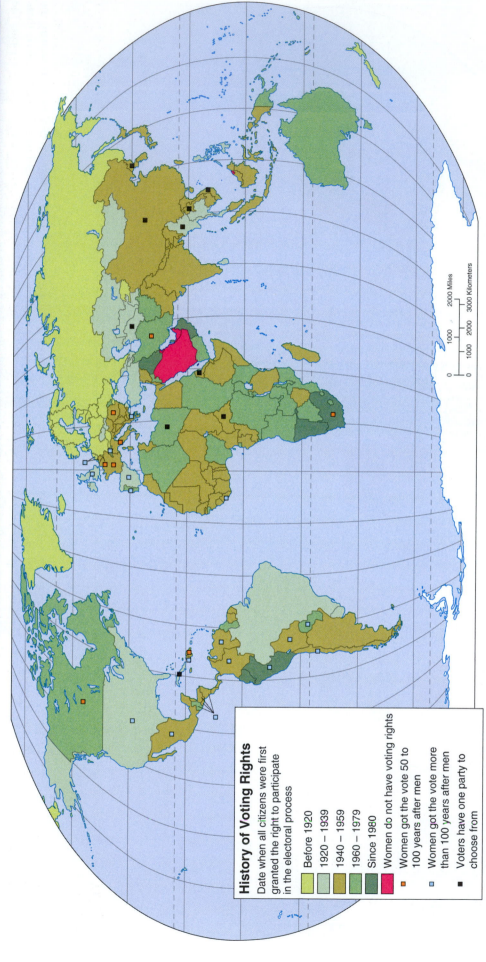

History of Voting Rights

Date when all citizens were first granted the right to participate in the electoral process

- Before 1920
- 1920 – 1939
- 1940 – 1959
- 1960 – 1979
- Since 1980
- Women do not have voting rights
- Women got the vote 50 to 100 years after men
- Women got the vote more than 100 years after men
- Voters have one party to choose from

In 1800 most of the world's people were ruled by a monarch—sometimes with a constitution and sometimes with a parliament, but a monarch, nevertheless. Few monarchs remain and where they do, their role is largely that of a figurehead than of an actual ruler. For all the veneration given to, say, Queen Elizabeth II of the United Kingdom, whoever is prime minister in that government has enormously more power—and the prime minister is elected by a constituency. But elected leaders as opposed to hereditary rulers is not the only criterion for identifying a democracy. Suffrage or who possesses the right to vote ought to be universal (it is worth noting that in the United States—one of the world's oldest democracies—black men possessed the right to vote more than a half century before white women). Elections should be competitive: a true democracy can rarely exist without multiple political parties with equal access to the electorate. Elections must also be free from internal or external coercion or corruption. And, in true democracies, there

are a number of other identifying components such as a free press, an advanced legal system in which laws rather than men rule, and a respect for human rights. By these more rigorous standards perhaps only two-thirds of the countries on this map are true, western-style liberal democracies. But that is a great deal better (unless, of course, you are a monarchist) than the situation in 1800. What explains the rise of democratic systems? Most experts point to such things as higher levels of economic development and greater literacy among populations. Even more important, however, is globalization and the shrinking of the world through increased communication systems. People who live in democracies tend to have certain advantages not possessed by those who do not. In a world of low or no barriers to interpersonal and interregional information flows, this fact is hard to hide. Isolated countries (North Korea is a case-in-point) are more likely to be autocratic than democratic.

Map 120 Nations with Nuclear Weapons

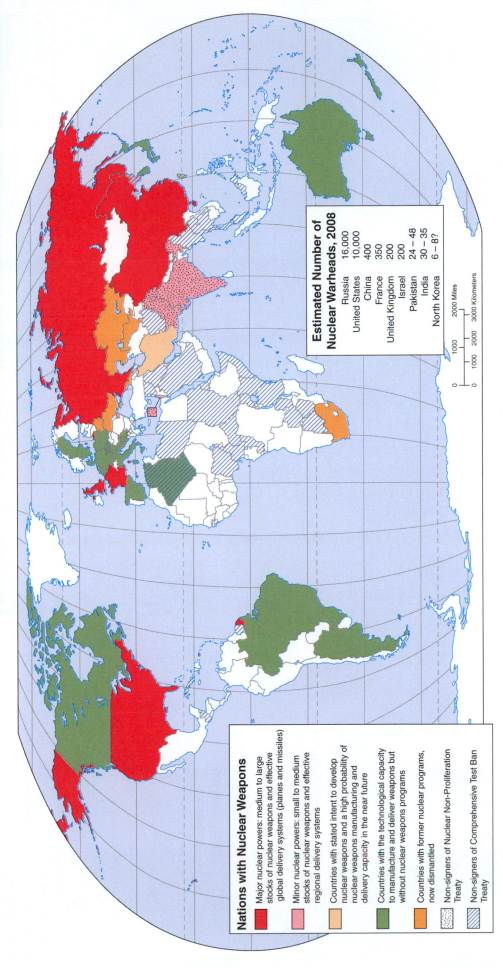

Nations with Nuclear Weapons

- Major nuclear powers: medium to large stocks of nuclear weapons and effective global delivery systems (planes and missiles)
- Minor nuclear powers: small to medium stocks of nuclear weapons and effective regional delivery systems
- Countries with stated intent to develop nuclear weapons and a high probability of nuclear weapons manufacturing and delivery capacity in the near future
- Countries with the technological capacity to manufacture and deliver weapons but without nuclear weapons programs
- Countries with former nuclear programs, now dismantled
- Non-signers of Nuclear Non-Proliferation Treaty
- Non-signers of Comprehensive Test Ban Treaty

Estimated Number of Nuclear Warheads, 2008

Russia	16,000
United States	10,000
China	400
France	350
United Kingdom	200
Israel	200
Pakistan	24 – 48
India	30 – 35
North Korea	6 – 8?

Since 1980, the number of countries possessing the capacity to manufacture and deliver nuclear weapons has grown dramatically, increasing the chances of accidental or intentional nuclear exchanges. In addition to the traditional nuclear powers of the United States, Russia, China, the United Kingdom, and France, must now be added Israel, India, and Pakistan as countries that, without possessing the large stocks of weapons of the major powers, nor the extensive delivery systems of the United States and Russia, still have effective regional (and possibly global) delivery systems and medium stocks of warheads. Countries such as Kazakhstan, Ukraine, Georgia, and Belarus that were created out of what had been the Soviet Union did have some nuclear capacity in the 1991–1995 period but have since had all nuclear weapons removed from their territories. North Korea has recently announced the re-suspension of its nuclear weapons programs, although it may possess a small stock of nuclear warheads along with the capacity to deliver those weapons regionally. Iran has announced nuclear ambitions and recently tested delivery systems. Until the overthrow of the Baathist regime of Saddam Hussein by a U.S.-led military coalition in 2003, Iraq also had nuclear ambitions. The proliferation of nuclear states threatens global security, and the objective of the Nuclear Non-Proliferation Treaty was to reduce the chances for expanding nuclear arsenals worldwide. This treaty has been partially successful in that a number of countries in the developed world certainly have the capacity to manufacture and deliver nuclear weapons but have chosen not to do so. These countries include Canada, European countries other than the United Kingdom and France, South Korea, Japan, Australia, and New Zealand, and Brazil and Argentina in South America. On the other side of the coin, the still-possible intent of North Korea to emerge as a nuclear power may force countries such as South Korea and Japan to re-think their positions as non-nuclear countries.

-143-

Map 121 Military Expenditures as a Percentage of Gross National Product

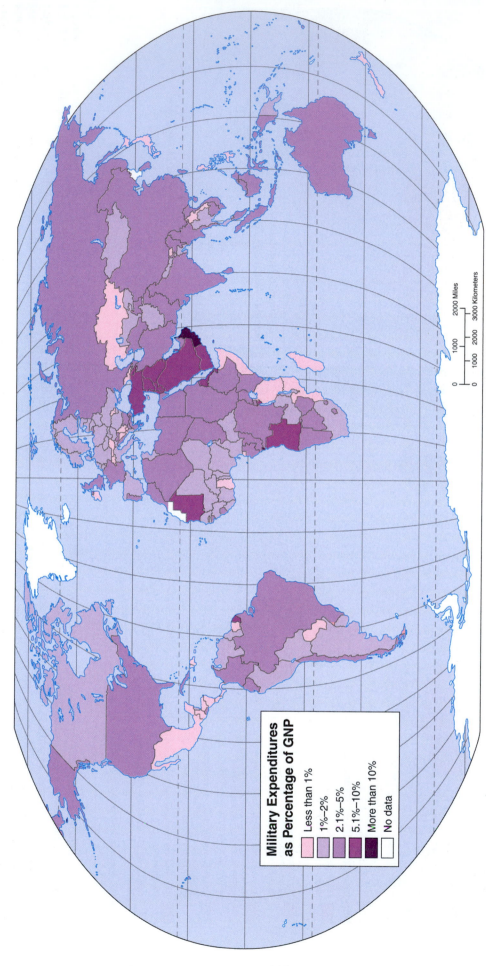

Military Expenditures as Percentage of GNP

- Less than 1%
- 1%–2%
- 2.1%–5%
- 5.1%–10%
- More than 10%
- No data

2000 Miles

3000 Kilometers

0   1000   2000

0   1000   2000

Many countries devote a significant proportion of their total central governmental expenditures to defense: weapons, personnel, and research and development of military hardware. A glance at the map reveals that there are a number of regions in which defense expenditures are particularly high, reflecting the degree of past and present political tension between countries. The clearest example is the Middle East. The steady increase in military expenditures by developing countries is one of the most alarming (and least well-known) worldwide defense issues. Where the end of the cold war has meant a substantial reduction of military expenditures for the countries in North America and Europe and for Russia, in many of the world's developing countries military expenditures have risen between 15 percent and 20 percent per year for the past few years, averaging out to 7.5 percent per year for the past quarter century. Even though many developing countries still spend less than 5 percent of their gross national product on defense, these funds could be put to different uses in such human development areas as housing, land reform, health care, and education.

Map 122   Distribution of Minority Populations

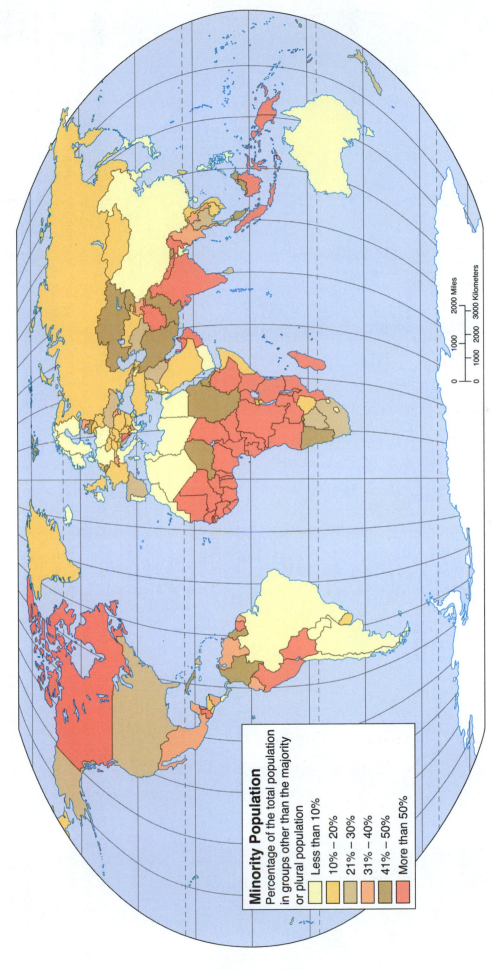

Minority Population

Percentage of the total population in groups other than the majority or plural population

- Less than 10%
- 10% – 20%
- 21% – 30%
- 31% – 40%
- 41% – 50%
- More than 50%

The presence of minority ethnic, national, or racial groups within a country's population can add a vibrant and dynamic mix to the whole. Plural societies with a high degree of cultural and ethnic diversity should, according to some social theorists, be among the world's most healthy. Unfortunately, the reality of the situation is quite different from theory or expectation. The presence of significant minority populations played an important role in the disintegration of the Soviet Union; the continuing existence of minority populations within the new states formed from former Soviet republics threatens the viability and stability of those young political units. In Africa, national boundaries were drawn by colonial powers without regard for the geographical distribution of ethnic groups, and the continuing tribal conflicts that have resulted hamper both economic and political development. Even in the most highly developed regions of the world, the presence of minority ethnic populations poses significant problems: witness the separatist movement in Canada, driven by the desire of some French-Canadians to be independent of the English majority, and the continuing ethnic conflict between Flemish-speaking and Walloon-speaking Belgians. This map, by arraying states on a scale of homogeneity to heterogeneity, indicates areas of existing and potential social and political strife.

Map 123 Marginalized Minorities: Declining Indigenous Populations

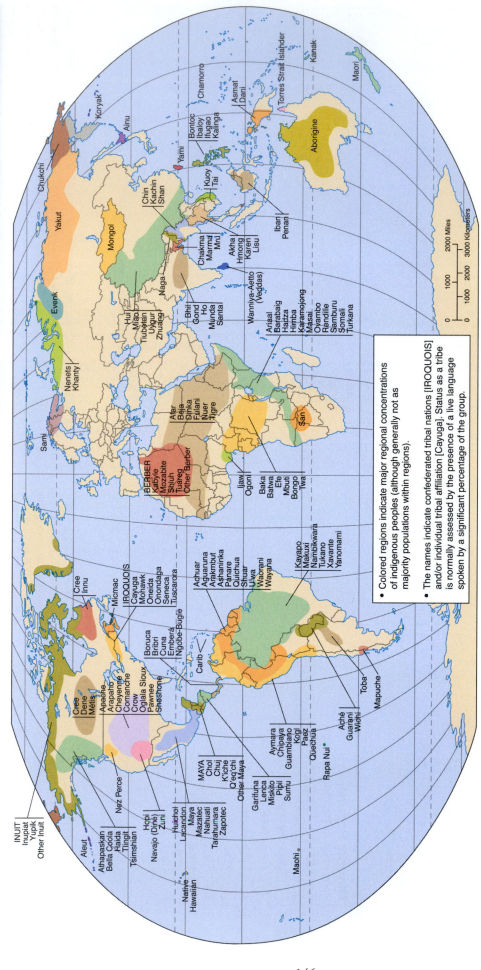

- Colored regions indicate major regional concentrations of indigenous peoples (although generally not as majority populations within regions).
- The names indicate confederated tribal nations [IROQUOIS] and/or individual tribal affiliation [Cayuga]. Status as a tribe is normally assessed by the presence of a live language spoken by a significant percentage of the group.

The world's indigenous (native) populations, those with the longest historical association with a geographic region, have nearly always been marginalized by colonialism, migration, economic development, or other external forces. Often confined to specific territories such as reservations or preserves, indigenous peoples have strived to preserve their languages, their cultures, their very identities—and often have sought to regain control of their ancestral lands as they have defined them. It has been estimated that, of the approximately 5,000 indigenous cultures remaining, fewer than half will survive through the end of the twenty-first century. There are many reasons for this: departure from the traditional life and homeland for urban areas and jobs; populations and population growth rates that have dropped below critical mass for maintenance; the scourge of substance abuse, particularly alcohol; disease rates that are and have been marginal, they may locationally marginal populations; because they are and have been marginal, they may not have had the same levels of exposure to certain diseases as more densely-crowded

populations. The best historical example of this is the massive decline in American Indian populations (in both North and South America) after European colonization. American Indians simply did not have the built-in immunities against diseases such as smallpox, measles, chicken pox, or even the common cold as did Europeans and Africans and their languages were enormous among many populations—often exceeding 90%. A similar process occurred in the latter half of the twentieth century as the Amazon Basin was rapidly occupied by representatives of mixed Old World populations (European and African) and Old World diseases took a heavy toll among indigenous peoples of the Amazon. The loss of indigenous culture is not just an academic issue but one that bears upon the survival of humanity. For it is the indigenous cultures, living close to their natural environment, who may have the knowledge of local biology that could aid in the identification of substances that could assist in the eradication of major diseases among non-indigenous populations.

Map 124

The Political Geography of a Global Religion: The Islamic World

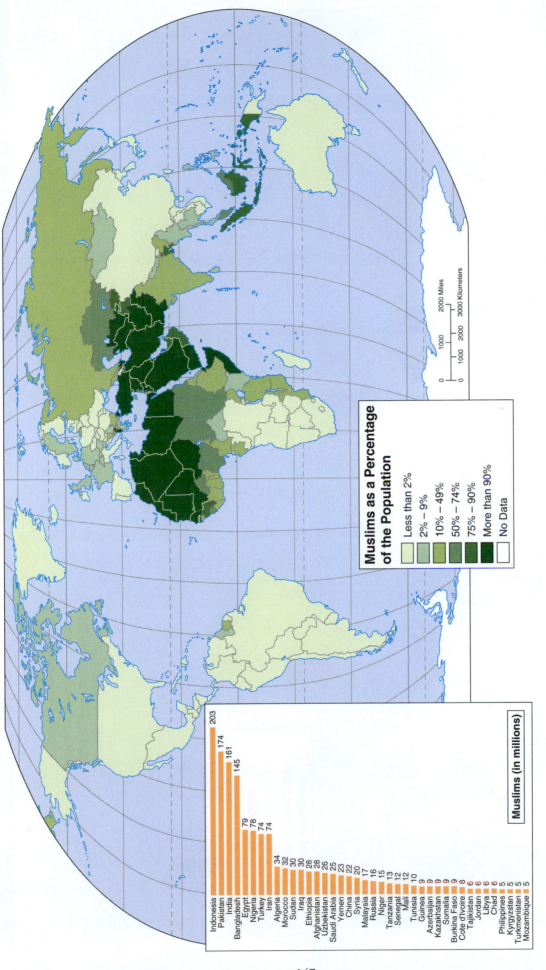

Muslims as a Percentage of the Population

- Less than 2%
- 2% – 9%
- 10% – 49%
- 50% – 74%
- 75% – 90%
- More than 90%
- No Data

Muslims (in millions)

Country	Muslims (millions)
Indonesia	203
Pakistan	174
India	161
Bangladesh	145
Egypt	79
Nigeria	78
Turkey	74
Iran	74
Algeria	34
Morocco	32
Sudan	30
Iraq	30
Ethiopia	28
Afghanistan	28
Uzbekistan	26
Saudi Arabia	25
Yemen	23
China	22
Syria	20
Malaysia	17
Russia	16
Niger	15
Tanzania	13
Senegal	12
Mali	12
Tunisia	10
Guinea	9
Azerbaijan	9
Kazakhstan	9
Somalia	9
Burkina Faso	9
Cote d'Ivoire	8
Tajikistan	6
Jordan	6
Libya	6
Chad	6
Philippines	5
Kyrgyzstan	5
Turkmenistan	5
Mozambique	5

Islam, as a religion, does not promote conflict. The term *jihad*, often mistranslated to mean "holy war," in fact refers to the struggle to find God and to promote the faith. In spite of the beneficent nature of Islamic teachings, the tensions between Muslims and adherents of other faiths often flare into warfare. A comparison of this map with the map of international conflict will show a disproportionate number of wars in that portion of the world where Muslims are either majority or significant minority populations. The reasons for this are based more in the nature of government, cultures, and social structure, than in the tenets of the faith of Islam. Nevertheless, the spatial correlations cannot be ignored. Similarly, terrorist incidents falling considerably short of open armed warfare are spatially consistent with the distribution of Islam and even more consistent with the presence of Islamic fundamentalism or "Islamism," which tends to be less tolerant and more aggressive than the mainstream of the religion. Terrorism is also consistent with those areas where the legacy of colonialism or the persistent presence of non-Islamic cultures intrude into the Islamic world.

Map 125 World Refugees: Country of Origin, 2009

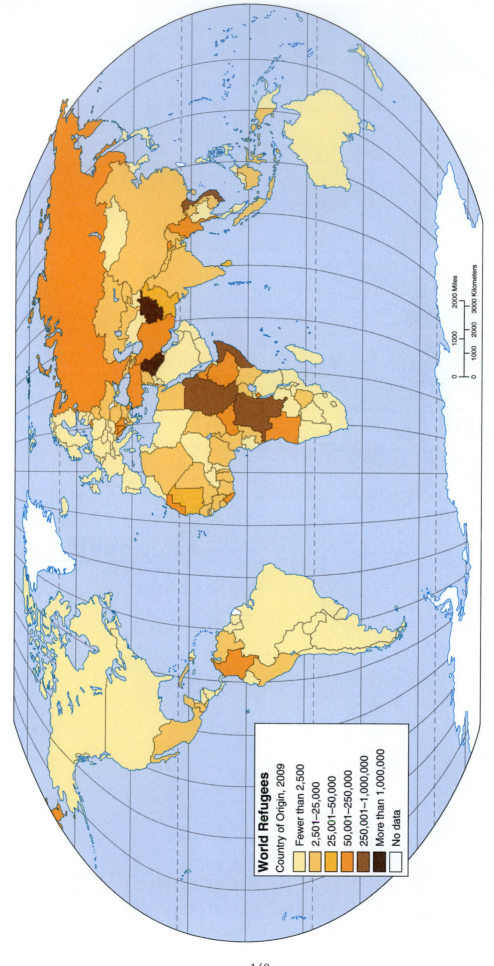

World Refugees
Country of Origin, 2009

- Fewer than 2,500
- 2,501–25,000
- 25,001–50,000
- 50,001–250,000
- 250,001–1,000,000
- More than 1,000,000
- No data

Refugees are persons who have been driven from their homes and seek refuge in another country. While there are many reasons why people flee their home country, the vast majority are fleeing armed conflict. In such cases, there may be a mass exodus from the country involving tens or perhaps hundreds of thousands of persons. Most refugees flee into neighboring countries. Because armed conflict is oftentimes short-lived, the number of refugees in the world changes from year to year, sometimes substantially. The refugee population is recognized by international agencies and is monitored by the United Nations High Commissioner for Refugees (UNHCR).

Map 126   World Refugees: Host Country, 2009

World Refugees

Host Country, 2009

- Fewer than 1,000
- 1,001–10,000
- 10,001–100,000
- 100,001–250,000
- 250,001–500,000
- 500,001–1,070,488
- No data

2000 Miles

1000   2000   3000 Kilometers

0   1000   2000

0   1000

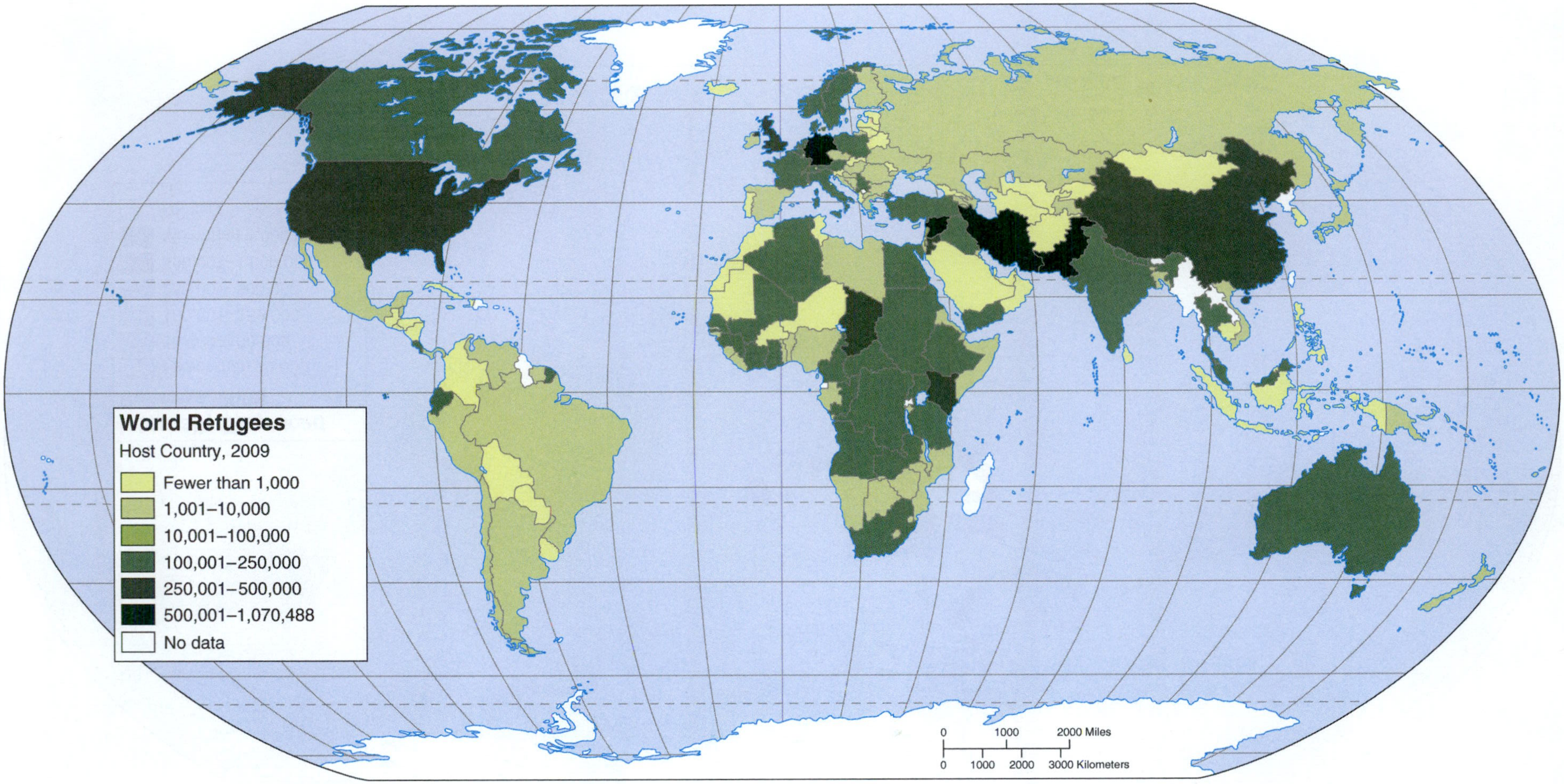

When refugees flee their country, they most commonly flee to a neighboring country, and not every country is equally equipped to handle such an influx of persons. During such times, international agencies often financially reward the countries of refuge for their willingness to take in externally displaced persons. For most of the host countries, the challenge of hosting a large refugee population is a short-term problem. As we have seen in recent decades, the burden of hosting massive numbers refugees can be a destabilizing force for the country of refuge.

Map 127 Internally Displaced Persons, 2009

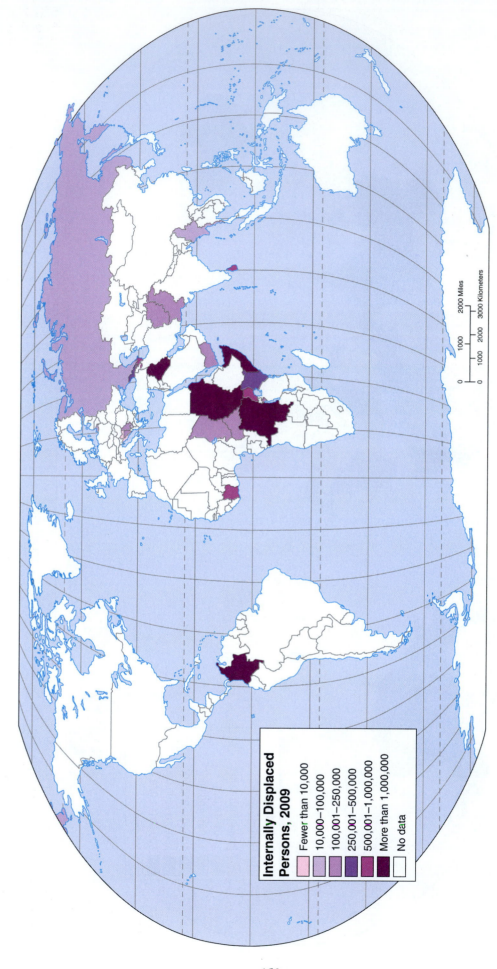

Internally Displaced Persons, 2009

- Fewer than 10,000
- 10,000–100,000
- 100,001–250,000
- 250,001–500,000
- 500,001–1,000,000
- More than 1,000,000
- No data

Internally displaced persons (IDPs) are those who flee their homes because of conflict, persecution, or disaster and seek refuge in another location within their country. With the exception of not leaving their country, they are essentially the same as refugees. International organizations offer the same assistance to internally displaced persons as is provided to refugees. The actual number of IDPs is more difficult to assess than refugee data. Not only do IDP populations fluctuate, there likely are a large number of displaced persons who flee to the larger cities in the countries rather than to the camps established by international relief organizations. In early 2009, the countries with the greatest number of IPDs were Colombia, Iraq, Democratic Republic of the Congo, Somalia, and Sudan.

Map 128 Abuse of Public Trust

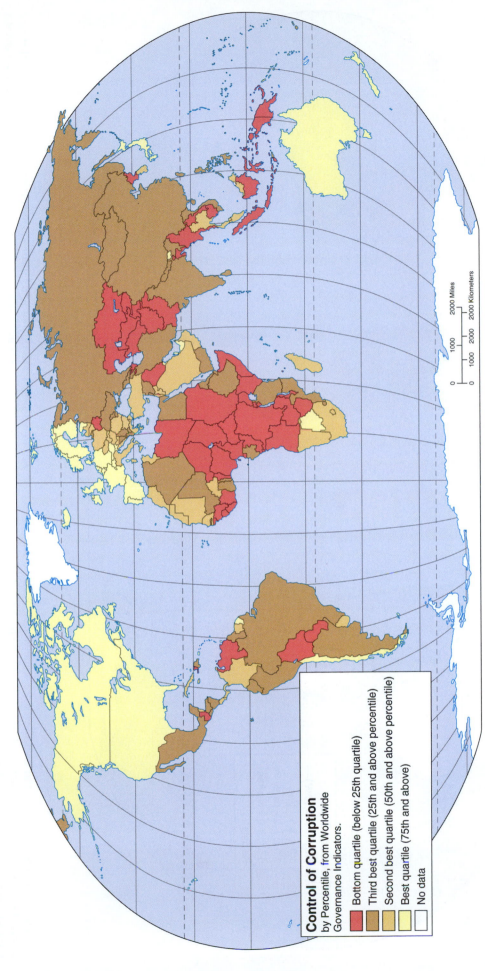

Control of Corruption
by Percentile, from Worldwide Governance Indicators.

- Bottom quartile (below 25th quartile)
- Third best quartile (25th and above percentile)
- Second best quartile (50th and above percentile)
- Best quartile (75th and above)
- No data

Abusing the public trust is simply another way of saying "corruption in government." In many parts of the world, corruption in the government is not an aberration but a way of life. Normally, although not always, governmental corruption is an indication of a weak and ineffective government, one that negatively affects such public welfare issues as public health, sanitation, education, and the provision of social services. It also tends to impact the cost of doing business and, thereby, drives away the foreign capital so badly needed in many African and Asian countries for economic development. Corruption is not automatic in poor countries, nor are rich countries free from it. But there is a general correlation between abuse of the public trust and lower levels of per capita income—excepting such countries as the Baltic states and Chile that have reached high standards of governance without joining the ranks of the wealthy countries. Studies by the World Bank have shown that countries that address issues of corruption and clean up the operations of their governments increase national incomes as much as four or five times. In those countries striving to attain governments that function according to a rule of law—rather than a rule of abusing the public trust—such important demographic measures as child mortality drop by as much as 75%. Clearly, good government and good business and higher incomes and better living conditions for the general public all go hand-in-hand.

Map 129 Political and Civil Liberties, 2008

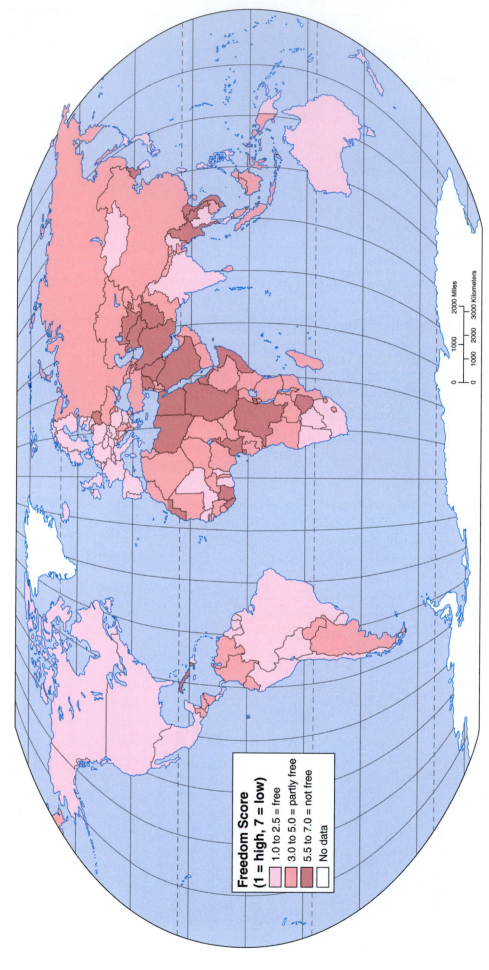

**Freedom Score
(1 = high, 7 = low)**

- 1.0 to 2.5 = free
- 3.0 to 5.0 = partly free
- 5.5 to 7.0 = not free
- No data

Although measures of political and civil liberty are somewhat difficult to obtain and assess, there are some generally accepted standards that can be evaluated: open elections and competitive political parties, the rule of law, freedoms of speech and press, judicial systems separate from other branches of government, and limits on the power of elected or appointed governmental officials. Interstingly, there appear to be correlations between "degrees of freedom" and such other characteristics of a state as per capita wealth, environmental quality, and healthy economic growth—characteristics that may be mutually contradictory. There is no empirical evidence of a causal link between democratic institutions and consumption; on the other hand, there is clear evidence of a positive relationship between wealth and consumption. Therefore, the three variables are closely correlated and should be used in assessing the nature of the state in any part of the world.

Map 130   Human Rights Abuse

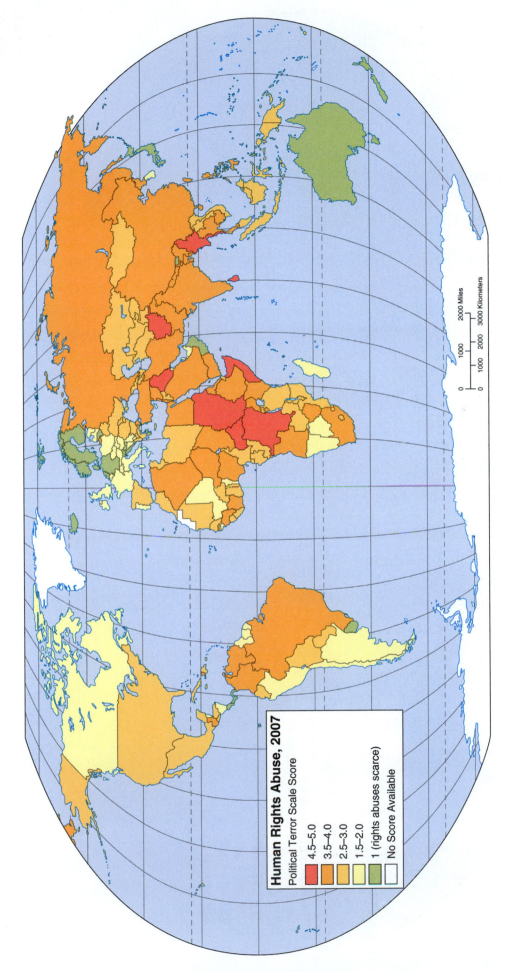

The Political Terror Scale measures levels of political violence and terror that a country experiences in a particular year. The score is calculated using data from the U.S. State Department Country Reports on Human Rights and yearly country reports from Amnesty International. At the lowest end of the scale, torture or political murder are scarce and the rule of law dominates. Higher scores indicate increasing pervasiveness of human rights abuses. At the highest level, a country's entire population is affected by political rights abuses. Countries with high scores typically are dictatorships or totalitarian states whose leaders oftentimes carry out political terror through secret police forces or death squads. In many politically unstable countries, the Political Terror Scale may fluctuate from year to year. In the early 2000s countries like Liberia, Colombia, and Rwanda ranked much higher than in 2007. Conversely, Central African Republic, Thailand, and Kenya have seen increases in the score since the early 2000s. Afghanistan, Democratic Republic of the Congo, Somalia, and Iraq have had very high scores since the turn of the century.

Map 131 Women's Rights

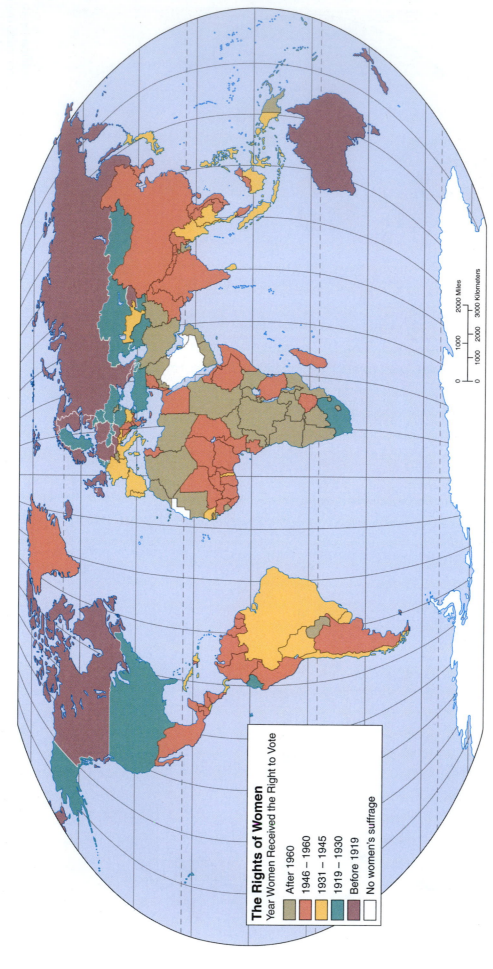

The Rights of Women
Year Women Received the Right to Vote

- After 1960
- 1946 – 1960
- 1931 – 1945
- 1919 – 1930
- Before 1919
- No women's suffrage

The "rights" referred to in this map refer primarily to the right to vote. But where women have the right to vote in free elections, the other fundamental rights tend to become available as well: the right to own property, the right to an education, the right to leave a domestic alliance without fear of retribution, or the right to be treated as a human being rather than property. But the time lag between women receiving the right to vote and their attainment of other fundamental human rights does not occur immediately or, in many cases, even relatively quickly. On the map, the most recent countries to grant suffrage to women are in Africa and Southwest Asia. In these regions, women still do not have access to many of the basic rights of what we would consider to be a civilized life. And, of course, there are still areas where women cannot vote: Kuwait, for example. Neither men nor women are allowed to vote in Brunei, Saudi Arabia, United Arab Emirates, or Western Sahara.

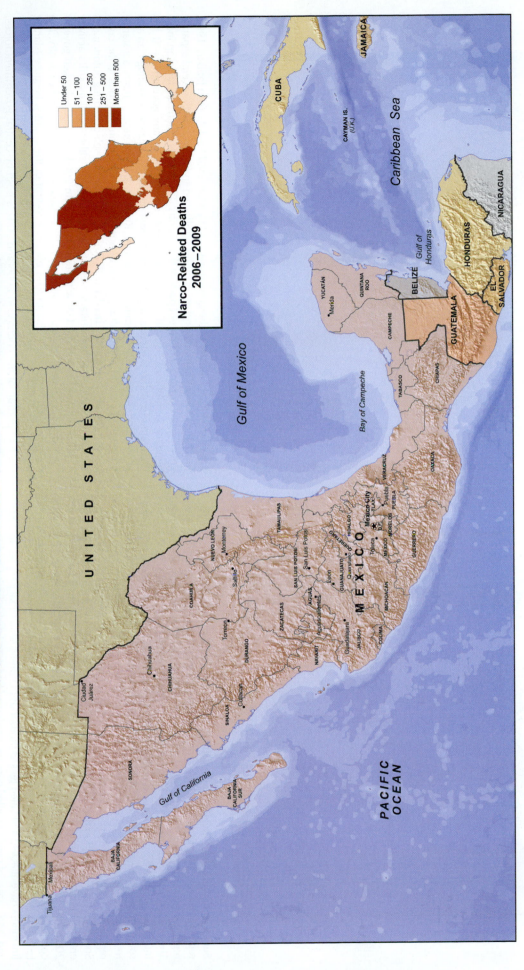

Map 132

Map 132   Flashpoints, 2011

Mexico: Although drug-related crime has been present in Mexico for decades, the last few years has seen a dramatic increase in narco-related violence. The surge in violence came as a result of Mexican drug cartels supplanting Colombian drug cartels for control of illegal drug trafficking into the United States. These cartels have become heavily armed and not only are capable of battling government troops, but also have targeted police officers, politicians, journalists, and entertainers for kidnapping and murder. Although there have been many drug cartels operating in Mexico, power recently has begun to consolidate among those operating in the cities of Ciudad Juaréz and Tijuana as well as the states of Michoacán, Sinaloa, and Tamaulipas. Mexico ranks in the top ten most dangerous countries for journalists to work in and, in 2006, it ranked behind only Iraq for this dubious honor. *The Los Angeles Times* reported in 2010 that nearly 23,000 persons have died in drug-related violence in Mexico since the start of 2007. All of this presents problems both for Mexico and for the United States. At nearly 2,000 miles in length, the border between Mexico and the United States is the most frequently crossed international border in the world, and the border cities of Ciudad Juaréz and Tijuana are the most violent. Given the likelihood of continued demand for foreign narcotics in the United States in the near future, narco-related violence, and the ever-increasing attention to U.S.-Mexican border issues, Mexico will likely continue to be a flashpoint for the foreseeable future.

Map 132c

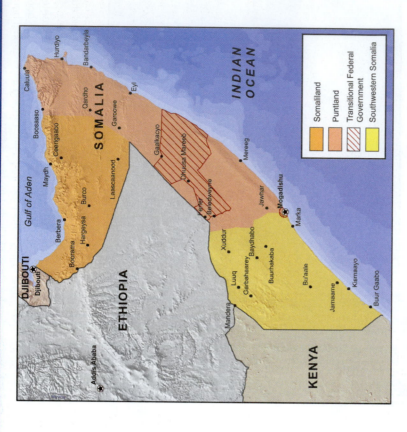

Somalia: With the ouster of the government led by Mohamed Said Barre in January 1991, turmoil, factional fighting, and anarchy have followed in Somalia, with several separate governments arising in different parts of the country. The northern clans declared an independent Republic of Somaliland. Although not recognized by any government, it has maintained a stable existence, aided by the overwhelming dominance of a ruling clan and economic infrastructure left behind by British, Russian, and American military assistance programs. Puntland, the central portion of Somalia, from the Horn of Africa to the coast of the Indian Ocean and the border with Ethiopia, has been a self-governing autonomous state since 1998. UN humanitarian efforts were unable to either quell guerrilla activity or alleviate famine, and the UN withdrew in 1995. In 2004 a new UN-backed government, the Transitional Federal Government (TFG), was created for the entire country. The government has not been able to gain effective control of the country, because Islamist insurgencies such as al-Shabaab have been able to gain control over much of the southwestern portion of the country. In 2006 and 2007 a push by central government forces, backed up by units of the Ethiopian regular army, succeeded in driving Islamic extremist forces out of the Mogadishu region. However, Ethiopian troops pulled out of Somalia in early 2009, and there is still no unification of the country. In fact, the lack of effective governmental control has led to Somali pirates becoming a major threat to shipping in the Indian Ocean. By any definition, Somalia is a "disordered" or "failed" state.

Map 132b

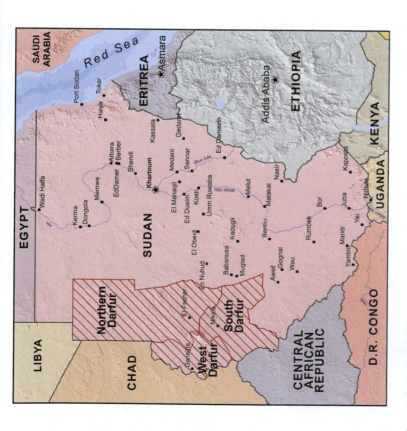

Sudan and the Darfur Region: Since Sudan achieved independence in 1956, military regimes favoring Islamic-oriented governments have dominated national politics. These regimes have embroiled the country in a civil war for nearly all of the past half-century. These wars have been rooted in the attempts of northern economic, political, and social interests dominated by Muslims to control territories occupied by non-Muslim, non-Arab southern Sudanese such as the Dinka tribal groups. Since 1983, war and famine have resulted in more than 2 million deaths and over 4 million people displaced. The current regime is a mixture of military elite and an Islamist party that came to power in a 1989 coup. Some northern opposition parties have made common cause with the southern rebels and entered the war as part of an anti-government alliance. Peace talks gained momentum in 2002–2003 with the signing of several accords, including a cease-fire agreement. However, conflicts have continued to persist in the three Darfur provinces of western Sudan adjacent to the border with Chad and the Central African Republic, where government-backed Muslim militia have attacked and killed tens of thousands of non-Muslim tribal peoples. In 2005 and 2006, areas of conflict spilled over the borders of Sudan to involve both Chad and the Central African Republic. The Darfur region is relatively water-rich and forested in a country that is chiefly desert and is therefore desired by Muslim pastoral groups from the north for settlement purposes. Another cease-fire was agreed to in February 2010, but it was short-lived; the Sudanese army launched raids and air strikes later in the year.

Map 132e

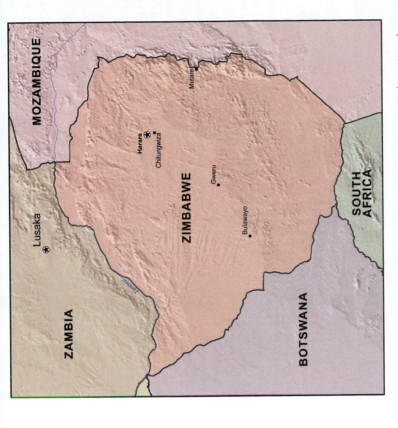

Map 132d

Dem. Rep. Congo: The war in the Democratic Republic of the Congo (formerly Zaire) has preoccupied United Nations and African diplomats since 1999. Troops from Zimbabwe, Angola, Sudan, Chad, and Namibia joined with president of the Congo Laurent Kabila against his former allies Rwanda, Burundi, and Uganda, who each backed several separate Congolese rebel groups. The origins of the conflict lie in the overthrow of longtime dictator Mobutu Sese Seko by Kabila's army in 1997 after a year of civil war. Kabila's failure to call elections or stabilize the country's economy led to further rounds of rebellion in the huge but fractious nation—rebellion supported by the economic and military assistance of neighboring Rwanda, Burundi, and Uganda. After Kabila's assassination in 2001 and the succession of his son, Joseph, to the presidency, accord seemed to have been reached, and the various conflicting parties agreed to withdraw troops in 2002. But in early 2003, new fighting flared along the country's eastern border, threatening a new and broadened war and the addition of more deaths to the 3.9 million since 1998. Diplomats called the conflict "Africa's first world war." With fighting continuing in the east, fears are that the Congo conflict could destabilize the entire southern half of the continent, leading to massive refugee flows and abject poverty.

Zimbabwe: When Zimbabwe achieved independence in 1980, Robert Mugabe assumed leadership of the country—a position he has held ever since. In 2000, Mugabe instituted a highly controversial land reform program, appropriating white-owned farms and giving them to tribal leaders. A country that had Africa's highest literacy rate and was one of the continent's leaders in agricultural production, Zimbabwe quickly deteriorated into conditions of abject poverty and famine. Hyperinflation was rampant in the early 2000s, peaking at a rate of over 11 million percent and bringing the country to the brink of economic collapse. Elections held in 2008 left Mugabe's chief rival, Morgan Tsvangirai, with the largest number of votes, but he did not receive enough votes to prevent a run-off election. Before that election could be held, Tsvangirai withdrew from the race, claiming (quite probably with some justification) that his supporters had been threatened with beatings, imprisonment, murder, and torture and that he did not wish to subject them to those dangers. In 2009, Tsvangirai was sworn in as Prime Minister, part of a power-sharing agreement with Mugabe—at least on paper—in which he and Mugabe would combine their forces to try to lead the country out of its economic chaos and medical crisis created by a cholera epidemic. Although Zimbabwe does not have the tribal conflicts that beset so many African nations (most Zimbabweans are Shona), the feelings between the supporters of the two rival political factions run deeply enough that, if the power-sharing arrangement does not work and the economy continues to deteriorate, the country could be plunged into a civil war.

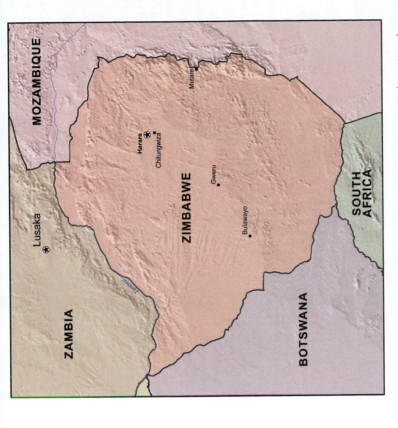

Map 132g

Map 132f

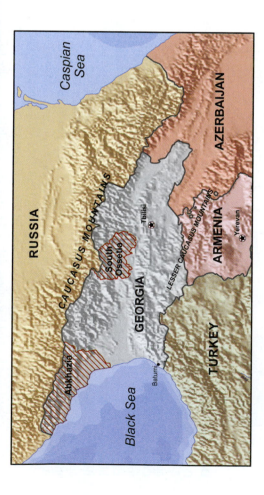

Georgia: Along with Azerbaijan and Armenia, Georgia is one of the trans-Caucasus states that were former Soviet Socialist Republics, attaining their independence after the breakup of the U.S.S.R. Unlike Azerbaijan, which is fairly homogenous in ethnicity and religion (Azerbaijani and Muslim), and Armenia, which is also an ethnically and religiously uniform area (Armenian and Christian), Georgia is a crazy quilt of ethnic, linguistic, and religious groups. Such a lack of national uniformity does not bode well for the stability of a political unit, and almost from its independence Georgia has been beset by internal turmoil. Much of this has been centered in the north, in the province of South Ossetia, and in the west, in the province of Abkhazia. Both provinces have wanted, almost from the creation of Georgian independence, to achieve independence on their own. Matters came to a head in the summer of 2008 when the Russian Federation announced it was formally recognizing the independence of Abkhazia and South Ossetia and sent troops to the breakaway provinces to augment already existing "peace-keeping" forces. The Russians have no particular religious or ethnic links with populations in either South Ossetia or Abkhazia; neither province's language is related to Russian, and in religion the populations are far from uniformly Russian Orthodox Christian. So why did Russia "invade" Georgia? Partly to demonstrate that it could, perhaps, using an ill-advised Georgian military thrust into South Ossetia as an excuse. More important, the increased Russian occupation of areas of the Republic of Georgia signaled to the Georgians that the Russians would not take it kindly if Georgia continued its overtures to the West, particularly to the NATO alliance. Although probably not a significant enough flashpoint to precipitate a regional trans-Caucasian war, the Russian/Georgian conflict signals the beginning of what could be a period of unrest and conflict for the entire region. It may also signal the beginning of a period of increased military tension between the Russian Federation and the United States that some political and military experts are already calling "a new Cold War."

Kyrgyzstan: Kyrgyzstan was one of the Central Asian republics of the USSR that achieved independence in 1991 following the collapse of the Soviet Union. Conflict in Kyrgyzstan gained little international attention until an outbreak of ethnic unrest in 2010 and the subsequent resignation and expulsion of President Kurmanbek Bakiev. There have been several clashes between ethnic Kyrgyz and Uzbeks over the years, and much of the animosity seems to be centered on land ownership. Uzbeks comprise about fifteen percent of the population, and most live in the fertile Fergana Valley near the border with Uzbekistan. Early in 2010, demonstrations organized by opposition leaders protesting corruption and inflation turned violent. Subsequent clashes between ethnic Kyrgyz and Uzbeks in the city of Osh prompted Bakiev to request Russian military assistance to quell a potential civil war—a request denied by the Russians. Bakiev resigned the presidency in April 2010 and was replaced by Roza Otunbayeva, who called for elections in 2011. In addition to continuing concerns of corruption and ethnic violence, Kyrgyzstan also has had occasional outbreaks of violence connected to jihadist groups in the south. The United States has had a military presence in the country since 2001, when it established an air base near the capital city of Bishkek in support of the war in Afghanistan.

Map 132i

Iraq: Prior to the 1990–91 invasion of Kuwait by Iraq and the subsequent United Nations coalition's military expulsion of Iraq from its neighbor, Iraq was one of the most prosperous countries in the Middle East and the only one with full capacity to feed itself, even without the vast oil revenues generated by the country's immense reserves. Despite the inefficiencies of the Baathist dictatorship of Saddam Hussein, the country had a solid agricultural base and a burgeoning industry. The combination of military adventurism and conflict, in the form of a lengthy war with Iran and the ill-advised invasion of Kuwait, limited further economic development, however. Development was also problematic given the country's internal tensions between Arabic Sunni Muslims and Arabic Shiite Muslims, and between Arabs and Kurds and a few other minority populations in the northern parts of the country. Seven years after the U.S.-led invasion, Iraq is still in critical condition, although important strides were made following the 2007 U.S. troop surge. By 2008, violence had dropped substantially and the U.S. began troop draw-downs. In 2009, U.S. troops withdrew from urban areas, transitioning security to Iraqi forces. President Obama announced that the U.S. combat mission would end in 2010 and that all troops are to be gone by the end of 2011.

Map 132h

Israel and Its Neighbors: The modern state of Israel was created out of the former British Protectorate of Palestine, inhabited primarily by Muslim Arabs, after World War II. Conflict between Arabs and Israeli Jews has been a constant ever since. Much of the present tension revolves around the West Bank area, not part of the original Israeli state but taken from Jordan, an Arab country, in the Six-Day War of 1967. Many Palestinians had settled this part of Jordan after the creation of Israel and remain as a majority population in the West Bank region today. Israel has established many agricultural settlements within the region since 1967, angering Palestinian Arabs. For Israel, the West Bank is the region of ancient Judea and this region, won in battle, will not be ceded back to Palestinian Arabs without protracted or severe military action. The West Bank, inhabited by nearly 400,000 Israeli settlers and 4 million Palestinians, is also the location of most of the suicide bombings carried out by Islamic militant groups from 2001 to 2009. By early 2008, the Gaza Strip had emerged as the most critical flashpoint in the area. The Israeli government and the Palestinian Authority had agreed to resume peace talks with the goal being a peace agreement by the end of the year. But in late 2008 and early 2009, Israeli troops responded to rocket attacks by Hamas, a leading Palestinian political party, by attacking Gaza in force. By January 2009, over 1,000 Palestinians and Israelis had died in this latest conflict.

-159-

Map 132k

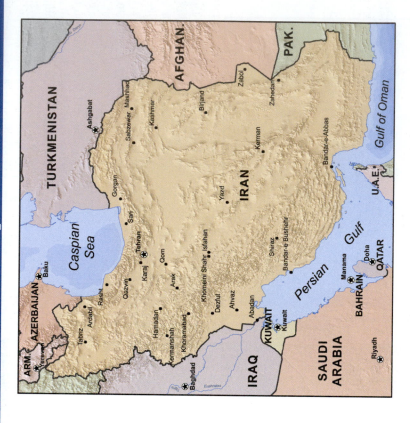

Iran: Iran has been a bone in the throat of the United States and Israel since the overthrow of the pro-Western government of the Shah and the installation of a fundamentalist Islamic republic in 1979. The Iranians took a large number of American hostages following the political revolution and held them for over a year. During an eight-year war between Iran and its nearest neighbor, Iraq, the United States provided military aid to Iraq, further straining the relations between Iran and the West. Iran's promise to continue development of nuclear facilities that could lead to the development of nuclear weapons has heightened distrust of Iran in the West. More likely, however, is that once Iran has developed facilities capable of producing weapons-grade plutonium, Israel will carry out the same type of preemptive strikes it has previously used on Iraq and Syria. Iran is a large and important country, poorly understood by the United States. In 2009, hard-line President Ahmadinejad won re-election in a highly contested and controversial vote. Millions of Iranians took to the street to protest, but Supreme Leader Ayatollah Khamenei endorsed Ahmadinejad as the winner and declared the protests illegal. Ahmadinejad has pushed construction of an atomic power station, declaring production of nuclear fuel to be an "inalienable right." In response, the UN imposed sanctions on Iran in 2010. With an ancient imperial tradition, Iran is the historical core of Shiah Islam (nearly 90% of Iranians are Shiite Muslims), and it possesses enormous reserves of oil and natural gas.

Map 132j

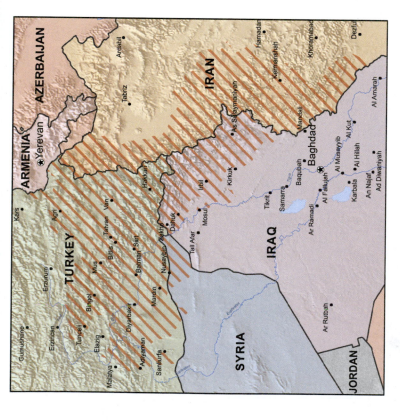

Kurdistan: Where Turkey, Iran, and Iraq meet in the high mountain region of the Tauros and Zagros mountains, a nation of 25 million people exists. This nation is "Kurdistan," but the Kurds, the occupants of this area for over 3,000 years, have no state, and receive much less attention than other stateless nations like the Palestinians. Following the 1991 Gulf War between Iraq and a U.S.-led coalition of European and Arabic states, the United Nations demarcated a Kurdish "security zone" in northern Iraq. From 1991 to 2003 the Security Zone was anything but secure as Iraqi militants from the south and Turks from the north infringed on Kurdish territory, and the internal militant extremist groups, such as the Kurdish Workers' Party, staged periodic attacks on rival villages. During the 2003 U.S.-led invasion of Iraq that eliminated the Baathist regime of Saddam Hussein, the Kurds played an important role in securing the northern portions of Iraq for the U.S.-British coalition and fought alongside American troops in expelling elements of the Iraqi army from cities like Mosul and Kirkuk. Rich in oil and history, Kurdistan will probably remain as a nation without a state, shared by Iraq, Turkey, and Iran—none of which is likely to give up substantial portions of territory for the establishment of a Kurdish state. In 2011, the portion of northern Iraq under Kurdish control was the most stable of that war-torn country.

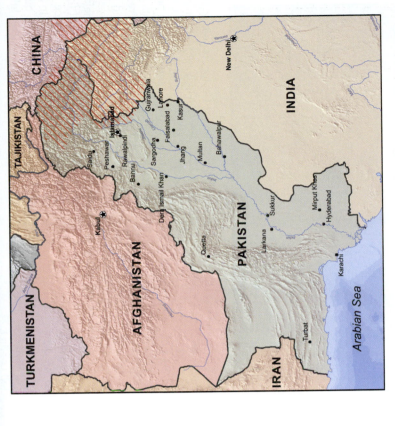

Pakistan: Pakistan's importance to the West and its potential as a flashpoint represent one of the most critical threats to world peace in the opening years of the twenty-first century. Pakistan is one of the 10 most populous countries in the world; it possesses both nuclear weapons and a delivery system (as does its nearest neighbor and chief antagonist, India); and it teeters on the brink of being either a western-style representative democracy or an Islamic fundamentalist state. Strategically, Pakistan lies at the western end of the core of the Muslim world (although large Muslim populations exist to the south and east in India, Malaysia, and Indonesia) and is immediately adjacent to U.S. military operations against the Taliban in Afghanistan. Physically, Pakistan is an incredibly rugged country mixing a large river floodplain (the Indus) with high mountain country with peaks in excess of 25,000 feet in elevation in the northwest. Culturally, the country is a mixture of different linguistic and ethnic groups. The government has tried to encourage the use of Urdu as the national language, but less than 10% of Pakistanis speak Urdu as their primary language. Almost the only source of unity in Pakistan is Sunni Islam, but even here, nearly 20% of the Pakistani population adheres to Shi'a Islam. The long-serving president, Musharraf, who resigned in 2008, was a purported ally of the United States, although his actions indicated otherwise. Asif Ali Zardari, a more West-leaning political figure, was elected president in a landslide but almost immediately was confronted with Islamist militants, mainly Taliban, expanding their control in the Northwest Frontier Province toward the capital city of Islamabad.

Afghanistan: In the aftermath of the tragic September 11, 2001 terrorist attacks on the World Trade Center and the Pentagon, the United States (backed to varying degrees by its allies) declared a massive and global "war on terrorism" targeting terrorist groups and states that provide "safe harbor" to them. Front and center in this war was the Taliban regime of Islamic extremists who controlled about 95 percent of Afghanistan and harbored the al-Qaeda terrorist network of the beleaguered nation of Afghanistan. U.S. and British forces, aided by members of the Northern Alliance of Afghan rebels, expelled the Taliban government in 2002, and in 2003, Hamid Karzai became the first democratically elected president of the country. Despite the imposition of democracy, Afghanistan still is plagued by warlords in remote areas of the country who refuse to recognize the legally constituted government. In addition, significant pockets of resistance from remnants of the former Taliban regime and from al-Qaeda forces are engaged in ongoing military conflict with American and Pakistani troops along the Afghanistan-Pakistan border. Taliban resurgence, particularly that based in Pakistan, grew throughout 2008 and into 2009, partly as the result of an ineffective and corrupt central government that, for all practical purposes, controls only the region of the capital city, Kabul. And that control is tenuous. President Obama ordered an additional 30,000 extra U.S. military personnel into the country to "disrupt, dismantle and defeat" al-Qaeda.

Map 132o

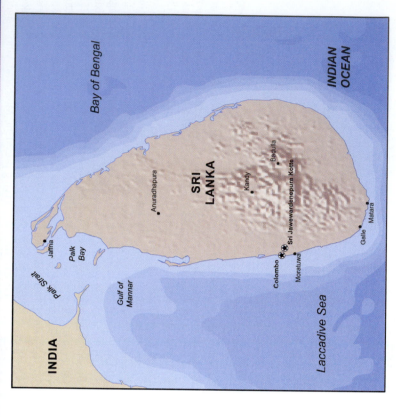

Bay of Bengal

INDIAN OCEAN

SRI LANKA

Jaffna

Palk Strait

Palk Bay

Gulf of Mannar

Anuradhapura

Kandy

Badulla

Sri Jawewardenepura Kotte

Colombo

Moratuwa

Galle

Matara

Laccadive Sea

INDIA

Sri Lanka: The island state of Sri Lanka, historically known as Ceylon, is potentially one of the most agriculturally productive regions of Asia. Unfortunately for plans related to agricultural development, two quite different peoples have occupied the island country: the Buddhist Sinhalese, originally from northern India and long the dominant population in Sri Lanka, and the minority Hindu Tamil, a Dravidian people from south India. Since independence from Britain, Sri Lankan governments have sought to "resettle" the Tamil population in south India, actions that finally precipitated an armed rebellion by Tamils against the Sinhalese-dominated government. The Tamils at present are demanding a complete separation of the state into two parts, with a Tamil homeland in the north and along the east coast. A cease-fire between Sinhalese and Tamil fighters was brokered in 2001 but fell apart in late 2003 with the resumption of violence. 2006 and 2007 brought a renewal of terrorist attacks instigated by the Tamil rebels. In 2009, government forces eliminated the last Tamil stronghold, seemingly ending a quarter-century of violence. Time will tell if this is a lasting peace or simply another lull in a protracted conflict.

Map 132n

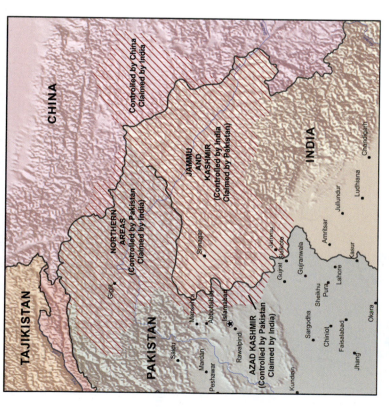

CHINA

TAJIKISTAN

Controlled by China Claimed by India

NORTHERN AREAS (Controlled by Pakistan) Claimed by India

JAMMU AND KASHMIR (Controlled by India Claimed by Pakistan)

Gilgit

Skardu

Srinagar

Jammu

Sialkot

INDIA

Chandigarh

Mardan

Peshawar

Abbottabad

Islamabad

Rawalpindi

AZAD KASHMIR Controlled by Pakistan Claimed by India

PAKISTAN

Gujrat

Gujranwala

Amritsar

Jullundur

Ludhiana

Sheikhu Pura

Lahore

Kasur

Sargodha

Chiniot

Faisalabad

Jhang

Okara

Kundian

Jammu and Kashmir: When Britain withdrew from South Asia in 1947, the former states of British India were asked to decide whether they wanted to become part of a new Hindu India or a Muslim Pakistan. In the state of Jammu and Kashmir, the rulers were Hindu and the majority population was Muslim. The maharajah (prince) of Kashmir opted to join India, but an uprising of the Muslim majority precipitated a war between India and Pakistan over control of this high mountain region. In 1949 a cease-fire line was established by the UN, leaving most of the territory of Jammu and Kashmir in Indian hands. Since then Pakistan and India have waged intermittent skirmishes over the disputed territory that holds the headwaters of the Indus River, a life-giving stream to desert Pakistan. While Jammu and Kashmir refers specifically to the state in northern India, "Kashmir" is used to describe the larger area of contention that includes Pakistan's northern areas and Azad Kashmir as well as territory controlled by China. In 1999 extremist Muslim groups demanding independence escalated the periodic battles into a full-fledged, if small, war between two of Asia's major powers—both possessing nuclear weapons. The specter of nuclear exchange caused both Pakistan and India to back down, and while the area remains disputed, military activity has quieted somewhat. Given that there has been no resolution to the situation, Kashmir will continue to be at the leading edge of the simmering feud between Hindu and Muslim that has been part of South Asian politics since independence from Great Britain and partition into separate states in 1947.

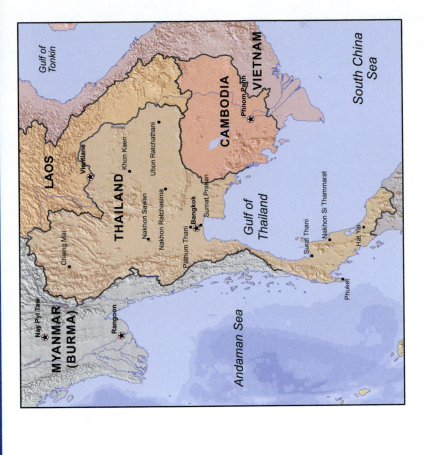

Korean Peninsula: Although active military conflict has not existed since the 1950s, the Korean peninsula remains an important flashpoint. Since the end of the "Korean War," South Korea has flourished economically. North Korea, on the other hand, adopted a policy of diplomatic and economic "self-reliance," becoming one of the world's most authoritarian and isolated states. North Korea molded its political, economic, and military policies around the core ideological objective of eventual unification of Korea and retains that objective. Yet, North Korea so mismanaged and misallocated its resources that, by the mid-1990s, the country was unable to feed itself. An estimated two million North Koreans have died in the past decade as a result of severe food shortages. It continues to expend resources to maintain one of the world's largest armies. In 2006, North Korea announced the development of nuclear weapons and delivery systems designed to "protect" against American aggression. In 2009, North Korea conducted a nuclear test, which was followed by the test firing of several short-range missiles and a long-range rocket. The United Nations imposed sanctions that were supported by longtime North Korean allies China and Russia. Peninsular tensions reached another high in 2010 when the South Korean warship *Cheonan* exploded and sank. South Korea blamed North Korea, then ceased all cross-border trade. North Korea denied attacking the ship and responded to the trade embargo by severing all ties with South Korea.

Thailand: Thailand has the distinction of being the only Southeast Asian country never to have been controlled by a European power. A constitutional monarchy since the early 1930s, Thailand enjoyed relative peace and prosperity until the 2000s. The economic boom that had propelled much of Southeast Asia collapsed in 1997 and led to political unrest in Thailand as people became increasingly disillusioned with government policies. A military coup in 2006 ousted the then-Prime Minister. At the same time, an ongoing ethnic separatist insurgency flared up again in southern Thailand along the border of Malaysia. The junta drafted a new constitution and held general elections in December 2007, which resulted in a coalition government. There has been ongoing conflict between the two main parties in the coalition government: the People's Power Party and the People's Alliance for Democracy. As a result, opposition to the government grew, with the National United front of Democracy Against Dictatorship, also known as the "Red Shirts," being one of the more visible opposition elements. In spring 2010, widespread protests culminated in clashes with government forces. Dozens were killed and several hundred were injured as the Thai army engaged Red Shirt protest camps. Time will tell if this is a short-lived period of instability in the country or a prelude to greater internal conflict.

Unit VII

World Regions

Map 133 North America: Political Divisions

Map 134 North America: Physical Features

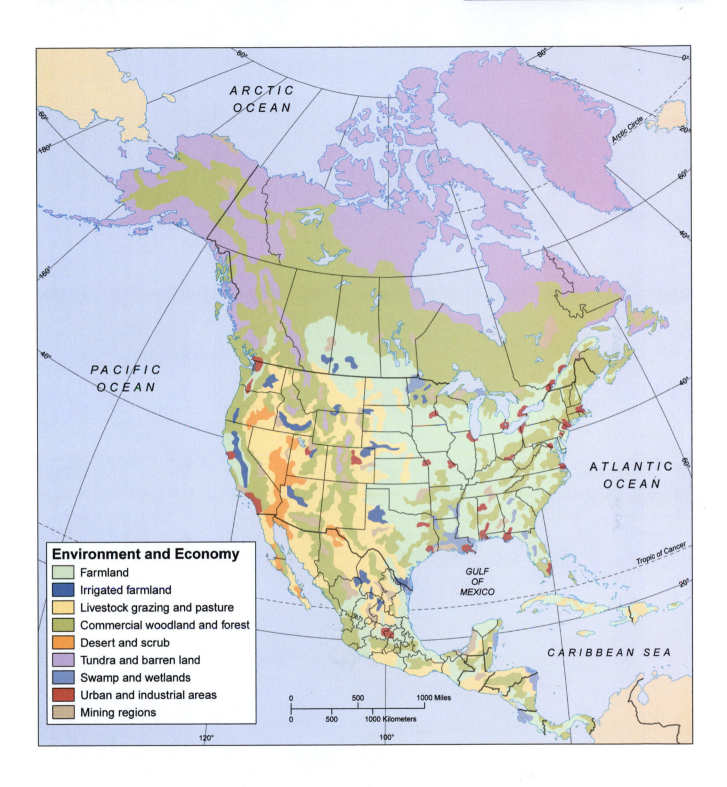

Environment and Economy

- Farmland
- Irrigated farmland
- Livestock grazing and pasture
- Commercial woodland and forest
- Desert and scrub
- Tundra and barren land
- Swamp and wetlands
- Urban and industrial areas
- Mining regions

ARCTIC OCEAN

PACIFIC OCEAN

ATLANTIC OCEAN

GULF OF MEXICO

CARIBBEAN SEA

Arctic Circle

Tropic of Cancer

The use of land in North America represents a balance between agriculture, resource extraction, and manufacturing that is unmatched. The United States, as the world's leading industrial power, is also the world's leader in commercial agricultural production. Canada, despite its small population, is a ranking producer of both agricultural and industrial products and Mexico has begun to emerge from its developing nation status to become an important industrial and agricultural nation as well. The countries of Middle America and the Caribbean are just beginning the transition from agriculture to modern industrial economies. Part of the basis for the high levels of economic productivity in North America is environmental: a superb blend of soil, climate, and raw materials. But just as important is the cultural and social mix of the plural societies of North America, a mix that historically aided the growth of the economic diversity necessary for developed economies.

Map 135c North America: Population Distribution

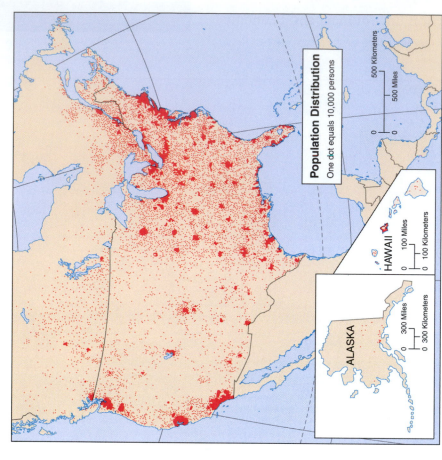

Population Distribution
One dot equals 10,000 persons

HAWAII

ALASKA

Although population clustering is characteristic of highly economically developed regions, the Anglo-American population exhibits a remarkably clustered pattern. The primary reasons for this remarkable development are agricultural technology and affluence. Highly developed agricultural technology allows a small number of farmers using sophisticated machinery to grow and harvest enormous quantities of agricultural produce on large farms. In the United States, the world's leader in commercial agricultural production, only 2 percent of the population are farmers and that population is thinly distributed over wide areas. The vast majority of Americans live and work in city-suburb systems in which the widespread availability of private automobiles allows people to live considerable distances from where they work, keeping overall urban population densities relatively low but allowing for extensive urbanization—with cities large enough in area to be visible as population clusters on maps at this scale.

Map 135b North America: Population Density

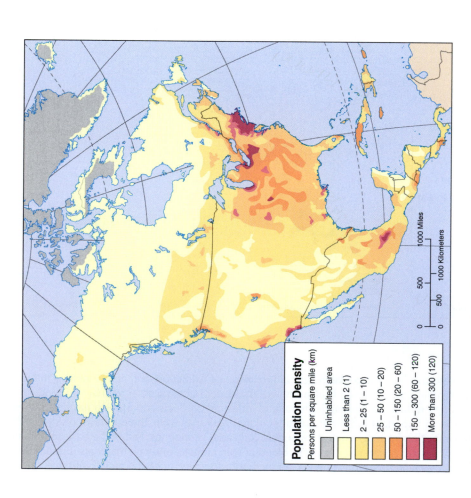

Population Density
Persons per square mile (km)

- Uninhabited area
- Less than 2 (1)
- 2 – 25 (1 – 10)
- 25 – 50 (10 – 20)
- 50 – 150 (20 – 60)
- 150 – 300 (60 – 120)
- More than 300 (120)

North America contains over 530 million people and the United States, with nearly 310 million inhabitants, is the third most populous country in the world after China and India. Most of the present North American population has roots in the Old World. The native populations of the Americas had little or no resistance to Old World diseases in 1500 and within a couple of centuries of their first contact with Europeans, most of the native peoples had either died out or preserved their genetic heritage by mixing with disease-resistant Old World populations. This left a North American population that is largely European, but with significant minorities resulting from the mixture of native Americans with Europeans and/or Africans. The density of that population is largely the consequence of environmental factors (good soil, the availability of water, the presence of other resources) and cultural/economic ones (agricultural production, urbanization, industrialization).

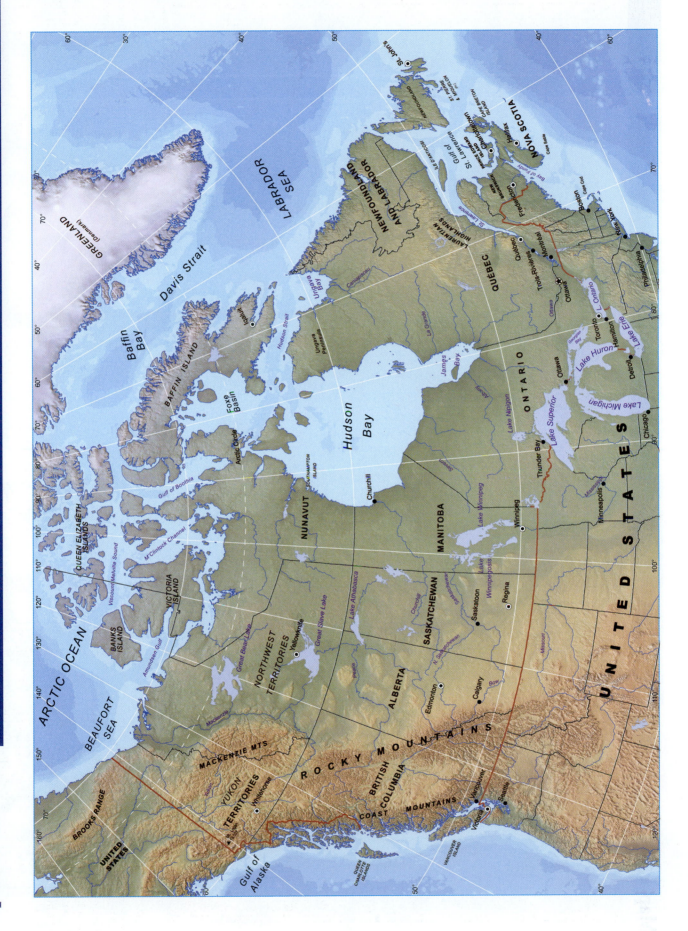

Map 136   Canada

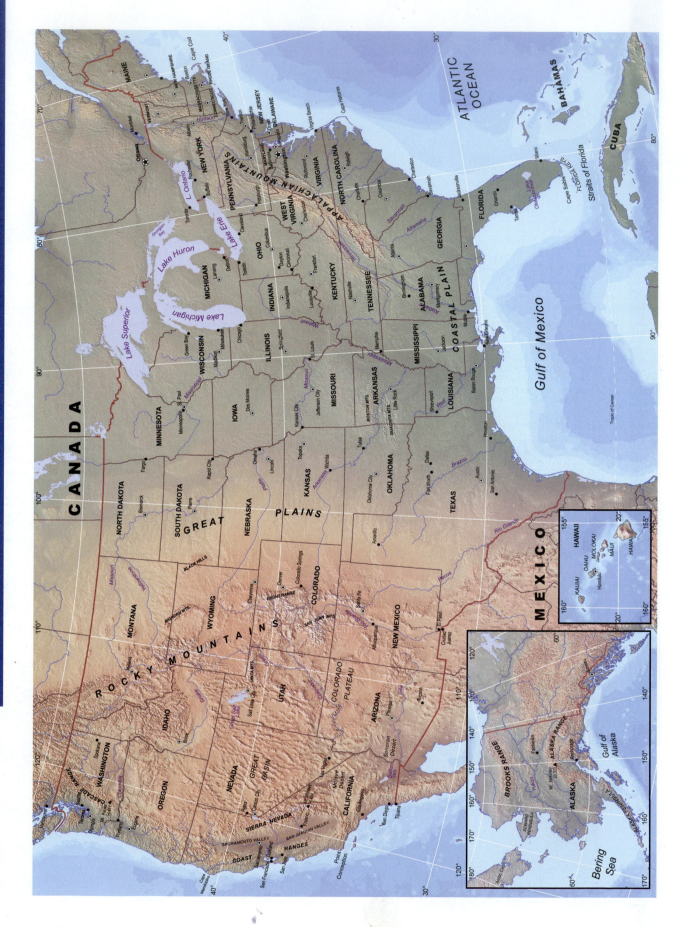

Map **137** United States

ATLANTIC OCEAN

BAHAMAS

CUBA

MAINE

NEW HAMPSHIRE

Cape Cod

VERMONT

MASSACHUSETTS

CONNECTICUT

RHODE ISLAND

NEW YORK

PENNSYLVANIA

NEW JERSEY

DELAWARE

MARYLAND

WEST VIRGINIA

VIRGINIA

Virginia Beach

Cape Hatteras

NORTH CAROLINA

Raleigh

Charlotte

Columbia

SOUTH CAROLINA

Charleston

Savannah

GEORGIA

Jacksonville

FLORIDA

Orlando

Tampa

Miami

Cape Sable

FLORIDA KEYS

Straits of Florida

Tropic of Cancer

L. Ontario

Lake Erie

Lake Huron

Georgian Bay

Lake Michigan

Lake Superior

MICHIGAN

OHIO

INDIANA

ILLINOIS

WISCONSIN

MINNESOTA

IOWA

KENTUCKY

TENNESSEE

ALABAMA

MISSISSIPPI

LOUISIANA

ARKANSAS

MISSOURI

APPALACHIAN MOUNTAINS

COASTAL PLAIN

Gulf of Mexico

CANADA

NORTH DAKOTA

SOUTH DAKOTA

NEBRASKA

KANSAS

OKLAHOMA

TEXAS

GREAT PLAINS

BLACK HILLS

BIGHORN MTS.

WYOMING

MONTANA

COLORADO

FRONT RANGE

SAN JUAN MTS.

NEW MEXICO

ARIZONA

UTAH

COLORADO PLATEAU

IDAHO

NEVADA

GREAT BASIN

CALIFORNIA

OREGON

WASHINGTON

CASCADE RANGE

SIERRA NEVADA

COAST RANGES

SACRAMENTO VALLEY

SAN JOAQUIN VALLEY

Point Conception

Mojave Desert

Sonoran Desert

ROCKY MOUNTAINS

MEXICO

Rio Grande

HAWAII

KAUAI

OAHU

MOLOKAI

MAUI

HAWAII

Honolulu

BROOKS RANGE

ALASKA RANGE

ALASKA

Mt. McKinley 20,320

Fairbanks

Anchorage

Juneau

SEWARD PENINSULA

ALASKA PENINSULA

Gulf of Alaska

Bering Sea

Map **138** Middle America

ATLANTIC
OCEAN

CARIBBEAN SEA

COLOMBIA

BAHAMAS

Kingston

JAMAICA

CUBA

ISTHMUS OF PANAMA

PANAMA

Panama
City

Gulf of
Panamá

Miami

Havana

Yucatán Channel

COZUMEL

Cancún

San José

COSTA RICA

NICARAGUA

Lago de Nicaragua

Managua

Tampa

Gulf of Mexico

Chetumal

YUCATÁN
PENINSULA

QUINTANA
ROO

BELIZE

Belmopan

HONDURAS

Tegucigalpa

New Orleans

Mérida

YUCATÁN

CAMPECHE

GUATEMALA

EL SALVADOR

San Salvador

Houston

UNITED STATES

Campeche

Bay of
Campeche

TABASCO

Villahermosa

Guatemala City

Dallas

Fort Worth

Austin

San Antonio

Ciudad Victoria

TAMAULIPAS

NUEVO
LEON

Monterrey

Saltillo

SIERRA MADRE ORIENTAL

SAN LUIS
POTOSI

San Luis Potosí

Veracruz

VERACRUZ

Xalapa

Orizaba

ISTHMUS OF TEHUANTEPEC

Tuxtla
Gutiérrez

CHIAPAS

Gulf of
Tehuantepc

COAHUILA

ZACATECAS

Zacatecas

GUANAJUATO

Guanajuato

HIDALGO

Pachuca

Mexico City

MÉXICO

PUEBLA

Puebla

OAXACA

Oaxaca

SIERRA MADRE DEL SUR

Chihuahua

CHIHUAHUA

Chihuahuan
Desert

El Paso

Ciudad
Juárez

DURANGO

Durango

Aguascalientes

Querétaro

Morelia

MICHOACAN

GUERRERO

Chilpancingo

Acapulco

SINALOA

Culiacán

NAYARIT

Tepic

Guadalajara

JALISCO

Colima

COLIMA

SIERRA MADRE OCCIDENTAL

Tucson

Phoenix

SONORA

Hermosillo

La Paz

BAJA
CALIFORNIA
SUR

Gulf of California

Cabo San Lucas

BAJA CALIFORNIA

PACIFIC
OCEAN

San Diego

Tijuana

Mexicali

BAJA
CALIFORNIA

Map 139   The Caribbean

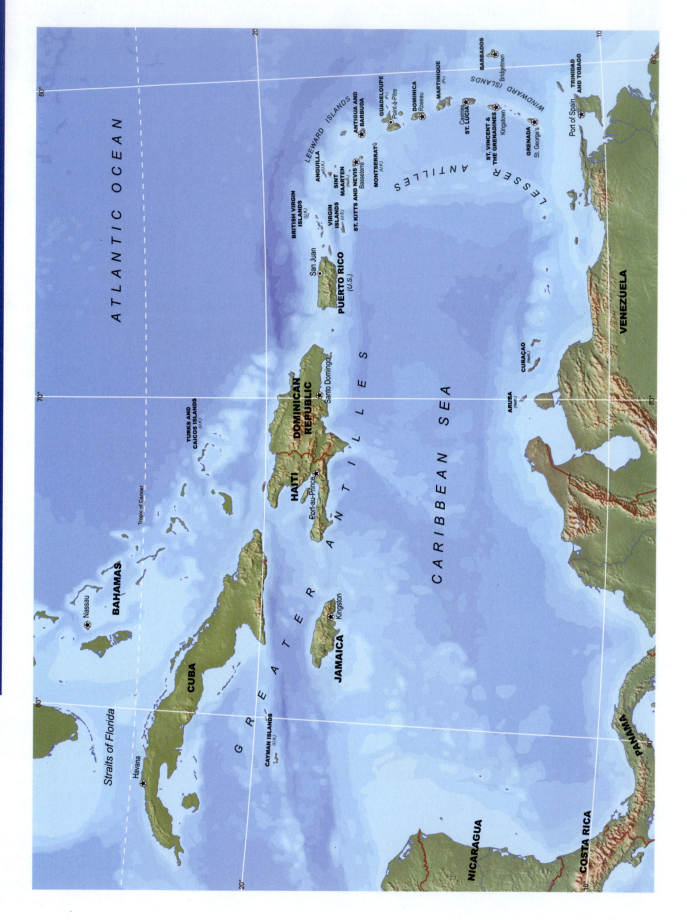

ATLANTIC OCEAN

20°

60°

TURKS AND
CAICOS ISLANDS
(U.K.)

BAHAMAS

Nassau

Tropic of Cancer

70°

BRITISH VIRGIN
ISLANDS
(U.K.)

San Juan

PUERTO RICO
(U.S.)

VIRGIN
ISLANDS
(U.S.)

ANGUILLA
(U.K.)

SINT
MAARTEN
(Neth.)

ST. KITTS AND NEVIS
Basseterre

MONTSERRAT
(U.K.)

LEEWARD ISLANDS

ANTIGUA AND
BARBUDA

GUADELOUPE
(Fr.)
Point-à-Pitre

DOMINICA
Roseau

MARTINIQUE
(Fr.)

BARBADOS
Bridgetown

WINDWARD ISLANDS

Castries
ST. LUCIA

ST. VINCENT &
THE GRENADINES
Kingstown

GRENADA
St. George's

Port of Spain

TRINIDAD
AND TOBAGO

10°

60°

LESSER ANTILLES

DOMINICAN
REPUBLIC
Santo Domingo

HAITI
Port-au-Prince

G R E A T E R   A N T I L L E S

CURAÇAO
(Neth.)

ARUBA
(Neth.)

VENEZUELA

70°

CARIBBEAN SEA

Straits of Florida

Havana

CUBA

CAYMAN ISLANDS
(U.K.)

JAMAICA
Kingston

80°

NICARAGUA

COSTA RICA

PANAMA

20°

Map 140

National capital
State capital
City

Map 141 South America: Physical Features

Map 142a   South America: Environment and Economy

Environment and Economy
- Farmland
- Irrigated farmland
- Livestock grazing and pasture
- Commercial woodland and forest
- Desert and scrub
- Tundra and barren land
- Swamp and wetlands
- Urban and industrial areas
- Mining regions

South America is a region just beginning to emerge from a colonial-dependency economy in which raw materials flowed from the continent to more highly developed economic regions. With the exception of Brazil, Argentina, Chile, and Uruguay, most of the continent's countries still operate under the traditional mode of exporting raw materials in exchange for capital that tends to accumulate in the pockets of a small percentage of the population. The land use patterns of the continent are, therefore, still dominated by resource extraction and agriculture. A problem posed by these patterns is that little of the continent's land area is actually suitable for either commercial forestry or commercial crop agriculture without extremely high environmental costs. Much of the agriculture, then, is based on high value tropical crops that can be grown in small areas profitably, or on extensive livestock grazing. Even within the forested areas of the Amazon Basin where forest clearance is taking place at unprecedented rates, much of the land use that replaces forest is grazing.

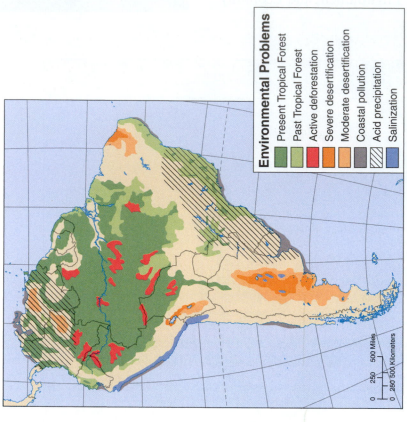

Environmental Problems

- Present Tropical Forest
- Past Tropical Forest
- Active deforestation
- Severe desertification
- Moderate desertification
- Coastal pollution
- Acid precipitation
- Salinization

The drainage basin of the Amazon River and its tributaries, along with adjacent regions, is the world's largest remaining area of tropical forest. Much of the periphery of this vast forested region has already been cleared for farming and grazing, and the ax and chainsaw and flames are working their way steadily toward the interior. Tropical deforestation produces a loss of the biological diversity represented by the world's most biologically productive ecosystem, along with changes in soil and soil-water systems. South America has other environmental problems: *desertification* in which grassland and/or scrub vegetation is converted to desert through overgrazing or other unwise agricultural practices; over-soil *salinization* in which soils become increasingly salty as the consequence of the over-application of irrigation water; *coastal and estuarine pollution* resulting from unregulated or unchecked use of coastal waters for industrial, commercial, and transportation purposes; and *acid precipitation* resulting from the combination of airborne industrial wastes and automobile/truck exhausts with water vapor to produce dry or wet acidic fallout.

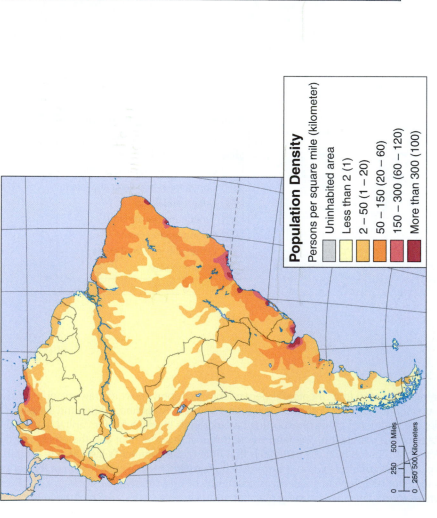

Population Density

Persons per square mile (kilometer)

- Uninhabited area
- Less than 2 (1)
- 2 – 50 (1 – 20)
- 50 – 150 (20 – 60)
- 150 – 300 (60 – 120)
- More than 300 (100)

Since so much of interior South America is uninhabitable (the high Andes) or only sparsely populated (the interior of the Amazon Basin), the continent's population tends to be peripheral—approximately 90 percent of the continent's nearly 400 million people live within 150 miles of the sea. This population also tends to be heavily urbanized. Over 80 percent of South America's population lives in cities and the continent has three of the world's 15 largest cities—Rio de Janeiro, São Paulo, and Buenos Aires. São Paulo is the world's third largest urban agglomeration after Tokyo and New York. As in North America, most of the population of South America can trace at least part of its ancestry to the Old World. Throughout the Spanish-speaking parts of the continent the population is predominantly *mestizo* or mixed European and native South American. In Portuguese-speaking Brazil, in addition to a *mestizo* population, there is a significant admixture of African blood, the result of a large slave labor force imported from Africa to work the sugar, cotton, and other plantations of the colonial period.

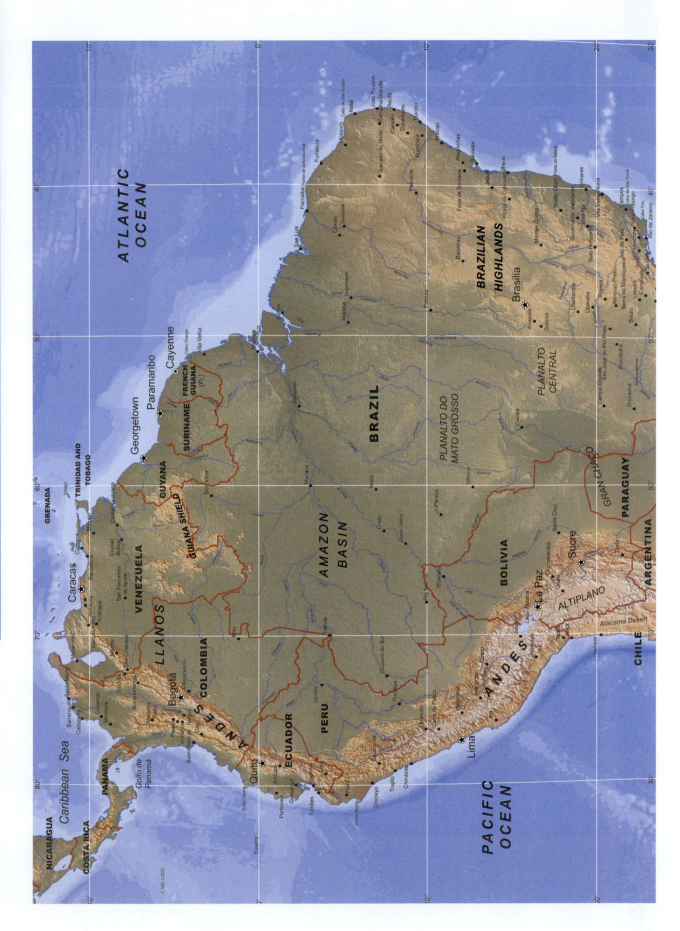

Map **143** Northern South America

ATLANTIC OCEAN

PACIFIC OCEAN

Caribbean Sea

NICARAGUA

COSTA RICA

PANAMA

Golfo de Panamá

CABO CORRIENTES

COLOMBIA

ECUADOR

PERU

VENEZUELA

LLANOS

ANDES

Bogotá

Quito

Lima

GRENADA

TRINIDAD AND TOBAGO

Caracas

Georgetown

Paramaribo

Cayenne

GUYANA

SURINAME

FRENCH GUIANA (Fr.)

GUIANA SHIELD

AMAZON BASIN

BRAZIL

BRAZILIAN HIGHLANDS

Brasília

PLANALTO CENTRAL

PLANALTO DO MATO GROSSO

BOLIVIA

La Paz

Sucre

ALTIPLANO

Atacama Desert

GRAN CHACO

PARAGUAY

ARGENTINA

CHILE

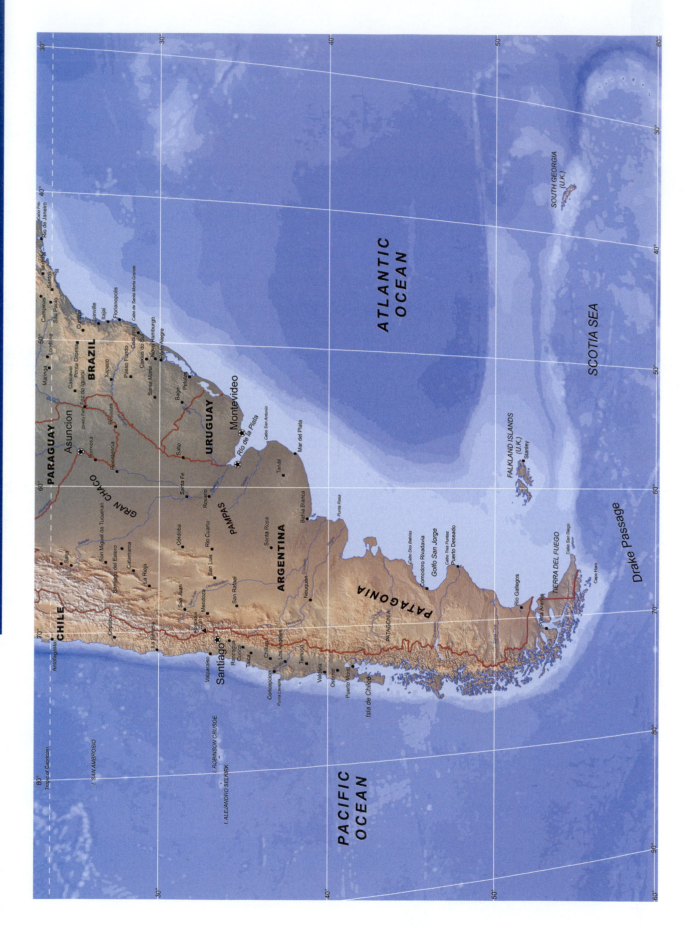

Map 144 Southern South America

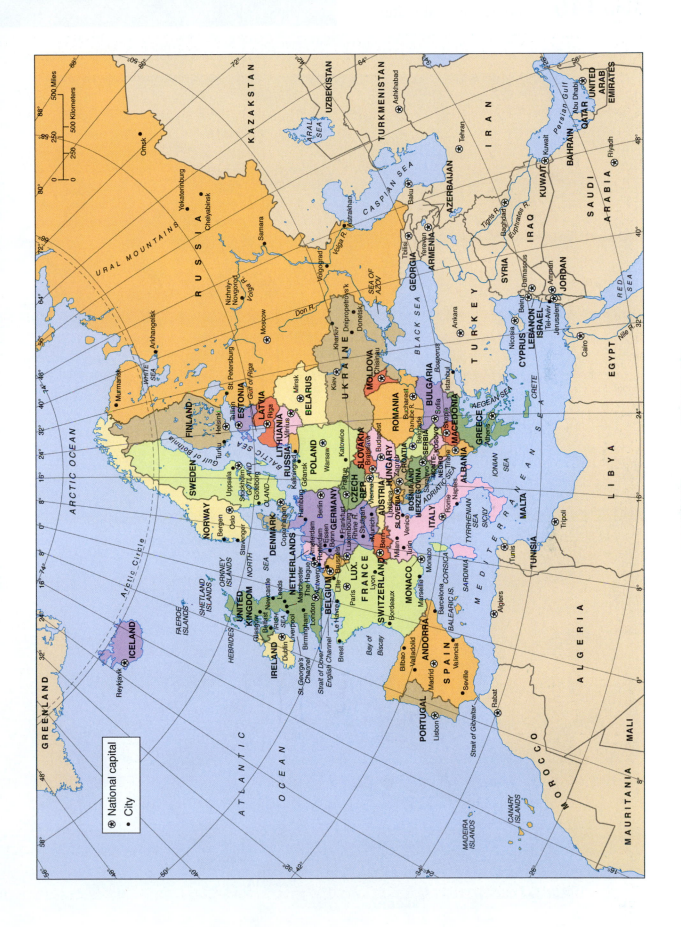

Map 145 Europe: Political Divisions

Map 146 Europe: Physical Features

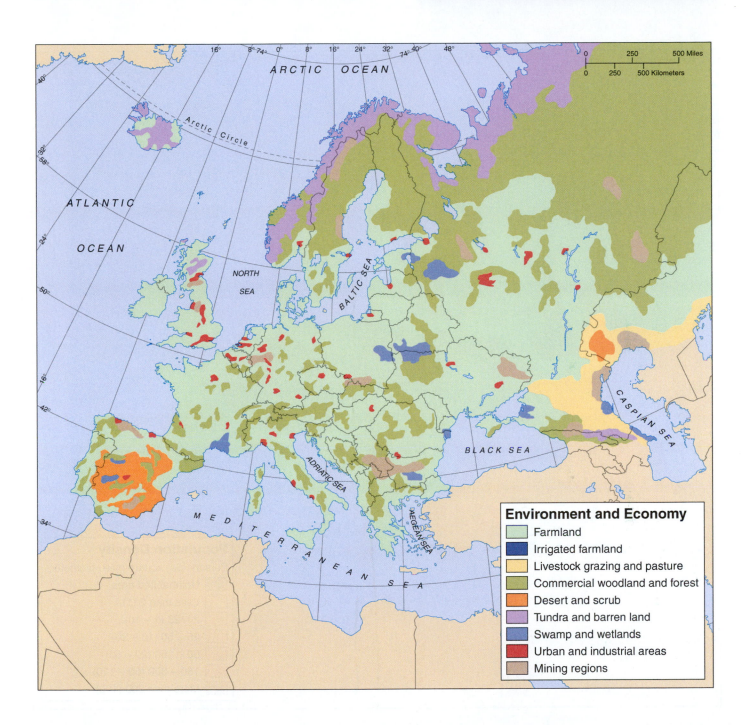

Environment and Economy

- Farmland
- Irrigated farmland
- Livestock grazing and pasture
- Commercial woodland and forest
- Desert and scrub
- Tundra and barren land
- Swamp and wetlands
- Urban and industrial areas
- Mining regions

More than any other continent, Europe bears the imprint of human activity—mining, forestry, agriculture, industry, and urbanization. Virtually all of western and central Europe's natural forest vegetation is gone, lost to clearing for agriculture beginning in prehistory, to lumbering that began in earnest during the Middle Ages, or more recently, to disease and destruction brought about by acid precipitation. Only in the far north and the east do some natural stands remain. The region is the world's most heavily industrialized and the industrial areas on the map represent only the largest and most significant. Not shown are the industries that are found in virtually every small town and village and smaller city throughout the industrial countries for Europe. Europe also possesses abundant raw materials and a very productive agricultural base. The mineral resources have long been in a state of active exploitation and the mining regions shown on the map are, for the most part, old regions in upland areas that are somewhat less significant now than they may have been in the past. Agriculturally, the northern European plain is one of the world's great agricultural regions but most of Europe contains decent land for agriculture.

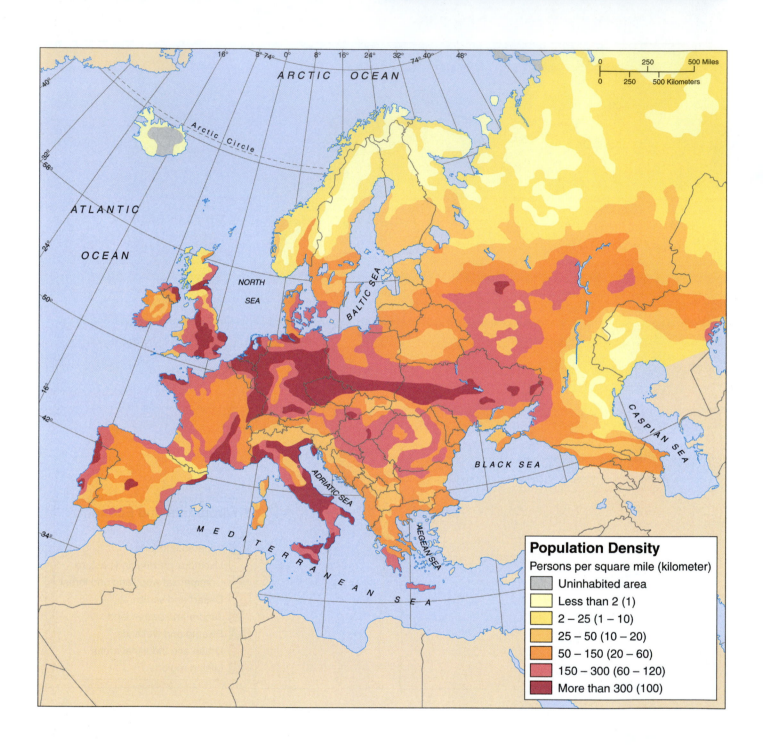

Population Density
Persons per square mile (kilometer)

- Uninhabited area
- Less than 2 (1)
- 2 – 25 (1 – 10)
- 25 – 50 (10 – 20)
- 50 – 150 (20 – 60)
- 150 – 300 (60 – 120)
- More than 300 (100)

Europe is one of the most densely settled regions of the world with an overall population density nearing 200 persons per square mile (80 per square kilometer), the consequence of a high level of urbanization and an economic system that is heavily industrialized. Even in agricultural regions, the population density is high. Beyond high density, the two chief identifying marks of the European population are remarkable diversity and unusual dynamics. For a small part of the world, Europe has cultural and ethnic diversity that is rarely matched elsewhere; more than 60 languages are spoken in an area not much larger than the United States. The population dynamics of Europe show a mature population that has passed through the "Demographic Transition"—a remarkable increase and then decline in growth rates resulting from the rise of an urban-industrial society. Only in other heavily industrialized regions of the world are found the very small overall growth rates characteristic of Europe.

Map 147c   European Political Boundaries, 1914–1948

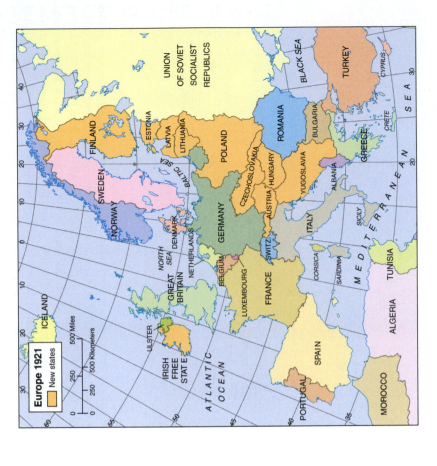

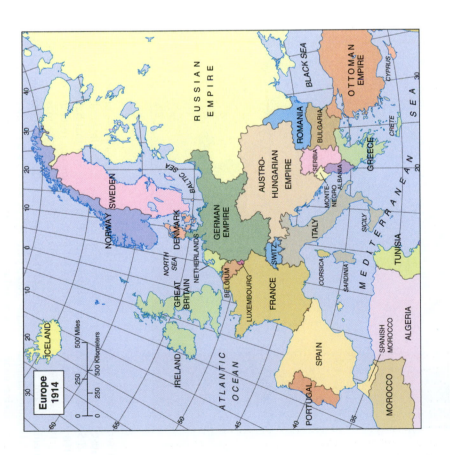

In 1914, on the eve of the First World War, Europe was dominated by the United Kingdom and France in the west, the German Empire and the Austro-Hungarian Empire in central Europe, and the Russian Empire in the east. Battle lines for the conflict that began in 1914 were drawn when the United Kingdom, France, and the Russian Empire joined together as the Triple Entente. In the view of the Germans, this coalition was designed to encircle Germany and its Austrian ally, which, along with Italy, made up the Triple Alliance. The German and Austrian fears were heightened from 1912–14 when a Russian-sponsored "Balkan League" pushed the Ottoman Turkish Empire from Europe, leaving behind the weak and mutually antagonistic Balkan states Serbia and Montenegro. In August 1914, Germany and Austria-Hungary attacked in several directions and World War I began. Four years later, after massive loss of life and destruction, the central European empires were defeated. The victorious French, English, and Americans (who had entered the war in 1917) restructured the map of Europe in 1919, carving nine new states out of the remains of the German and Austro-Hungarian empires and the westernmost portions of the Russian Empire which, by the end of the war, was deep in the Revolution that deposed the czar and brought the Communists to power in a new Union of Soviet Socialist Republics.

-185-

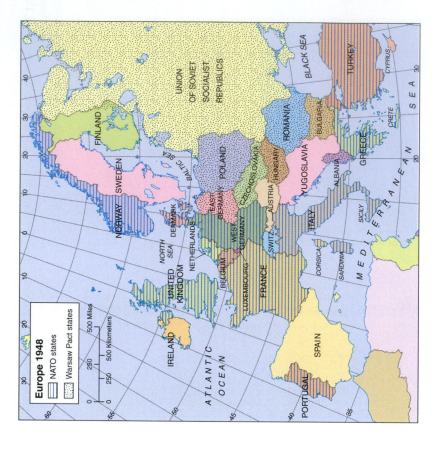

Europe 1948
☐ NATO states
⋯ Warsaw Pact states

500 Miles
500 Kilometers

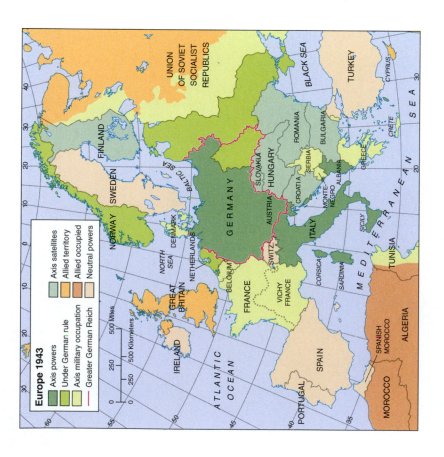

Europe 1943
■ Axis powers
■ Under German rule
□ Axis military occupation
— Greater German Reich
■ Axis satellites
□ Allied territory
□ Allied occupied
□ Neutral powers

500 Miles
500 Kilometers

When the victorious Allies redrew the map of central and eastern Europe in 1919, they caused as many problems as they were trying to solve. The interval between the First and Second World Wars was really just a lull in a long war that halted temporarily in 1918 and erupted once again in 1939. Defeated Germany, resentful of the terms of the 1918 armistice and 1919 Treaty of Versailles and beset by massive inflation and unemployment at home, overthrew the Weimar republican government in 1933 and installed the National Socialist (Nazi) party led by Adolf Hitler in Berlin. Hitler quickly began making good on his promises to create a "thousand year realm" of German influence by annexing Austria and the Czech region of Czechoslovakia and allying Germany with a fellow fascist state in Mussolini's Italy. In September 1939 Germany launched the lightning-quick combined infantry, artillery, and armor attack known as **der Blitzkrieg** and took Poland to the east and, in quick succession, the Netherlands, Belgium, and France to the west. By 1943 the greater German Reich extended from the

Russian Plain to the Atlantic and from the Black Sea to the Baltic. But the Axis powers of Germany and Italy could not withstand the greater resources and manpower of the combined United Kingdom–United States–USSR–led Allies and, in 1945, Allied armies occupied Germany. Once again, the lines of the central and eastern European map were redrawn. This time, a strengthened Soviet Union took back most of the territory the Russian Empire had lost at the end of the First World War. Germany was partitioned into four occupied sectors (English, French, American, and Russian) and later into two independent countries, the Federal Republic of Germany (West Germany) and the German Democratic Republic (East Germany). Although the Soviet Union's territory stopped at the Polish, Hungarian, Czechoslovakian, and Romanian borders, the eastern European countries (Poland, East Germany, Czechoslovakia, Hungary, Romania, Yugoslavia, Albania, and Bulgaria) became Communist between 1945 and 1948 and were separated from the West by the Iron Curtain.

-186-

Map 147d   Europe: Political Changes, 1989–2009

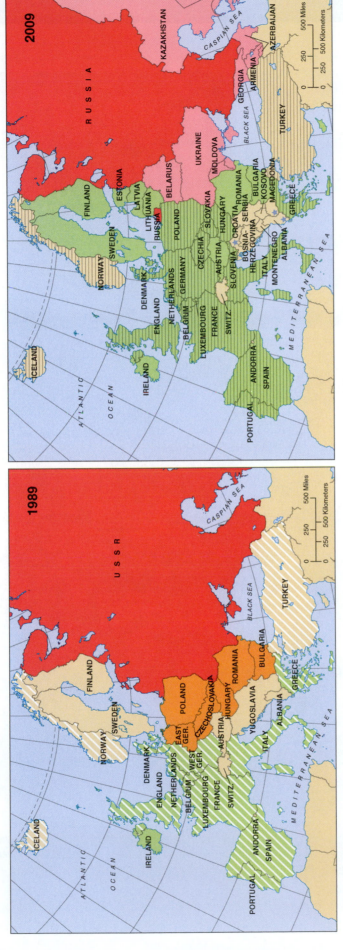

1989

■ (red)	Union of Soviet Socialist Republics
■ (orange)	Warsaw Pact Countries (excluding the USSR)
▨	North Atlantic Treaty Organization Countries
▨ (green)	European Community (formerly the EEC)

Europe: Political Changes 1989–2009

2009

■ (pink)	Former Republics of the USSR, now independent countries
■ (red)	Russian Federation
▥	NATO Countries, 2005
★	Associated with NATO / petitioned for entry
■ (green)	European Union (formerly the EC)

During the last decade of the twentieth century, one of the most remarkable series of political geographic changes of the last 500 years took place. The bipolar East-West structure that had characterized Europe's political geography since the end of the Second World War altered in the space of a very few years. In the mid-1980s, as Soviet influence over eastern and central Europe weakened, those countries began to turn to the capitalist West. Between 1989, when the country of Hungary was the first Soviet satellite to open its borders to travel, and 1991, when the Soviet Union dissolved into 15 independent countries, abrupt change in political systems occurred. The result is a new map of Europe that includes a number of countries not present on the map of 1989. These countries have emerged as the result of reunification, separation, or independence from the former Soviet Union and Yugoslavia. The new political structure has been accompanied by growing economic cooperation.

-187-

Map 148   Western Europe

Map 149   Eastern Europe

Map 150 Northern Europe

BARENTS SEA

White Sea

RUSSIA

Lake Onega

Lake Ladoga

FINLAND

Gulf of Finland

Helsinki

Tallinn

ESTONIA

Gulf of Riga

Riga

LATVIA

LITHUANIA

Vilnius

Minsk

BELARUS

NORTHERN EUROPEAN PLAIN

POLAND

Gulf of Bothnia

Stockholm

Gotland

BALTIC SEA

Gdansk

SWEDEN

SCANDINAVIAN PENINSULA

Oslo

Copenhagen

GERMANY

NORWAY

KJØLEN MOUNTAINS

KATTEGAT

DENMARK

JUTLAND PENINSULA

SKAGERRAK

NORTH SEA

NORWEGIAN SEA

Arctic Circle

SHETLAND IS. (U.K.)

UNITED KINGDOM

Newcastle

Middlesbrough

Edinburgh

JAN MAYEN (NOR.)

FAROE ISLANDS (Den.)

HEBRIDES IS. (U.K.)

Belfast

IRELAND

GREENLAND SEA

ATLANTIC OCEAN

Cape Brewster

GREENLAND (Den.)

ICELAND

Reykjavik

Map 151   Africa: Political Divisions

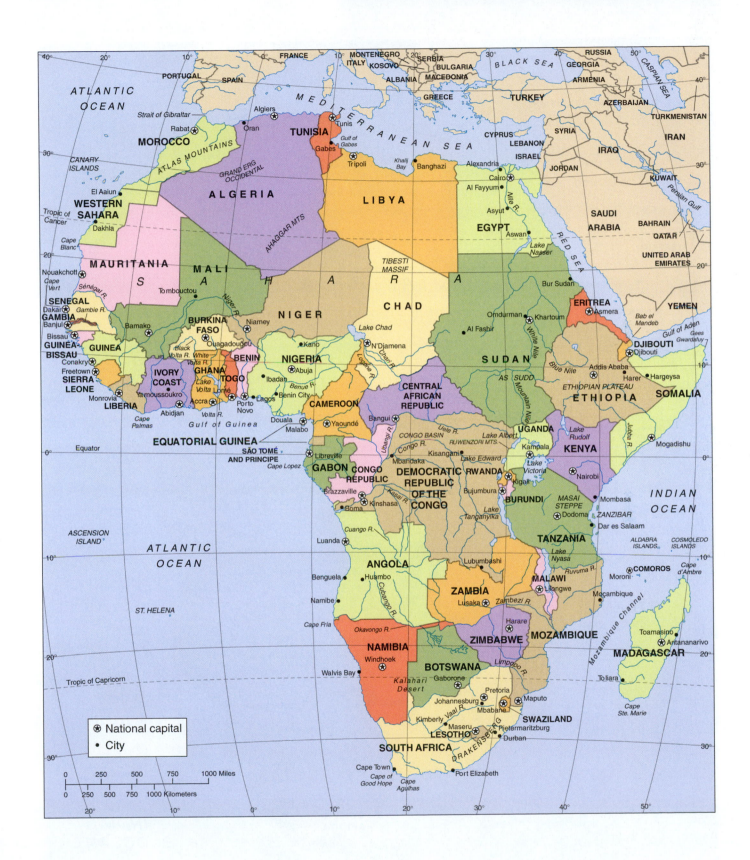

Map 152   Africa: Physical Features

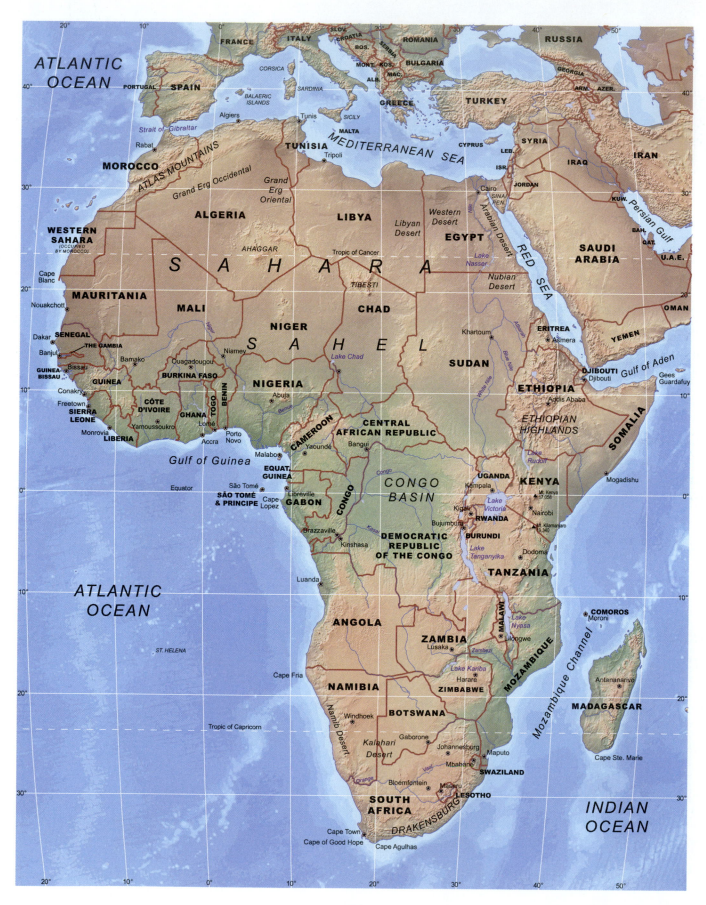

Map 153b Africa: Population Density

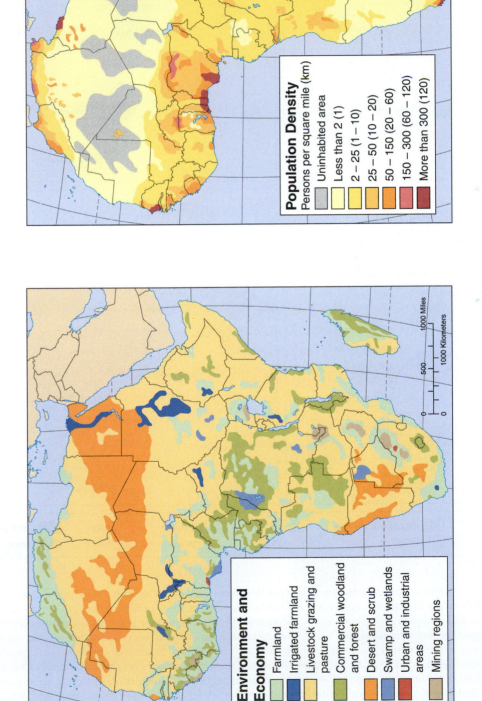

Population Density
Persons per square mile (km)

- Uninhabited area
- Less than 2 (1)
- 2 – 25 (1 – 10)
- 25 – 50 (10 – 20)
- 50 – 150 (20 – 60)
- 150 – 300 (60 – 120)
- More than 300 (120)

1000 Miles

1000 Kilometers

500

0

Over one billion people occupy the African continent, approximately one-seventh of the world's population. In general, this population has two chief characteristics: a low level of quality of life and growth rates that are among the world's highest. On a continent beset by poverty, recurrent internal civil and tribal war, and a host of environmental problems, the populations of many African countries are nevertheless increasing at a rate above 3 percent per year. The bulk of the population is concentrated in relatively small areas of the Mediterranean coastal regions of the north, the bulge of West Africa, and the eastern coastal and highland zone stretching from South Africa to Kenya. Where populations in other parts of the world tend to avoid highland locations, in Africa highlands often tend to be the most densely settled regions: moister with better soils, freer of insect pests, and somewhat cooler.

Map 153a Africa: Environment and Economy

Environment and Economy

- Farmland
- Irrigated farmland
- Livestock grazing and pasture
- Commercial woodland and forest
- Desert and scrub
- Swamp and wetlands
- Urban and industrial areas
- Mining regions

1000 Miles

1000 Kilometers

500

0

Africa's economic landscape is dominated by subsistence, or marginally-commercial agricultural activities and raw material extraction, engaging three-fourths of Africa's workers. Much of this grazing land is very poor desert scrub and bunch grass that is easily impacted by cattle, sheep, and goats. Growing human and livestock populations place enormous stress on this fragile support capacity and the result is desertification: the conversion of even the most minimal of grazing environments to a small quantity of land suitable for crop farming. Although the continent has approximately 20 percent of the world's total land area, the proportion of Africa's arable land is small. The agricultural environment is also uncertain; unpredictable precipitation and poor soils hamper crop agriculture.

-193-

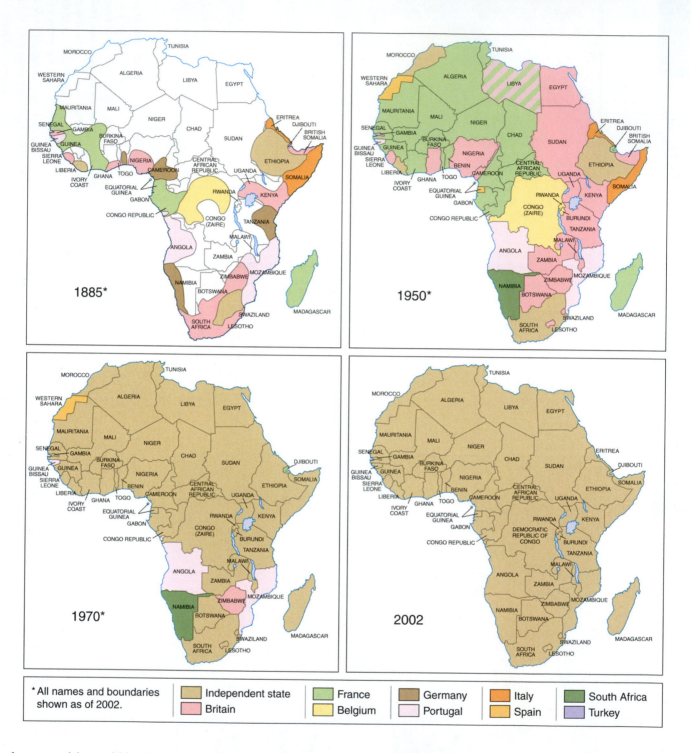

1885*

1950*

1970*

2002

* All names and boundaries shown as of 2002.

Independent state	France	Germany	Italy	South Africa
Britain	Belgium	Portugal	Spain	Turkey

In few parts of the world has the transition from colonialism to independence been as abrupt as on the African continent. Most African states did not become colonies until the nineteenth century and did not become independent until the twentieth, nearly all of them after World War II. Much of the colonial power in Africa is social and economic. The African colony provided the mother country with raw materials in exchange for marginal economic returns, and many African countries still exist in this colonial dependency relationship. An even more important component of the colonial legacy of Europe in Africa is geopolitical. When the world's colonial powers joined at the Conference of Berlin in 1884, they divided up Africa to fit their own needs, drawing boundary lines on maps without regard for terrain or drainage features, or for tribal/ethnic linguistic, cultural, economic, or political borders. Traditional Africa was enormously disrupted by this process. After independence, African countries retained boundaries that are legacies of the colonial past; and African countries today are beset by internal problems related to tribal and ethnic conflicts, the disruption of traditional migration patterns, and inefficient spatial structures of market and supply.

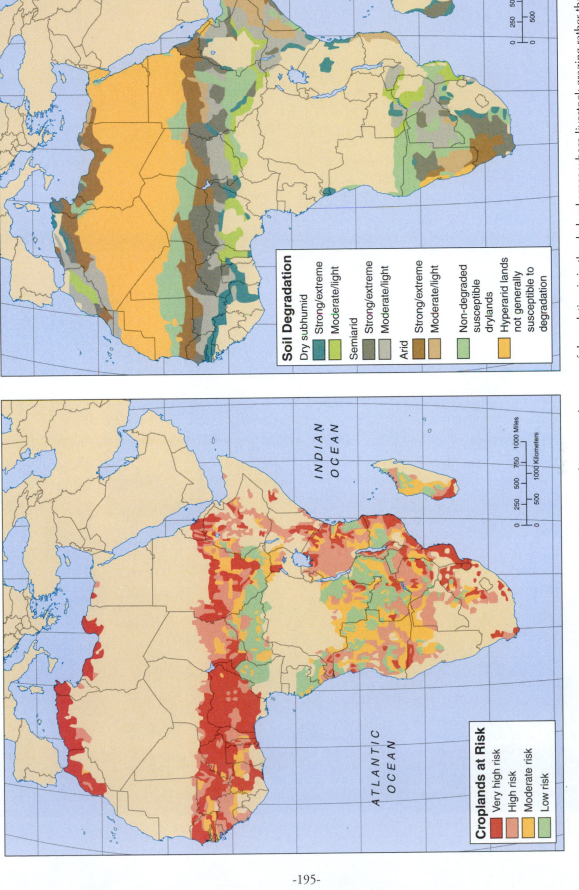

Map 153d   African Cropland and Dryland Degradation

Soil Degradation

Dry subhumid
- Strong/extreme
- Moderate/light

Semiarid
- Strong/extreme
- Moderate/light

Arid
- Strong/extreme
- Moderate/light

- Non-degraded susceptible drylands
- Hyperarid lands not generally susceptible to degradation

INDIAN OCEAN

0   250   500   750   1000 Miles
0   500   1000 Kilometers

Croplands at Risk
- Very high risk
- High risk
- Moderate risk
- Low risk

INDIAN OCEAN

ATLANTIC OCEAN

0   250   500   750   1000 Miles
0   500   1000 Kilometers

The economy of the African continent is largely agricultural and given an African population of over 1 billion people, much of the agricultural environment is degraded. Two forms of degradation exist. The first of these is in cropland areas where susceptible tropical soils and high population densities have produced major cropland degradation, largely in the form of loss of fertility. Irrigation and the advent of artificial fertilizer have not helped soils to restore their natural chemical balances after generations of misuse. The second form of degradation is in the dryland areas where livestock grazing rather than cropping is the dominant agricultural form. Populations of domesticated stock that are too large for the carrying capacity of the environment, exacerbated by the view of livestock as wealth and worsened by increasing human populations, have all contributed to the conversion of semi-arid grasslands into desert.

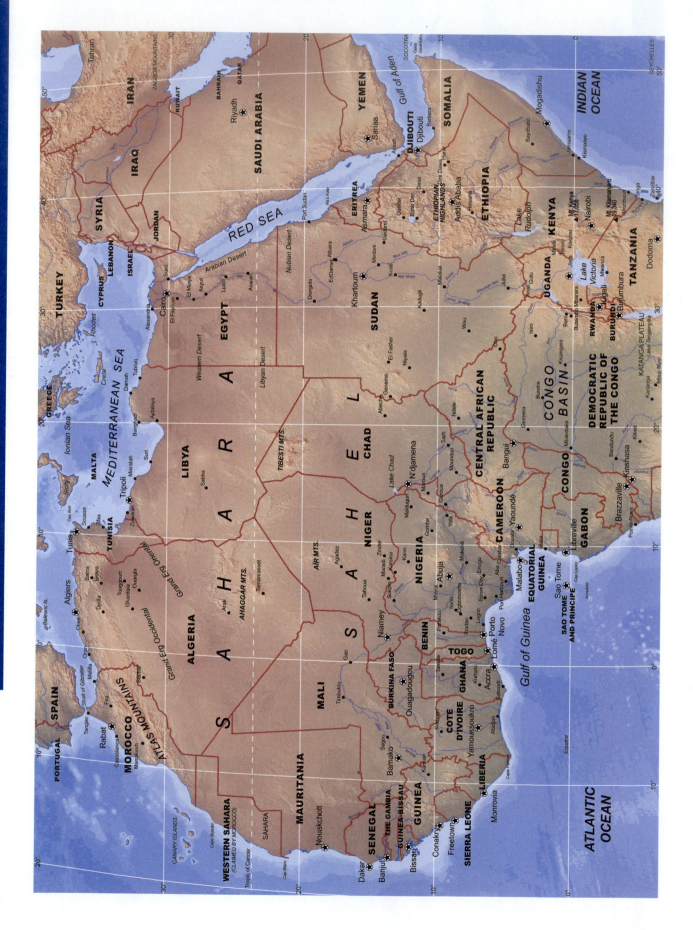

Map **154** Northern Africa

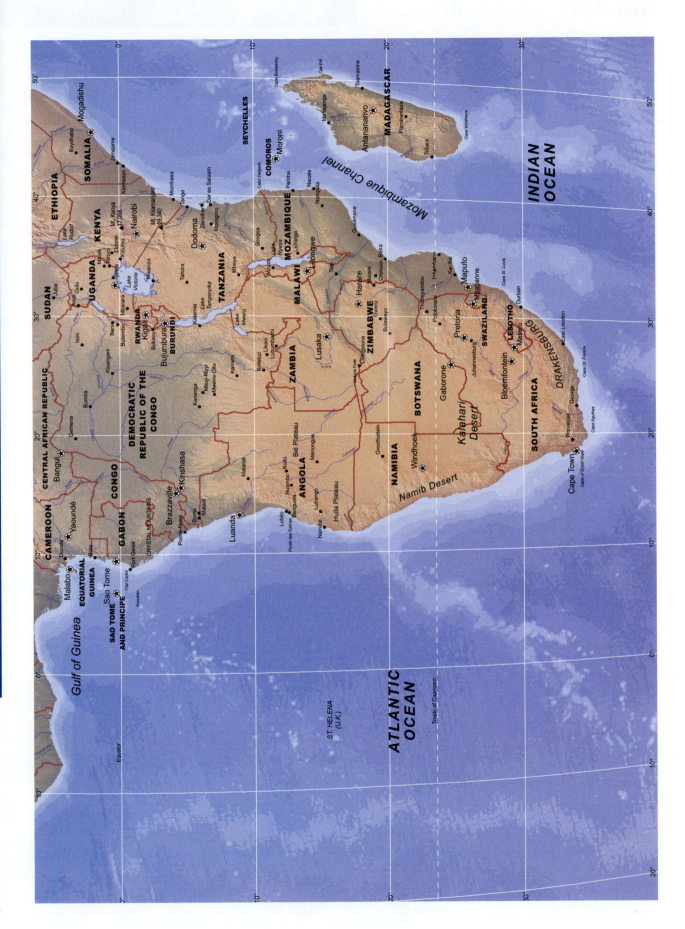

Map 155   Southern Africa

-197-

Map 156　Asia: Political Divisions

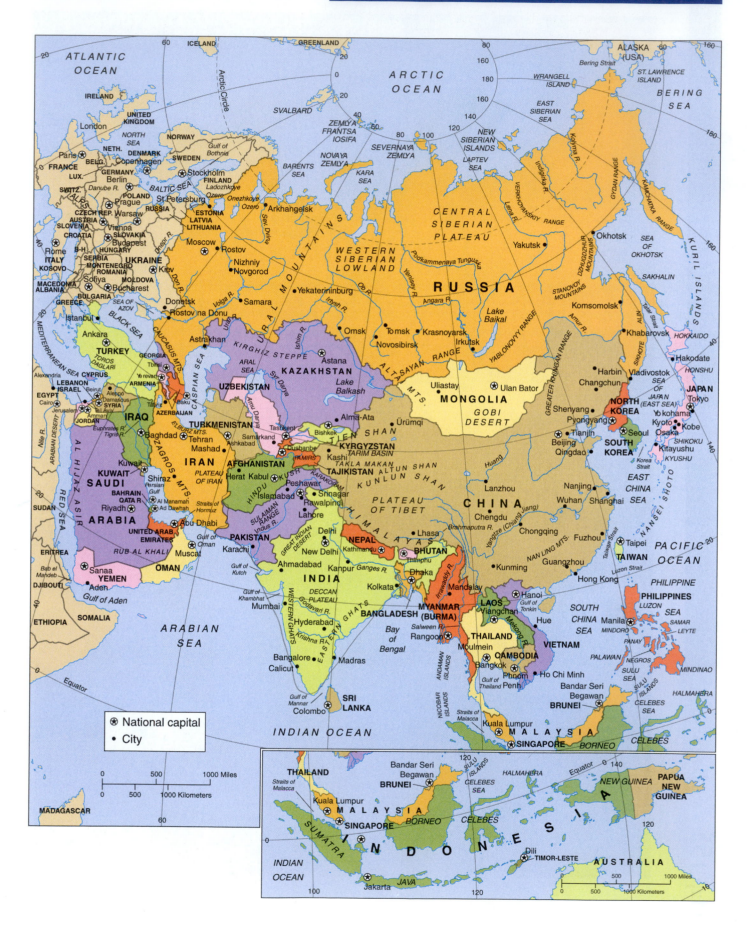

National capital

City

| 0 | 500 | 1000 Miles |
| 0 | 500 | 1000 Kilometers |

-198-

Map 157 Asia: Physical Features

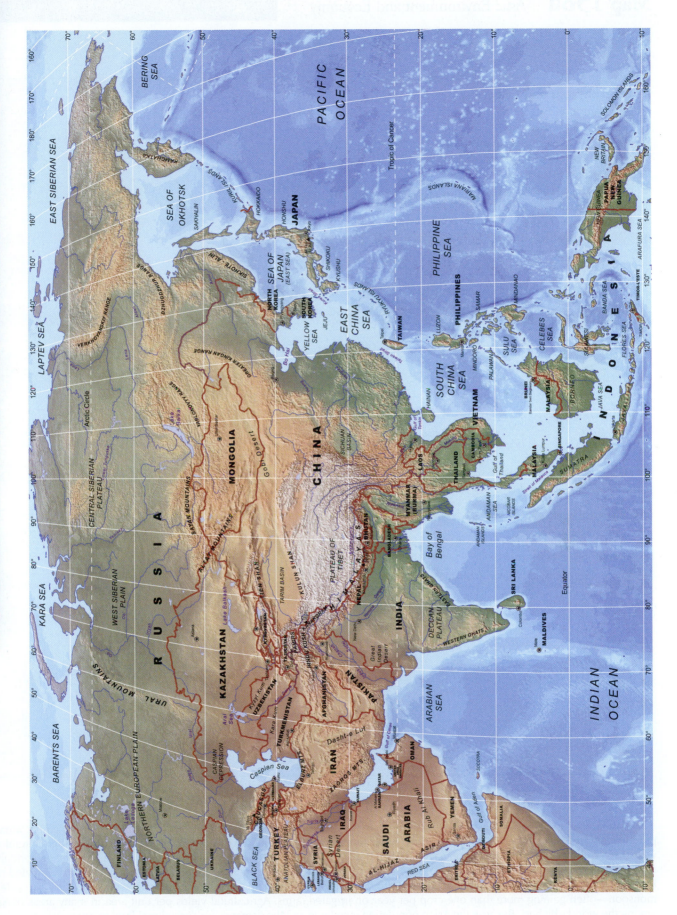

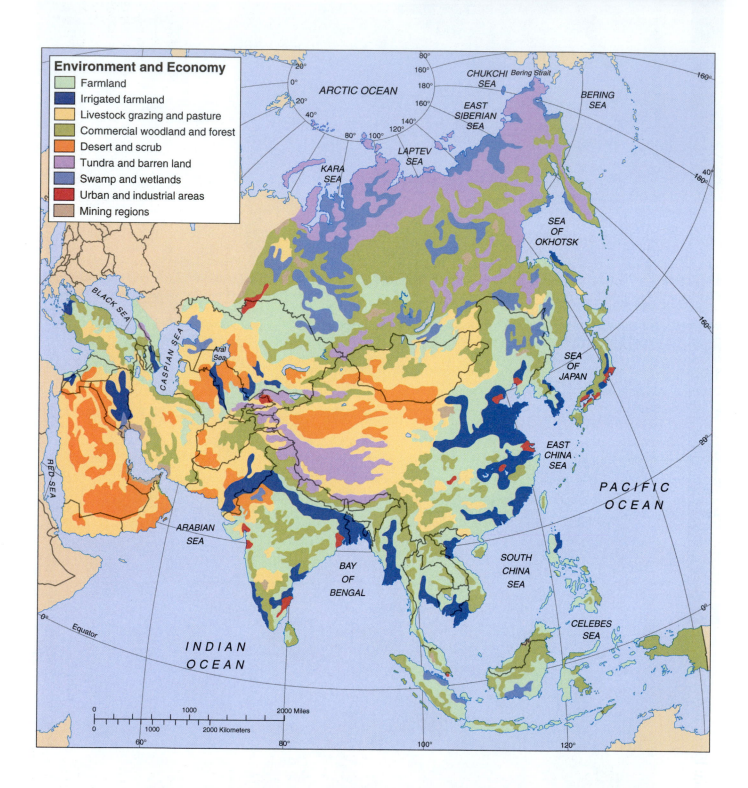

Environment and Economy
- Farmland
- Irrigated farmland
- Livestock grazing and pasture
- Commercial woodland and forest
- Desert and scrub
- Tundra and barren land
- Swamp and wetlands
- Urban and industrial areas
- Mining regions

Asia is a land of extremes of land use with some of the world's most heavily industrialized regions, barren and empty areas, and productive and densely populated farm regions. Asia is a region of rapid industrial growth. Yet Asia remains an agricultural region with three out of every four workers engaged in agriculture. Asian commercial agriculture and intensive subsistence agriculture is characterized by irrigation. Some of Asia's irrigated lands are desert requiring additional water. But most of the Asian irrigated regions have sufficient precipitation for crop agriculture and irrigation is a way of coping with seasonal drought—the wet-and-dry cycle of the monsoon—often gaining more than one crop per year on irrigated farms. Agricultural yields per unit area in many areas of Asia are among the world's highest. Because the Asian population is so large and the demands for agricultural land so great, Asia is undergoing rapid deforestation and some areas of the continent have only small remnants of a once-abundant forest reserve.

Map 158b   Asia: Population Density

Population Density

Persons per square mile (km)

- Uninhabited area
- Less than 2 (1)
- 2 – 25 (1 – 10)
- 25 – 50 (10 – 20)
- 50 – 150 (20 – 60)
- 150 – 300 (60 – 120)
- More than 300 (120)

1000 Miles

1000 Kilometers

0

With one-third of the world's land area and nearly two-thirds of the world's population, Asia is more densely settled than any other region of the world. In some of the continent's farming regions, agricultural population density exceeds 2,000 persons per square mile. In some portions of the continent, particularly the Islamic areas of Central and Southwest Asia, this already large population is growing very rapidly with some countries having population growth rates above 3 percent per year and doubling times between 20 and 25 years. The populations of neither China nor India are growing particularly rapidly but since both countries have population bases that are enormous, the absolute number of Indians and Chinese added to the world's population each year is staggering. In spite of these massive populations, Asia also contains areas that are either completely uninhabited or have population densities that are as low as any on earth.

Map 158c   Asia: Industrialization: Manufacturing and Resources

Manufacturing and Resources

- Major manufacturing region
- Minor manufacturing region
- Iron producing region
- Producing coal field
- Producing oil and gas field
- Railroad

1000 Miles

1000 Kilometers

0
0

International economists have predicted that the next century will be one of Asian economic dominance as other Asian nations approach the economic levels of industrial giant Japan. Asian countries have begun to industrialize rapidly and have increased industrial production more than 100 percent in the last decade. But nothing guarantees industrial output; it must master the great distances separating critical raw material and production sites from the locations of the markets for them. Such a mastery is achieved only by the development of efficient transportation systems. Usually that means water travel and here Asia is remarkably deficient when compared with Europe and North America. Nevertheless, the mixed prospects for Asian economic growth into the next century are a great deal better than would have been predicted a decade ago.

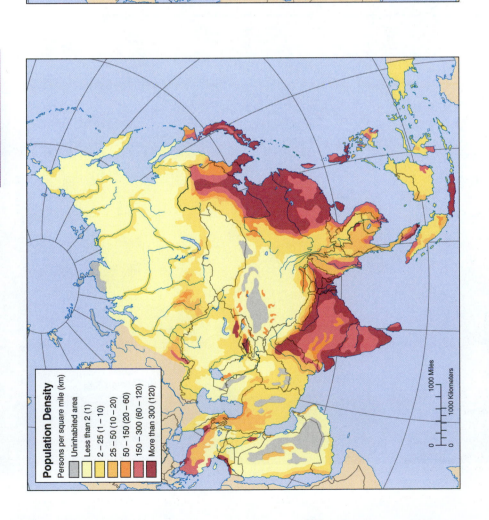

-201-

Map 159 Northern Asia

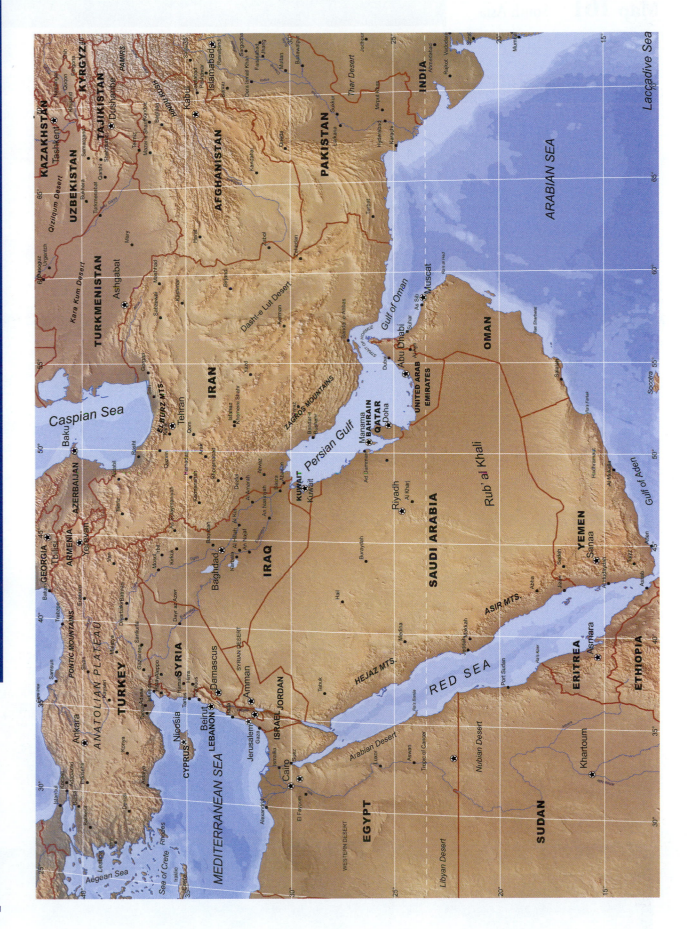

Map 160 Southwestern Asia

KAZAKHSTAN

KYRGYZSTAN

UZBEKISTAN

TAJIKISTAN

TURKMENISTAN

Ashgabat

Caspian Sea

Baku

GEORGIA
Tbilisi
ARMENIA
AZERBAIJAN
Yerevan

PONTIC MOUNTAINS

TURKEY

ANATOLIAN PLATEAU

Ankara

NICOSIA

CYPRUS

MEDITERRANEAN SEA

Aegean Sea

Sea of Crete

SYRIA
Damascus

LEBANON
Beirut

ISRAEL
Jerusalem
JORDAN
Amman

Cairo

EGYPT

WESTERN DESERT

Libyan Desert

Arabian Desert

Nubian Desert

SUDAN

Khartoum

AFGHANISTAN

HINDU KUSH

PAMIRS

Kabul

Islamabad

PAKISTAN

INDIA

Thar Desert

ELBURZ MTS.

Tehran

IRAN

Dasht-e Lut Desert

ZAGROS MOUNTAINS

Persian Gulf

KUWAIT
Kuwait

IRAQ

Baghdad

SYRIAN DESERT

HEJAZ MTS.

Riyadh

SAUDI ARABIA

Rub' al Khali

QATAR
Doha
BAHRAIN
Manama

UNITED ARAB
EMIRATES

Abu Dhabi

Muscat

OMAN

Gulf of Oman

STRAIT OF HORMUZ

ARABIAN SEA

Laccadive Sea

Medina

RED SEA

ASIR MTS.

YEMEN
Sanaa

Aden

Gulf of Aden

ERITREA
Asmara

ETHIOPIA

Map 161   South Asia

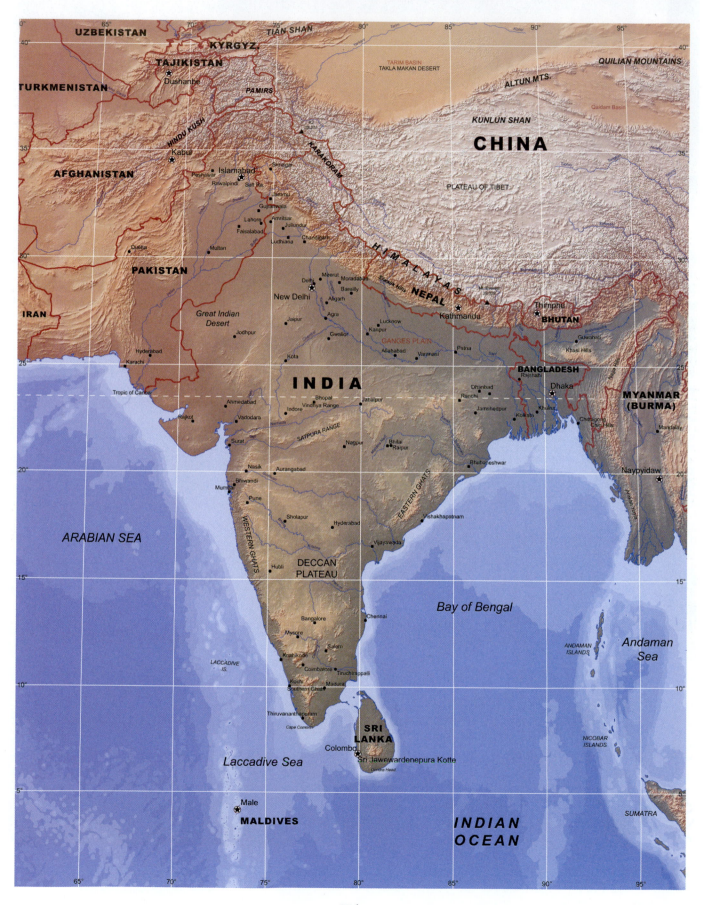

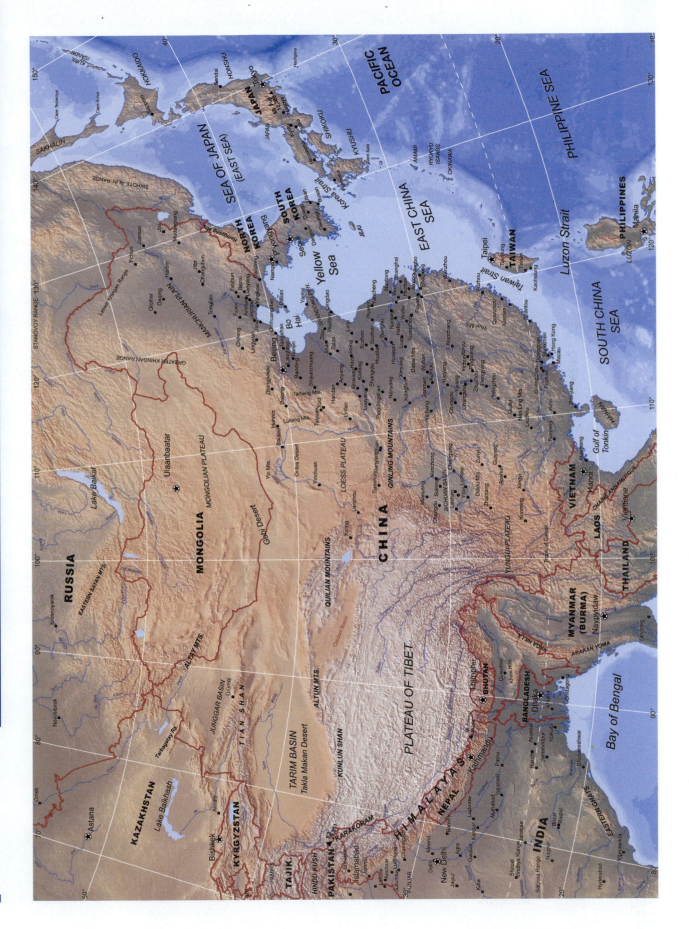

Map 162 East Asia

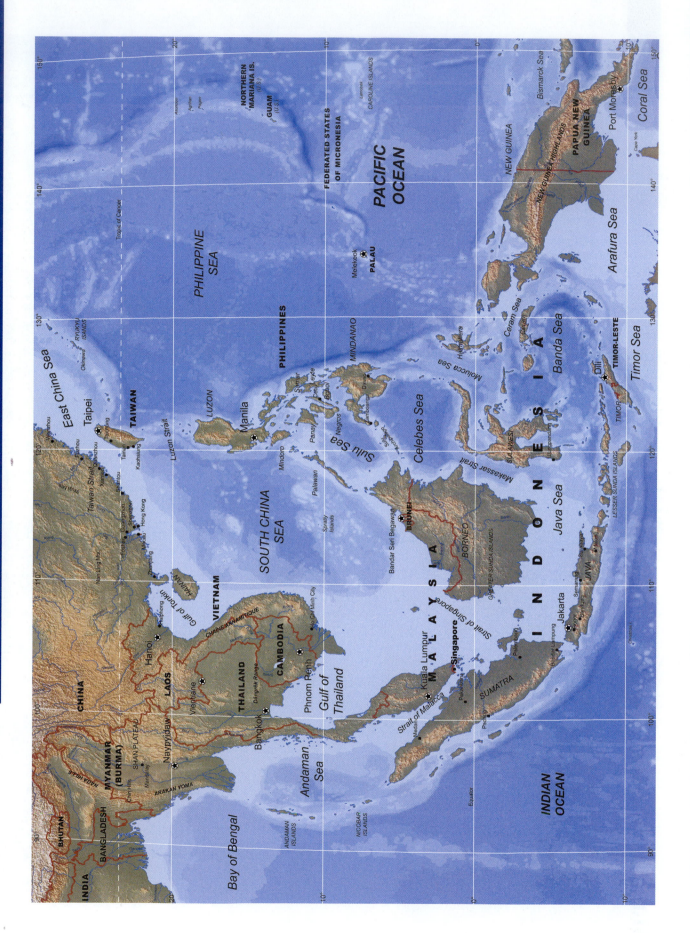

Map 163 Southeast Asia

CHINA

INDIA

BANGLADESH

BHUTAN

MYANMAR (BURMA)

SHAN PLATEAU

ARAKAN YOMA

Mandalay

Naypyidaw

LAOS

Vientiane

THAILAND

Bangkok

CAMBODIA

Phnom Penh

VIETNAM

Hanoi

Haiphong

Ho Chi Minh City

CHAINE ANNAMITIQUE

Dangrek Range

HAINAN

Gulf of Tonkin

Gulf of Thailand

Andaman Sea

Bay of Bengal

ANDAMAN ISLANDS

NICOBAR ISLANDS

TAIWAN

Taipei

Taichung

Tainan

Kaohsiung

Taiwan Strait

Luzon Strait

East China Sea

RYUKYU ISLANDS

Okinawa

Wenzhou

Fuzhou

Quanzhou

Guangzhou

Hong Kong

Macau

Shantou

Zhanjiang

Nanling Mts.

Tropic of Cancer

PHILIPPINE SEA

PHILIPPINES

LUZON

Manila

Mindoro

Panay

Negros

Cebu

Bohol

Leyte

Samar

MINDANAO

Davao

Zamboanga

Sulu Sea

Spratly Islands

SOUTH CHINA SEA

NORTHERN MARIANA IS.

Pagan

Aguijan

Saipan

GUAM (U.S.)

CAROLINE ISLANDS

FEDERATED STATES OF MICRONESIA

PACIFIC OCEAN

PALAU

Melekeok

MALAYSIA

Kuala Lumpur

Singapore

Strait of Malacca

Medan

Palembang

Strait of Singapore

BRUNEI

Bandar Seri Begawan

BORNEO

GREATER SUNDA ISLANDS

Celebes Sea

Makassar Strait

SULAWESI

Ujungpandang

Manado

Halmahera

Ceram Sea

Molucca Sea

Banda Sea

I N D O N E S I A

Java Sea

SUMATRA

Bandar Lampung

Padang

Pekanbaru

Jakarta

Bandung

Semarang

Surabaya

JAVA

LESSER SUNDA ISLANDS

Denpasar

TIMOR

Kupang

Dili

TIMOR-LESTE

Timor Sea

Arafura Sea

NEW GUINEA

NEW GUINEA HIGHLANDS

PAPUA NEW GUINEA

Port Moresby

Cape York

Bismarck Sea

Coral Sea

Equator

INDIAN OCEAN

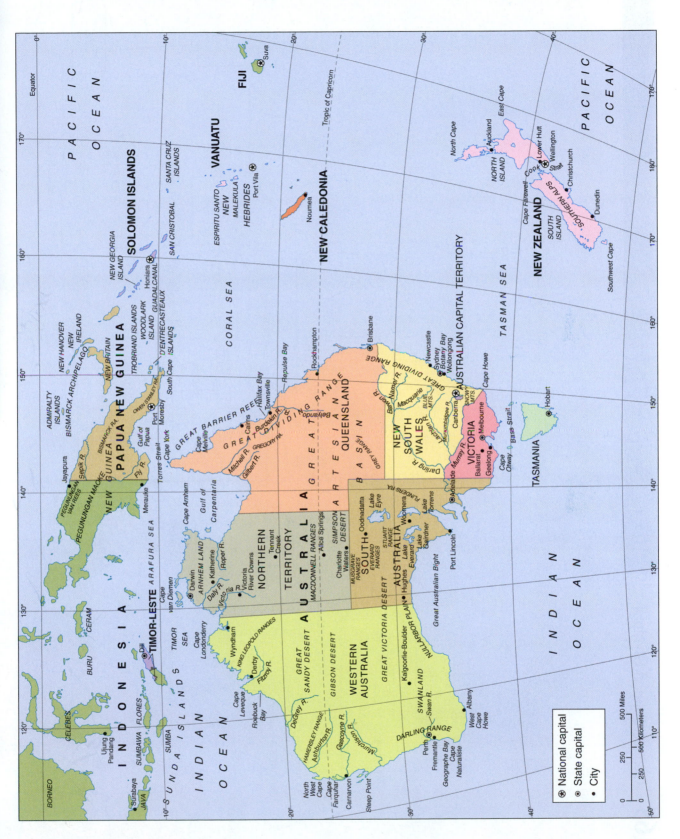

Map 164   Australasia and Western Oceania: Political Divisions

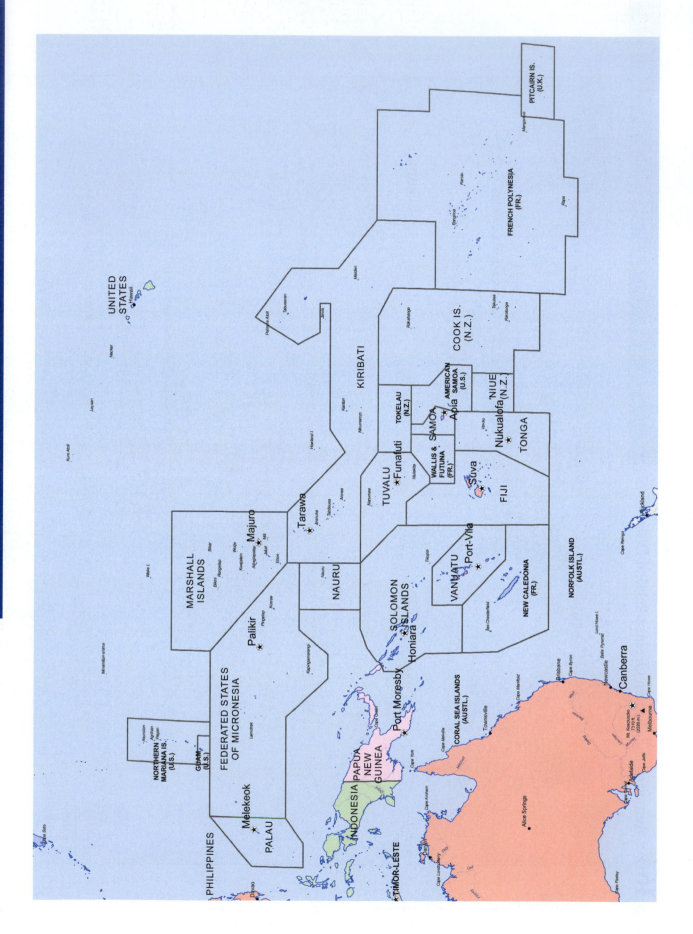

Map 165 Oceania: Political Divisions

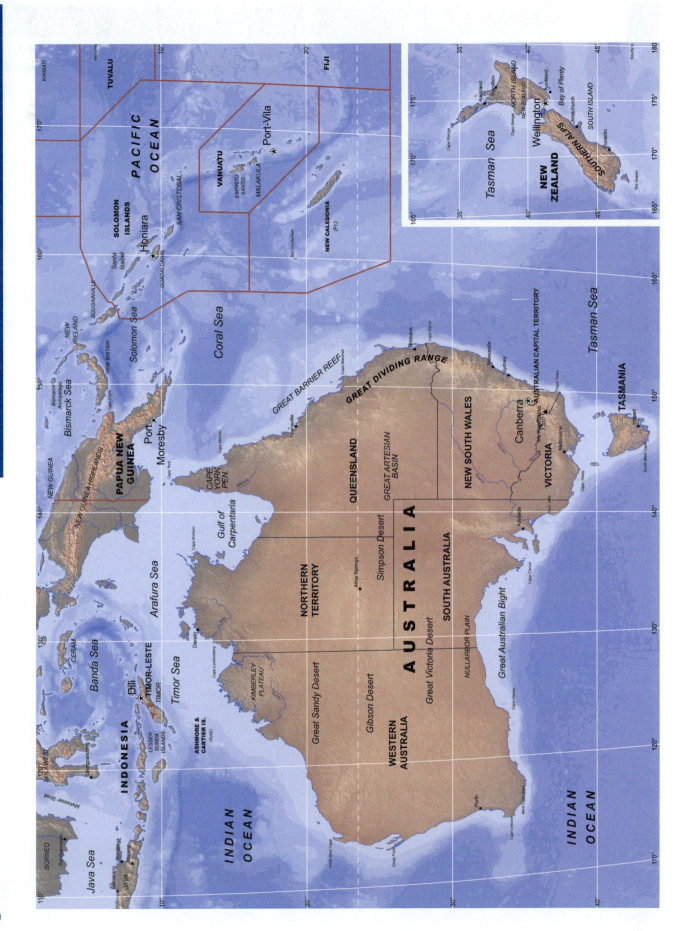

Map 166 Australasia and Western Oceania: Physical Features

KIRIBATI

TUVALU

PACIFIC OCEAN

FIJI

Port-Vila

VANUATU

ESPIRITU SANTO

MALAKULA

SOLOMON ISLANDS

Honiara

Santa Isabel

SAN CRISTOBAL

GUADALCANAL

NEW CALEDONIA (Fr.)

Iles Chesterfield

BOUGAINVILLE

NEW IRELAND

NEW BRITAIN

Bismarck Archipelago

Bismarck Sea

Solomon Sea

Coral Sea

GREAT BARRIER REEF

Cape Melville

GREAT DIVIDING RANGE

Brisbane

Cape Byron

Newcastle

Sydney

Cape Howe

Townsville

NEW GUINEA

NEW GUINEA HIGHLANDS

PAPUA NEW GUINEA

Port Moresby

CAPE YORK PEN.

Cape York

Mitchell

Cape Arnhem

Gulf of Carpentaria

QUEENSLAND

GREAT ARTESIAN BASIN

NEW SOUTH WALES

Cape Jervis

AUSTRALIAN CAPITAL TERRITORY

Canberra

Mt. Kosciuszko 7,310

Melbourne

VICTORIA

TASMANIA

Hobart

South West Cape

Tasman Sea

CERAM

Banda Sea

Arafura Sea

Cape Londonderry

Timor Sea

Dili

TIMOR-LESTE

TIMOR

Darwin

NORTHERN TERRITORY

Alice Springs

Simpson Desert

A U S T R A L I A

SOUTH AUSTRALIA

Adelaide

Cape Catastrophe

BORNEO

Banjarmasin

Java Sea

Surabaya

Malang

JAVA

Semarang

SULAWESI

Ujungpandang

Makassar Strait

INDONESIA

LESSER SUNDA ISLANDS

ASHMORE & CARTIER IS. (Aust.)

KIMBERLEY PLATEAU

Great Sandy Desert

Gibson Desert

WESTERN AUSTRALIA

Great Victoria Desert

NULLARBOR PLAIN

Great Australian Bight

Cape Pasley

Perth

West Cape Howe

Cape Leeuwin

Steep Point

North West Cape

INDIAN OCEAN

INDIAN OCEAN

NORTH ISLAND

Auckland

Hamilton

Cape Runaway

Bay of Plenty

Wellington

Christchurch

SOUTH ISLAND

SOUTHERN ALPS

Dunedin

The Snares

NEW ZEALAND

Cape Reinga

Napier

Tasman Sea

Bounty Is.

Map 167a   Australasia: Environment and Economy

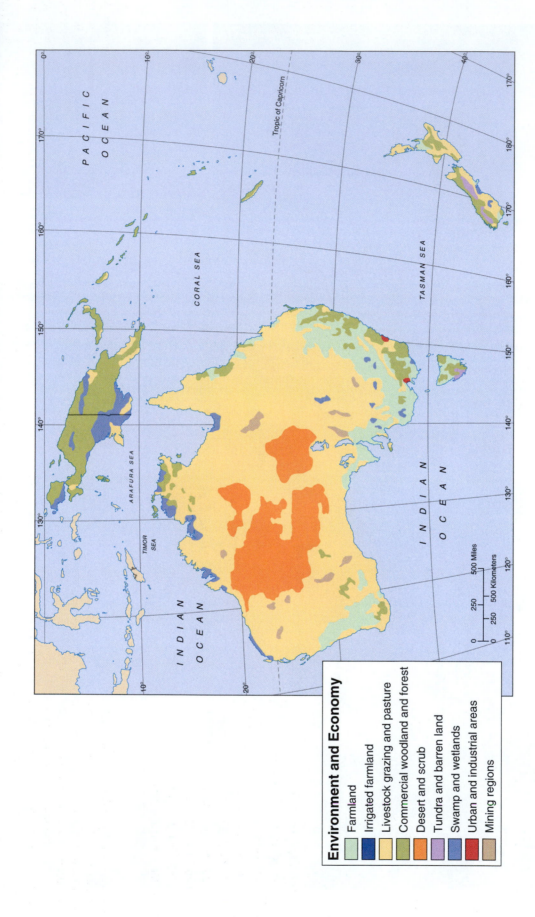

Environment and Economy

- Farmland
- Irrigated farmland
- Livestock grazing and pasture
- Commercial woodland and forest
- Desert and scrub
- Tundra and barren land
- Swamp and wetlands
- Urban and industrial areas
- Mining regions

PACIFIC OCEAN

Tropic of Capricorn

CORAL SEA

TASMAN SEA

TIMOR SEA

ARAFURA SEA

INDIAN OCEAN

INDIAN OCEAN

500 Miles
500 Kilometers

Australasia is dominated by the world's smallest and most uniform continent. Flat, dry and mostly hot, Australia has the simplest of land use patterns: where rainfall exists so does agricultural activity. Two agricultural patterns dominate the map: livestock grazing, primarily sheep, and wheat farming, although some sugar cane production exists in the north and some cotton is grown elsewhere. Only about 6 percent of the continent consists of arable land so the areas of wheat farming, dominant as they may be in the context of Australian agriculture, are small. Australia also supports a healthy mineral resource economy, with iron and copper and precious metals making up the bulk of the extraction. Elsewhere in the region, tropical forests dominate Papua New Guinea, with some subsistence agriculture and livestock. New Zealand's temperate climate with abundant precipitation supports a productive livestock industry and little else besides tourism— which is an important economic element throughout the remainder of the region as well.

-210-

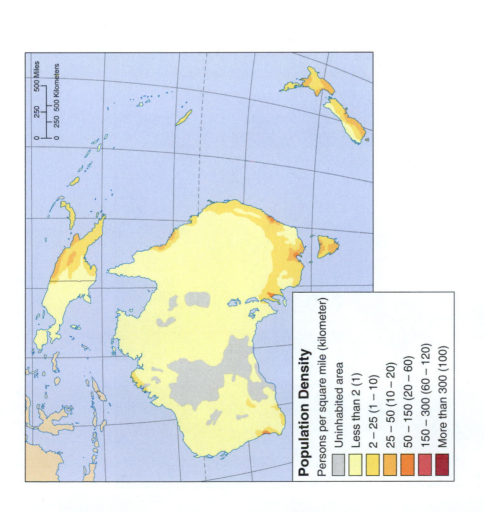

Climatic Patterns

- Tropical Rain Forest
- Tropical Savanna
- Tropical Steppe
- Tropical Desert
- Mediterranean Climate
- Humid Subtropical
- Marine West Coast
- Highland Climate

Population Density

Persons per square mile (kilometer)

- Uninhabited area
- Less than 2 (1)
- 2 – 25 (1 – 10)
- 25 – 50 (10 – 20)
- 50 – 150 (20 – 60)
- 150 – 300 (60 – 120)
- More than 300 (100)

The region's small population is remarkably diverse. To the north are the Melanesian New Guinea peoples, while Europeans dominate the populations of Australia and New Zealand, both of which have significant indigenous populations. Throughout most of the smaller island groups, the bulk of the population is Melanesian but with a scattering of Europeans. The distribution of population is extremely uneven with New Guinea, southeastern Australia, and New Zealand supporting the bulk of the region's people while the remainder of the region—meaning nearly all of Australia—is either sparsely populated or devoid of population altogether. Nowhere do population densities reach the levels they do in other major regions of the world and densities of 50 persons per square mile are the highest to be found. Population location is dependent on precipitation and population growth patterns are culturally variable.

Because of its nearly uniform surface, with only a few low and scattered uplands, Australian climate is a consequence only of the two great climate controls—latitudinal position and location relative to continental margin and interior. The continent bestrides the 30th parallel of latitude and its climatic pattern is dominated by the subtropical high pressure system with dry air masses that are responsible for the existence of great deserts. Toward the equator, the desert grades into steppe, savanna, and tropical forest as the subtropical high gives way to equatorial low pressure and abundant precipitation. Toward the pole, arid land fades into more well-watered steppe grasslands, the Mediterranean type climate of the southern margins of the continent, and the marine west coast climate of the Australian southeast and New Zealand. This latter climate is where most of the region's people live.

Map 169 Antarctica

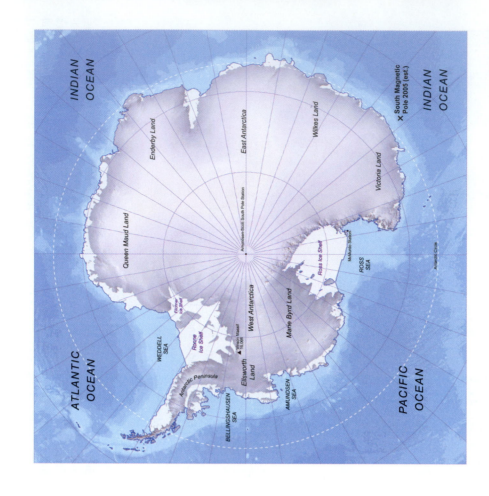

INDIAN OCEAN

Enderby Land

East Antarctica

Wilkes Land

Victoria Land

× South Magnetic Pole 2005 (est.)

INDIAN OCEAN

Queen Maud Land

Amundsen-Scott South Pole Station

McMurdo Station

Ross Ice Shelf

ROSS SEA

Antarctic Circle

ATLANTIC OCEAN

Filchner Ice Shelf

Ronne Ice Shelf

WEDDELL SEA

West Antarctica

Marie Byrd Land

▲ Vinson Massif 16,066

Ellsworth Land

Antarctic Peninsula

BELLINGSHAUSEN SEA

AMUNDSEN SEA

PACIFIC OCEAN

Map 168 The Arctic

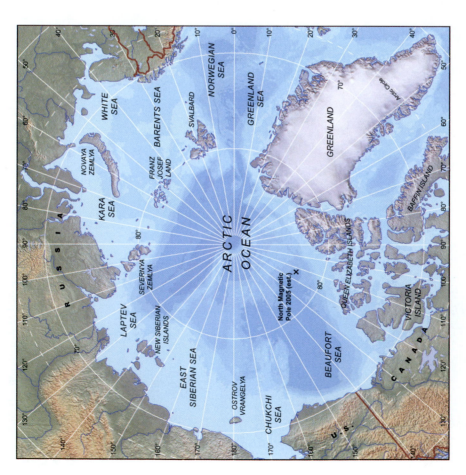

40° 30° 20° 10° 0° 10° 20° 30° 40°

50°

60°

70°

NORWEGIAN SEA

WHITE SEA

BARENTS SEA

SVALBARD

GREENLAND SEA

GREENLAND

Arctic Circle

50°

60°

NOVAYA ZEMLYA

FRANZ JOSEF LAND

70°

KARA SEA

R U S S I A

80°

SEVERNYA ZEMLYA

ARCTIC OCEAN

BAFFIN ISLAND

80°

LAPTEV SEA

NEW SIBERIAN ISLANDS

North Magnetic Pole 2005 (est.) ×

80°

QUEEN ELIZABETH ISLANDS

90°

100°

EAST SIBERIAN SEA

OSTROV VRANGELYA

CHUKCHI SEA

BEAUFORT SEA

VICTORIA ISLAND

C A N A D A

110°

U. S.

130° 140° 150° 160° 170° 180° 170° 160° 150° 140° 130°

Unit VIII

Country and Dependency Profiles

Country and Dependency Profiles

Afghanistan

Afghanestan

Official Name: Islamic Republic of Afghanistan
Capital: Kabul
Area: 250,001 sq mi (647,500 sq km)
Population: 28,395,716
Major Language(s): Dari, Pashto
Major Religion(s): Sunni Islam
Currency: afghani

Andorra

Official Name: Principality of Andorra
Capital: Andorra la Vella
Area: 181 sq mi (468 sq km)
Population: 83,888
Major Language(s): Catalan, Spanish, French
Major Religion(s): Roman Catholic Christianity
Currency: Euro

Albania

Shqiperia

Official Name: Republic of Albania
Capital: Tirana
Area: 11,100 sq mi (28,748 sq km)
Population: 3,639,453
Major Language(s): Albanian
Major Religion(s): Sunni Islam
Currency: lek

Angola

Official Name: Republic of Angola
Capital: Luanda
Area: 481,354 sq mi (1,246,700 sq km)
Population: 12,799,293
Major Language(s): Portuguese, Bantu languages
Major Religion(s): Indigenous beliefs, Christianity
Currency: kwanza

Algeria

Al Jaza'ir

Official Name: People's Democratic Republic of Algeria
Capital: Algiers
Area: 919,595 sq mi (2,381,740 sq km)
Population: 34,178,188
Major Language(s): Arabic, French, Berber dialects
Major Religion(s): Sunni Islam
Currency: Algerian dinar

Anguilla

(overseas territory of the United Kingdom)

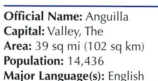

Official Name: Anguilla
Capital: Valley, The
Area: 39 sq mi (102 sq km)
Population: 14,436
Major Language(s): English
Major Religion(s): Protestant Christianity
Currency: East Caribbean dollar

American Samoa

(territory of the United States)

Official Name: Territory of American Samoa
Capital: Pago Pago
Area: 77 sq mi (199 sq km)
Population: 65,628
Major Language(s): Samoan
Major Religion(s): Protestant Christianity
Currency: US dollar

Antigua and Barbuda

Official Name: Antigua and Barbuda
Capital: Saint John's
Area: 171 sq mi (443 sq km)
Population: 85,632
Major Language(s): English
Major Religion(s): Protestant Christianity
Currency: East Caribbean dollar

Argentina

Official Name: Argentine Republic
Capital: Buenos Aires
Area: 1,068,302 sq mi (2,766,890 sq km)
Population: 40,913,584
Major Language(s): Spanish
Major Religion(s): Roman Catholic Christianity
Currency: Argentine peso

Austria
Oesterreich

Official Name: Republic of Austria
Capital: Vienna
Area: 32,382 sq mi (83,870 sq km)
Population: 8,210,281
Major Language(s): German
Major Religion(s): Roman Catholic Christianity
Currency: Euro

Armenia
Hayastan

Official Name: Republic of Armenia
Capital: Yerevan
Area: 11,484 sq mi (29,743 sq km)
Population: 2,967,004
Major Language(s): Armenian
Major Religion(s): Armenian Apostolic Christianity
Currency: dram

Azerbaijan
Azarbaycan

Official Name: Republic of Azerbaijan
Capital: Baku
Area: 33,436 sq mi (86,600 sq km)
Population: 8,238,672
Major Language(s): Azerbaijani
Major Religion(s): Shia Islam
Currency: manats

Aruba
(part of the Kingdom of the Netherlands)

Official Name: Aruba
Capital: Oranjestad
Area: 75 sq mi (193 sq km)
Population: 103,065
Major Language(s): Papiamento, Dutch
Major Religion(s): Roman Catholic Christianity
Currency: Arubian guilder/florin

Bahamas, The

Official Name: Commonwealth of the Bahamas
Capital: Nassau
Area: 5,382 sq mi (13,940 sq km)
Population: 307,552
Major Language(s): English
Major Religion(s): Protestant
Currency: Bahamian dollar

Australia

Official Name: Commonwealth of Australia
Capital: Canberra
Area: 2,967,909 sq mi (7,686,850 sq km)
Population: 21,262,641
Major Language(s): English, Aboriginal languages
Major Religion(s): Christianity
Currency: Australian dollar

Bahrain
Al Bahrayn

Official Name: Kingdom of Bahrain
Capital: Manama
Area: 257 sq mi (665 sq km)
Population: 728,709
Major Language(s): Arabic, English
Major Religion(s): Sunni Islam
Currency: Bahraini dollar

Country and Dependency Profiles

Bangladesh

Banladesh

Official Name: People's Republic of Bangladesh
Capital: Dhaka
Area: 55,599 sq mi (144,000 sq km)
Population: 156,050,883
Major Language(s): Bangla, English
Major Religion(s): Sunni Islam
Currency: taka

Belize

Official Name: Belize
Capital: Belmopan
Area: 8,867 sq mi (22,966 sq km)
Population: 307,899
Major Language(s): Spanish, Creole, English
Major Religion(s): Roman Catholic Christianity
Currency: Belizian dollar

Barbados

Official Name: Barbados
Capital: Bridgetown
Area: 166 sq mi (431 sq km)
Population: 284,589
Major Language(s): English
Major Religion(s): Protestant Christianity
Currency: Barbadian dollar

Benin

Official Name: Republic of Benin
Capital: Porto-Novo
Area: 43,483 sq mi (112,620 sq km)
Population: 8,791,832
Major Language(s): French, Fon, Yoruba, other African languages
Major Religion(s): Indigenous beliefs, Christianity
Currency: CFA franc

Belarus

Byelarus'

Official Name: Republic of Belarus
Capital: Minsk
Area: 80,155 sq mi (207,600 sq km)
Population: 9,648,533
Major Language(s): Belarusian, Russian
Major Religion(s): Eastern Orthodox Christianity
Currency: Belarusian ruble

Bermuda

(overseas territory of the United Kingdom)

Official Name: Bermuda
Capital: Hamilton
Area: 20 sq mi (53 sq km)
Population: 67,837
Major Language(s): English
Major Religion(s): Protestant Christianity
Currency: Bermudian dollar

Belgium

Belgique/Belgie

Official Name: Kingdom of Belgium
Capital: Brussels
Area: 11,787 sq mi (30,528 sq km)
Population: 10,414,336
Major Language(s): Dutch, French, German
Major Religion(s): Roman Catholic Christianity
Currency: Euro

Bhutan

Druk Yul

Official Name: Kingdom of Bhutan
Capital: Thimphu
Area: 18,147 sq mi (47,000 sq km)
Population: 691,141
Major Language(s): Dzongkha, Tibetan dialects, Nepalese dialects
Major Religion(s): Buddhism
Currency: ngultrum

Bolivia

Official Name: Plurinational State of Bolivia
Capital: La Paz (administrative), Sucre (constitutional)
Area: 424,164 sq mi (1,098,580 sq km)
Population: 9,775,246
Major Language(s): Spanish, Quechua, Aymara
Major Religion(s): Roman Catholic Christianity
Currency: boliviano

Brunei

Official Name: Brunei Darussalam
Capital: Bandar Seri Begawan
Area: 2,228 sq mi (5,770 sq km)
Population: 388,190
Major Language(s): Malay, English
Major Religion(s): Sunni Islam
Currency: Bruneian dollar

Bosnia and Herzegovina

Bosna i Hercegovina

Official Name: Bosnia and Herzegovina
Capital: Sarajevo
Area: 19,772 sq mi (51,209 sq km)
Population: 4,613,414
Major Language(s): Bosnian, Croatian, Serbian
Major Religion(s): Christianity, Islam
Currency: konvertibilna mark

Bulgaria

Balgariya

Official Name: Republic of Bulgaria
Capital: Sofia
Area: 42,823 sq mi (110,910 sq km)
Population: 7,204,687
Major Language(s): Bulgarian
Major Religion(s): Eastern Orthodox Christianity
Currency: leva

Botswana

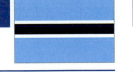

Official Name: Republic of Botswana
Capital: Gaborone
Area: 231,804 sq mi (600,370 sq km)
Population: 1,990,876
Major Language(s): Setswana, Kalanga, English
Major Religion(s): Christianity
Currency: pula

Burkina Faso

Official Name: Burkina Faso
Capital: Ouagadougou
Area: 105,869 sq mi (274,200 sq km)
Population: 15,746,232
Major Language(s): French, many African languages
Major Religion(s): Sunni Islam
Currency: CFA franc

Brazil

Brasil

Official Name: Federative Republic of Brazil
Capital: Brasilia
Area: 3,286,488 sq mi (8,511,965 sq km)
Population: 198,739,269
Major Language(s): Portuguese, many Amerindian languages
Major Religion(s): Roman Catholic Christianity
Currency: real

Burma (Myanmar)

Myanma Naingngandaw

Official Name: Union of Burma
Capital: Rangoon
Area: 261,970 sq mi (678,500 sq km)
Population: 48,137,741
Major Language(s): Burmese
Major Religion(s): Buddhism
Currency: kyat

Country and Dependency Profiles

Burundi

Official Name: Republic of Burundi
Capital: Bujumbura
Area: 10,745 sq mi (27,830 sq km)
Population: 9,511,330
Major Language(s): Kirundi, French, Swahili
Major Religion(s): Roman Catholic Christianity
Currency: Burundi franc

Cambodia

Kampuchea

Official Name: Kingdom of Cambodia
Capital: Phnom Penh
Area: 69,900 sq mi (181,040 sq km)
Population: 14,494,293
Major Language(s): Khmer
Major Religion(s): Buddhism
Currency: riel

Cameroon

Cameroun

Official Name: Republic of Cameroon
Capital: Yaounde
Area: 183,568 sq mi (475,440 sq km)
Population: 18,879,301
Major Language(s): English, French, many African languages
Major Religion(s): Indigenous beliefs, Christianity
Currency: CFA franc

Canada

Official Name: Canada
Capital: Ottawa
Area: 3,855,103 sq mi (9,984,670 sq km)
Population: 33,487,208
Major Language(s): English, French, Amerindian languages
Major Religion(s): Roman Catholic Christianity
Currency: Canadian dollar

Cape Verde

Cabo Verde

Official Name: Republic of Cape Verde
Capital: Praia
Area: 1,557 sq mi (4,033 sq km)
Population: 429,474
Major Language(s): Portuguese
Major Religion(s): Roman Catholic Christianity
Currency: escudo

Cayman Islands

(overseas territory of the United Kingdom)

Official Name: Cayman Islands
Capital: George Town
Area: 101 sq mi (262 sq km)
Population: 49,035
Major Language(s): English
Major Religion(s): Christianity
Currency: Caymanian dollar

Central African Republic

Official Name: Central African Republic
Capital: Bangui
Area: 240,535 sq mi (622,984 sq km)
Population: 4,511,488
Major Language(s): French, Sangho
Major Religion(s): Indigenous beliefs, Christianity
Currency: CFA franc

Chad

Tchad/Tshad

Official Name: Republic of Chad
Capital: N'Djamena
Area: 495,755 sq mi (1,284,000 sq km)
Population: 10,329,208
Major Language(s): French, Arabic, many African languages
Major Religion(s): Sunni Islam
Currency: CFA franc

Chile

Official Name: Republic of Chile
Capital: Santiago
Area: 292,260 sq mi (756,950 sq km)
Population: 16,601,707
Major Language(s): Spanish
Major Religion(s): Roman Catholic Christianity
Currency: Chilean peso

China

Zhongguo

Official Name: People's Republic of China
Capital: Beijing
Area: 3,705,407 sq mi (9,596,960 sq km)
Population: 1,338,612,968
Major Language(s): Chinese (Mandarin), Yue (Cantonese), many other dialects
Major Religion(s): Officially atheist
Currency: Renminbi yuan

Colombia

Official Name: Republic of Colombia
Capital: Bogotá
Area: 439,736 sq mi (1,138,910 sq km)
Population: 45,644,023
Major Language(s): Spanish
Major Religion(s): Roman Catholic Christianity
Currency: Colombian peso

Comoros

Komori/Comores/Juzur al Qamar

Official Name: Union of the Comoros
Capital: Moroni
Area: 838 sq mi (2,170 sq km)
Population: 752,438
Major Language(s): Arabic, French
Major Religion(s): Sunni Islam
Currency: Comoran franc

Congo, Democratic Republic of

Congo (Kinshasa)

Official Name: Democratic Republic of the Congo
Capital: Kinshasa
Area: 905,345 sq mi (2,344,858 sq km)
Population: 68,692,542
Major Language(s): French, Lingala, Kingwana
Major Religion(s): Christian, Muslim, Indigenous
Currency: Congolese franc

Congo, Republic of

Congo (Brazzaville)

Official Name: Republic of the Congo
Capital: Brazzaville
Area: 132,047 sq mi (342,000 sq km)
Population: 4,012,809
Major Language(s): French, Lingala, Monokutuba
Major Religion(s): Christian, Animist
Currency: Congolese franc

Costa Rica

Official Name: Republic of Costa Rica
Capital: San Jose
Area: 19,730 sq mi (51,100 sq km)
Population: 4,253,877
Major Language(s): Spanish
Major Religion(s): Roman Catholic Christianity
Currency: colón

Cote d'Ivoire

Official Name: Republic of Cote d'Ivoire
Capital: Yamoussoukro
Area: 124,503 sq mi (322,460 sq km)
Population: 20,617,068
Major Language(s): French, many African languages
Major Religion(s): Sunni Islam, Christianity
Currency: CFA franc

Country and Dependency Profiles

Croatia

Hrvatska

Official Name: Republic of Croatia
Capital: Zagreb
Area: 21,831 sq mi (56,542 sq km)
Population: 4,489,409
Major Language(s): Croatian
Major Religion(s): Roman Catholic Christianity
Currency: kuna

Cuba

Official Name: Republic of Cuba
Capital: Havana
Area: 42,803 sq mi (110,860 sq km)
Population: 11,451,652
Major Language(s): Spanish
Major Religion(s): Roman Catholic Christianity
Currency: Cuban peso

Cyprus

Kypros/Kibris

Official Name: Republic of Cyprus
Capital: Nicosia
Area: 3,571 sq mi (9,250 sq km)
Population: 796,740
Major Language(s): Greek, Turkish
Major Religion(s): Eastern Orthodox Christianity
Currency: Euro

Czechia (Czech Republic)

Cesko

Official Name: Czech Republic
Capital: Prague
Area: 30,450 sq mi (78,866 sq km)
Population: 10,211,904
Major Language(s): Czech
Major Religion(s): Roman Catholic Christianity
Currency: koruny

Denmark

Danmark

Official Name: Kingdom of Denmark
Capital: Copenhagen
Area: 16,639 sq mi (43,094 sq km)
Population: 5,500,510
Major Language(s): Danish
Major Religion(s): Protestant Christianity
Currency: Danish kroner

Djibouti

Jibuti

Official Name: Republic of Djibouti
Capital: Djibouti
Area: 8,880 sq mi (23,000 sq km)
Population: 516,055
Major Language(s): French, Arabic
Major Religion(s): Sunni Islam
Currency: Djiboutian franc

Dominica

Official Name: Commonwealth of Dominica
Capital: Roseau
Area: 291 sq mi (754 sq km)
Population: 72,660
Major Language(s): English
Major Religion(s): Roman Catholic Christianity
Currency: East Caribbean dollar

Dominican Republic

La Dominicana

Official Name: Dominican Republic
Capital: Santo Domingo
Area: 18,815 sq mi (48,730 sq km)
Population: 9,650,054
Major Language(s): Spanish
Major Religion(s): Roman Catholic Christianity
Currency: Dominican peso

Ecuador

Official Name: Republic of Ecuador
Capital: Quito
Area: 109,483 sq mi (283,560 sq km)
Population: 14,573,101
Major Language(s): Spanish, many Amerindian languages
Major Religion(s): Roman Catholic Christianity
Currency: US dollar

Egypt

Misr

Official Name: Arab Republic of Egypt
Capital: Cairo
Area: 386,662 sq mi (1,001,450 sq km)
Population: 83,082,869
Major Language(s): Arabic
Major Religion(s): Sunni Islam
Currency: Egyptian pound

El Salvador

Official Name: Republic of El Salvador
Capital: San Salvador
Area: 8,124 sq mi (21,040 sq km)
Population: 7,185,218
Major Language(s): Spanish
Major Religion(s): Roman Catholic Christianity
Currency: US dollar

Equatorial Guinea

Guinea Ecuatorial/Guinee equatoriale

Official Name: Republic of Equatorial Guinea
Capital: Malabo
Area: 10,831 sq mi (28,051 sq km)
Population: 633,441
Major Language(s): Spanish, French, Fang, Bubi
Major Religion(s): Roman Catholic Christianity
Currency: CFA franc

Eritrea

Ertra

Official Name: State of Eritrea
Capital: Asmara
Area: 46,842 sq mi (121,320 sq km)
Population: 5,647,168
Major Language(s): Tigrinya, Arabic
Major Religion(s): Sunni Islam, Christianity
Currency: nakfa

Estonia

Eesti

Official Name: Republic of Estonia
Capital: Tallinn
Area: 17,462 sq mi (45,226 sq km)
Population: 1,299,371
Major Language(s): Estonian, Russian
Major Religion(s): Protestant Christianity
Currency: Euro

Ethiopia

Ityop'iya

Official Name: Federal Democratic Republic of Ethiopia
Capital: Addis Ababa
Area: 435,186 sq mi (1,127,127 sq km)
Population: 85,237,338
Major Language(s): Amarigna, Oromigna, English
Major Religion(s): Ethiopian Orthodox Christianity, Islam
Currency: birr

Falkland Islands (Islas Malvinas)

(overseas territory of the United Kingdom, claimed by Argentina)

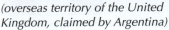

Official Name: Falkland Islands (Islas Malvinas)
Capital: Stanley
Area: 4,700 sq mi (12,173 sq km)
Population: 3,140
Major Language(s): English
Major Religion(s): Christianity
Currency: British pound

Country and Dependency Profiles

Faroe Islands

Foroyar
(part of the Kingdom of Denmark)

Official Name: Faroe Islands
Capital: Torshavn
Area: 540 sq mi (1,399 sq km)
Population: 48,856
Major Language(s): Faroese
Major Religion(s): Protestant Christianity
Currency: Danish kroner

Gabon

Official Name: Gabonese Republic
Capital: Libreville
Area: 103,347 sq mi (267,667 sq km)
Population: 1,514,993
Major Language(s): French, Fang
Major Religion(s): Christianity
Currency: CFA franc

Fiji

Fiji/Viti

Official Name: Republic of the Fiji Islands
Capital: Suva
Area: 7,054 sq mi (18,270 sq km)
Population: 944,720
Major Language(s): English, Fijian
Major Religion(s): Christianity
Currency: Fijian dollar

Gambia, The

Official Name: Republic of The Gambia
Capital: Banjul
Area: 4,363 sq mi (11,300 sq km)
Population: 1,782,893
Major Language(s): English, Mandinka, Wolof, Fula
Major Religion(s): Sunni Islam
Currency: dalasis

Finland

Suomi/Finland

Official Name: Republic of Finland
Capital: Helsinki
Area: 130,559 sq mi (338,145 sq km)
Population: 5,250,275
Major Language(s): Finnish
Major Religion(s): Protestant Christianity
Currency: Euro

Georgia

Sak'art'velo

Official Name: Georgia
Capital: T'bilisi
Area: 26,911 sq mi (69,700 sq km)
Population: 4,615,807
Major Language(s): Georgian, Russian
Major Religion(s): Eastern Orthodox Christianity
Currency: laris

France

Official Name: French Republic
Capital: Paris
Area: 248,429 sq mi (643,427 sq km)
Population: 64,057,792
Major Language(s): French
Major Religion(s): Roman Catholic Christianity
Currency: Euro

Germany

Deutschland

Official Name: Federal Republic of Germany
Capital: Berlin
Area: 137,847 sq mi (357,021 sq km)
Population: 82,329,758
Major Language(s): German
Major Religion(s): Christianity
Currency: Euro

Ghana

Official Name: Republic of Ghana
Capital: Accra
Area: 92,456 sq mi (239,460 sq km)
Population: 23,832,495
Major Language(s): English, many African languages
Major Religion(s): Christianity
Currency: cedi

Greece

Ellas or Ellada

Official Name: Hellenic Republic
Capital: Athens
Area: 50,942 sq mi (131,940 sq km)
Population: 10,737,428
Major Language(s): Greek
Major Religion(s): Eastern Orthodox Christianity
Currency: Euro

Greenland

Kalaallit Nunaat
(part of the Kingdom of Denmark)

Official Name: Greenland
Capital: Nuuk
Area: 836,330 sq mi (2,166,086 sq km)
Population: 57,600
Major Language(s): Greenlandic, Danish, English
Major Religion(s): Protestant Christianity
Currency: Danish kroner

Grenada

Official Name: Grenada
Capital: Saint George's
Area: 133 sq mi (344 sq km)
Population: 90,739
Major Language(s): English, French patois
Major Religion(s): Roman Catholic Christianity
Currency: East Caribbean dollar

Guatemala

Official Name: Republic of Guatemala
Capital: Guatemala
Area: 42,043 sq mi (108,890 sq km)
Population: 13,276,517
Major Language(s): Spanish, many Amerindian languages
Major Religion(s): Roman Catholic Christianity
Currency: quetzal

Guernsey

(British crown dependency)

Official Name: Bailiwick of Guernsey
Capital: Saint Peter Port
Area: 30 sq mi (78 sq km)
Population: 65,870
Major Language(s): English
Major Religion(s): Protestant Christianity
Currency: British pound

Guinea

Guinee

Official Name: Republic of Guinea
Capital: Conakry
Area: 94,926 sq mi (245,857 sq km)
Population: 10,057,975
Major Language(s): French, many African languages
Major Religion(s): Sunni Islam
Currency: Guinean franc

Guinea-Bissau

Guine-Bissau

Official Name: Republic of Guinea-Bissau
Capital: Bissau
Area: 13,946 sq mi (36,120 sq km)
Population: 1,533,964
Major Language(s): Portuguese, Crioulo
Major Religion(s): Sunni Islam, Indigenous beliefs
Currency: CFA franc

Country and Dependency Profiles

Guyana

Official Name: Cooperative Republic of Guyana
Capital: Georgetown
Area: 83,000 sq mi (214,970 sq km)
Population: 772,298
Major Language(s): English, Amerindian dialects
Major Religion(s): Hinduism, Christianity
Currency: Guyanese dollar

Haiti

Haiti/Ayiti

Official Name: Republic of Haiti
Capital: Port-au-Prince
Area: 10,714 sq mi (27,750 sq km)
Population: 9,035,536
Major Language(s): French, Creole
Major Religion(s): Roman Catholic Christianity
Currency: gourde

Holy See (Vatican City)

Santa Sede (Citta del Vaticano)

Official Name: The Holy See (State of the Vatican City)
Capital: Vatican City
Area: 0.4 sq mi (1.1 sq km)
Population: 826
Major Language(s): Italian, Latin
Major Religion(s): Roman Catholic Christianity
Currency: Euro

Honduras

Official Name: Republic of Honduras
Capital: Tegucigalpa
Area: 43,278 sq mi (112,090 sq km)
Population: 7,833,696
Major Language(s): Spanish, Amerindian dialects
Major Religion(s): Roman Catholic Christianity
Currency: lempira

Hungary

Magyarorszag

Official Name: Republic of Hungary
Capital: Budapest
Area: 35,919 sq mi (93,030 sq km)
Population: 9,905,596
Major Language(s): Hungarian
Major Religion(s): Roman Catholic Christianity
Currency: forint

Iceland

Island

Official Name: Republic of Iceland
Capital: Reykjavik
Area: 39,769 sq mi (103,000 sq km)
Population: 306,694
Major Language(s): Icelandic
Major Religion(s): Protestant Christianity
Currency: kronur

India

India/Bharat

Official Name: Republic of India
Capital: New Delhi
Area: 1,269,346 sq mi (3,287,590 sq km)
Population: 1,166,079,217
Major Language(s): Hindi, English, 17 other languages
Major Religion(s): Hinduism, Sunni Islam
Currency: Indian rupee

Indonesia

Official Name: Republic of Indonesia
Capital: Jakarta
Area: 741,100 sq mi (1,919,440 sq km)
Population: 240,271,522
Major Language(s): Bahasa Indonesia, many local dialects
Major Religion(s): Sunni Islam
Currency: rupiah

Iran

Official Name: Islamic Republic of Iran
Capital: Tehran
Area: 636,296 sq mi (1,648,000 sq km)
Population: 66,429,284
Major Language(s): Persian, Turkic dialects, Kurdish
Major Religion(s): Shia Islam
Currency: Iranian rial

Italy

Italia

Official Name: Italian Republic
Capital: Rome
Area: 116,306 sq mi (301,230 sq km)
Population: 58,126,212
Major Language(s): Italian, German, French
Major Religion(s): Roman Catholic Christianity
Currency: Euro

Iraq

Al Iraq

Official Name: Republic of Iraq
Capital: Baghdad
Area: 168,754 sq mi (437,072 sq km)
Population: 28,945,657
Major Language(s): Arabic, Kurdish
Major Religion(s): Shia and Sunni Islam
Currency: New Iraqi dinar

Jamaica

Official Name: Jamaica
Capital: Kingston
Area: 4,244 sq mi (10,991 sq km)
Population: 2,825,928
Major Language(s): English, English patois
Major Religion(s): Protestant Christianity
Currency: Jamaica dollar

Ireland

Eire

Official Name: Ireland
Capital: Dublin
Area: 27,135 sq mi (70,280 sq km)
Population: 4,203,200
Major Language(s): English, Irish
Major Religion(s): Roman Catholic Christianity
Currency: Euro

Japan

Nihon/Nippon

Official Name: Japan
Capital: Tokyo
Area: 145,883 sq mi (377,835 sq km)
Population: 127,078,679
Major Language(s): Japanese
Major Religion(s): Shintoism
Currency: yen

Israel

Yisra'el

Official Name: State of Israel
Capital: Jerusalem (proclaimed), Tel Aviv (de facto)
Area: 8,019 sq mi (20,770 sq km)
Population: 7,233,701
Major Language(s): Hebrew, Arabic
Major Religion(s): Judaism
Currency: Israeli new shekel

Jersey

(British crown dependency)

Official Name: Bailiwick of Jersey
Capital: Saint Helier
Area: 45 sq mi (116 sq km)
Population: 91,626
Major Language(s): English
Major Religion(s): Protestant Christianity
Currency: British pound

Country and Dependency Profiles

Jordan

Al Urdun

Official Name: Hashemite Kingdom of Jordan
Capital: Amman
Area: 35,637 sq mi (92,300 sq km)
Population: 6,342,948
Major Language(s): Arabic
Major Religion(s): Sunni Islam
Currency: Jordanian dinar

Korea, North

Choson

Official Name: Democratic People's Republic of Korea
Capital: P'yongyang
Area: 46,541 sq mi (120,540 sq km)
Population: 22,665,345
Major Language(s): Korean
Major Religion(s): Buddhism
Currency: North Korean won

Kazakhstan

Astana

Official Name: Republic of Kazakhstan
Capital: Almaty
Area: 1,049,155 sq mi (2,717,300 sq km)
Population: 15,399,437
Major Language(s): Kazakh, Russian
Major Religion(s): Sunni Islam, Eastern Orthodox Christianity
Currency: tenge

Korea, South

Han'guk

Official Name: Republic of Korea
Capital: Seoul
Area: 38,023 sq mi (98,480 sq km)
Population: 48,508,972
Major Language(s): Korean
Major Religion(s): Buddhism
Currency: South Korean won

Kenya

Official Name: Republic of Kenya
Capital: Nairobi
Area: 224,962 sq mi (582,650 sq km)
Population: 39,002,772
Major Language(s): English, Kiswahili
Major Religion(s): Christianity
Currency: Kenyan shilling

Kosovo

Kosova

Official Name: Republic of Kosovo
Capital: Pristina
Area: 4,203 sq mi (10,887 sq km)
Population: 1,804,838
Major Language(s): Albanian, Serbian
Major Religion(s): Sunni Islam
Currency: Euro

Kiribati

Official Name: Republic of Kiribati
Capital: Taraw
Area: 313 sq mi (811 sq km)
Population: 112,850
Major Language(s): I-Kiribati, English
Major Religion(s): Roman Catholic Christianity
Currency: Australian Dollar

Kuwait

Al Kuwayt

Official Name: State of Kuwait
Capital: Kuwait
Area: 6,880 sq mi (17,820 sq km)
Population: 2,691,158
Major Language(s): Arabic
Major Religion(s): Sunni Islam
Currency: Kuwaiti dinar

Kyrgyzstan

Official Name: Kyrgyz Republic
Capital: Bishkek
Area: 76,641 sq mi (198,500 sq km)
Population: 5,431,747
Major Language(s): Kyrgyz, Uzbek, Russian
Major Religion(s): Sunni Islam, Eastern Orthodox Christianity
Currency: som

Laos
Pathet Lao

Official Name: Lao People's Democratic Republic
Capital: Vientiane
Area: 91,429 sq mi (236,800 sq km)
Population: 6,834,942
Major Language(s): Lao, French, English
Major Religion(s): Buddhism
Currency: kip

Latvia
Latvija

Official Name: Republic of Latvia
Capital: Riga
Area: 24,938 sq mi (64,589 sq km)
Population: 2,231,503
Major Language(s): Latvian, Russian
Major Religion(s): Protestant Christianity
Currency: lati

Lebanon
Lubnan

Official Name: Lebanese Republic
Capital: Beirut
Area: 4,015 sq mi (10,400 sq km)
Population: 4,017,095
Major Language(s): Arabic, French, English, Armenian
Major Religion(s): Islam, Christianity
Currency: Lebanese pound

Lesotho

Official Name: Kingdom of Lesotho
Capital: Maseru
Area: 11,720 sq mi (30,355 sq km)
Population: 2,130,819
Major Language(s): Sesotho, English, Zulu, Xhosa
Major Religion(s): Christianity, Indigenous beliefs
Currency: maloti

Liberia

Official Name: Republic of Liberia
Capital: Monrovia
Area: 43,000 sq mi (111,370 sq km)
Population: 3,441,790
Major Language(s): English, many African languages
Major Religion(s): Christianity, Indigenous beliefs
Currency: Liberian dollar

Libya
Ar-Libya

Official Name: Great Socialist People's Libyan Arab Jamahiriya
Capital: Tripoli
Area: 679,362 sq mi (1,759,540 sq km)
Population: 6,310,434
Major Language(s): Arabic
Major Religion(s): Sunni Islam
Currency: Libyan dinar

Liechtenstein

Official Name: Principality of Liechtenstein
Capital: Vaduz
Area: 62 sq mi (160 sq km)
Population: 34,761
Major Language(s): German
Major Religion(s): Roman Catholic Christianity
Currency: Swiss franc

Country and Dependency Profiles

Lithuania
Lietuva

Official Name: Republic of Lithuania
Capital: Vilnius
Area: 25,212 sq mi (65,300 sq km)
Population: 3,555,179
Major Language(s): Lithuanian, Russian
Major Religion(s): Roman Catholic Christianity
Currency: litai

Malawi

Official Name: Republic of Malawi
Capital: Lilongwe
Area: 45,745 sq mi (118,480 sq km)
Population: 15,028,757
Major Language(s): Chichewa, Chinyanja, other African languages
Major Religion(s): Christianity, Islam
Currency: Malawian kwacha

Luxembourg

Official Name: Grand Duchy of Luxembourg
Capital: Luxembourg
Area: 998 sq mi (2,586 sq km)
Population: 491,775
Major Language(s): Luxembourgish, German, French
Major Religion(s): Roman Catholic Christianity
Currency: Euro

Malaysia

Official Name: Malaysia
Capital: Kuala Lumpur
Area: 127,317 sq mi (329,750 sq km)
Population: 25,715,819
Major Language(s): Bahasa Malaysia, Chinese, Tamil
Major Religion(s): Islam, Buddhism
Currency: ringgit

Macedonia
Makedonija

Official Name: Republic of Macedonia
Capital: Skopje
Area: 9,781 sq mi (25,333 sq km)
Population: 2,066,718
Major Language(s): Macedonian, Albanian
Major Religion(s): Eastern Orthodox Christianity, Sunni Islam
Currency: Macedonian denar

Maldives
Dhivehi Raajje

Official Name: Republic of Maldives
Capital: Male
Area: 116 sq mi (300 sq km)
Population: 396,334
Major Language(s): Maldivian Dhivehi
Major Religion(s): Sunni Islam
Currency: rufiyaa

Madagascar
Madagascar/Madagasikara

Official Name: Republic of Madagascar
Capital: Antananarivo
Area: 226,657 sq mi (587,040 sq km)
Population: 20,653,556
Major Language(s): English, French, Malagasy
Major Religion(s): Indigenous beliefs, Christianity
Currency: ariary

Mali

Official Name: Republic of Mali
Capital: Bamako
Area: 478,767 sq mi (1,240,000 sq km)
Population: 12,666,987
Major Language(s): Bambara, French, many African languages
Major Religion(s): Sunni Islam
Currency: CFA franc

Malta

Official Name: Republic of Malta
Capital: Valletta
Area: 122 sq mi (316 sq km)
Population: 405,165
Major Language(s): Maltese
Major Religion(s): Roman Catholic Christianity
Currency: Euro

Mayotte

(territorial overseas collectivity of France)

Official Name: Territorial Collectivity of Mayotte
Capital: Mamoutzou
Area: 144 sq mi (374 sq km)
Population: 223,765
Major Language(s): Mahorian, French
Major Religion(s): Sunni Islam
Currency: Euro

Marshall Islands

Official Name: Republic of the Marshall Islands
Capital: Majuro
Area: 70 sq mi (181 sq km)
Population: 64,522
Major Language(s): Marshallese
Major Religion(s): Protestant Christianity
Currency: US dollar

Mexico

Official Name: United Mexican States
Capital: Mexico City
Area: 761,606 sq mi (1,972,550 sq km)
Population: 111,211,789
Major Language(s): Spanish, Amerindian dialects
Major Religion(s): Roman Catholic Christianity
Currency: Mexican peso

Mauritania

Muritaniyah

Official Name: Islamic Republic of Mauritania
Capital: Nouakchott
Area: 397,955 sq mi (1,030,700 sq km)
Population: 3,129,486
Major Language(s): Arabic, Pulaar, Soninke, Wolof, French
Major Religion(s): Sunni Islam
Currency: ouguiya

Micronesia

Official Name: Federated States of Micronesia
Capital: Palikir
Area: 271 sq mi (702 sq km)
Population: 107,434
Major Language(s): English
Major Religion(s): Roman Catholic Christianity
Currency: US dollar

Mauritius

Official Name: Republic of Mauritius
Capital: Port Louis
Area: 788 sq mi (2,040 sq km)
Population: 1,284,264
Major Language(s): Creole, Bhojpuri, French
Major Religion(s): Hinduism
Currency: Mauritian rupee

Moldova

Official Name: Republic of Moldova
Capital: Chisinau
Area: 13,067 sq mi (33,843 sq km)
Population: 4,320,748
Major Language(s): Moldovan, Russian
Major Religion(s): Eastern Orthodox Christianity
Currency: lei

Country and Dependency Profiles

Monaco

Official Name: Principality of Monaco
Capital: Monaco
Area: 1 sq mi (2 sq km)
Population: 32,965
Major Language(s): French, English, Italian, Monegasque
Major Religion(s): Roman Catholic Christianity
Currency: Euro

Mongolia

Mongol Uls

Official Name: Mongolia
Capital: Ulaanbaatar
Area: 603,909 sq mi (1,564,116 sq km)
Population: 3,041,142
Major Language(s): Khalkha Mongol
Major Religion(s): Buddhism
Currency: tögrög

Montenegro

Crna Gora

Official Name: Montenegro
Capital: Cetinje
Area: 5,415 sq mi (14,026 sq km)
Population: 672,180
Major Language(s): Montenegrin, Serbian
Major Religion(s): Eastern Orthodox Christianity, Sunni Islam
Currency: Euro

Montserrat

(overseas territory of the United Kingdom)

Official Name: Montserrat
Capital: Plymouth
Area: 39 sq mi (102 sq km)
Population: 5,097
Major Language(s): English
Major Religion(s): Protestant Christianity
Currency: East Caribbean dollar

Morocco

Al Maghrib

Official Name: Kingdom of Morocco
Capital: Rabat
Area: 172,414 sq mi (446,550 sq km)
Population: 34,859,364
Major Language(s): Arabic, French, Berber dialects
Major Religion(s): Sunni Islam
Currency: Moroccan dirham

Mozambique

Moçambique

Official Name: Republic of Mozambique
Capital: Maputo
Area: 309,496 sq mi (801,590 sq km)
Population: 21,669,278
Major Language(s): Emakhuwa, Xichangana, Portuguese
Major Religion(s): Christianity, Sunni Islam
Currency: metical

Namibia

Official Name: Republic of Namibia
Capital: Windhoek
Area: 318,696 sq mi (825,418 sq km)
Population: 2,108,665
Major Language(s): Afrikaans, English, German
Major Religion(s): Christianity, Indigenous beliefs
Currency: Namibian dollar

Nepal

Official Name: Federal Democratic Republic of Nepal
Capital: Kathmandu
Area: 56,827 sq mi (147,181 sq km)
Population: 28,563,377
Major Language(s): Nepali, Maithali, Bhojpuri
Major Religion(s): Hinduism, Buddhism
Currency: Nepalese rupee

Netherlands

Nederland

Official Name: Kingdom of the Netherlands
Capital: Amsterdam
Area: 16,033 sq mi (41,526 sq km)
Population: 16,715,999
Major Language(s): Dutch, Frisian
Major Religion(s): Christianity
Currency: Euro

Netherlands Antilles

Nederlandse Antillen
(part of the Kingdom of the Netherlands)

Official Name: Netherlands Antilles
Capital: Willemstad
Area: 371 sq mi (960 sq km)
Population: 227,049
Major Language(s): Papiamento, English, Dutch
Major Religion(s): Roman Catholic Christianity
Currency: Netherlands Antilles guilder

New Caledonia

Nouvelle-Caledonie
(self-governing territory of France)

Official Name: Territory of New Caledonia and Dependencies
Capital: Noumea
Area: 7,359 sq mi (19,060 sq km)
Population: 227,436
Major Language(s): French, many Melanesian-Polynesian dialects
Major Religion(s): Roman Catholic Christianity
Currency: CFP franc

New Zealand

Official Name: New Zealand
Capital: Wellington
Area: 103,738 sq mi (268,680 sq km)
Population: 4,213,418
Major Language(s): English, Maori
Major Religion(s): Christianity
Currency: New Zealand dollar

Nicaragua

Official Name: Republic of Nicaragua
Capital: Managua
Area: 49,998 sq mi (129,494 sq km)
Population: 5,891,199
Major Language(s): Spanish
Major Religion(s): Roman Catholic Christianity
Currency: cordoba

Niger

Official Name: Republic of Niger
Capital: Niamey
Area: 489,191 sq mi (1,267,000 sq km)
Population: 15,306,252
Major Language(s): French, Hausa, Djerma
Major Religion(s): Sunni Islam
Currency: CFA franc

Nigeria

Official Name: Federal Republic of Nigeria
Capital: Abuja
Area: 356,669 sq mi (923,768 sq km)
Population: 149,229,090
Major Language(s): English, Hausa, Yoruba, Igbo, Fulani
Major Religion(s): Sunni Islam, Christianity
Currency: naira

Niue

*(self-governing territory in
free association with New Zealand)*

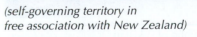

Official Name: Niue
Capital: Alofi
Area: 100 sq mi (260 sq km)
Population: 1,398
Major Language(s): Niuean, English
Major Religion(s): Protestant Christianity
Currency: New Zealand dollar

Country and Dependency Profiles

Norfolk Island

(self-governing territory of Australia)

Official Name: Territory of Norfolk Island
Capital: Kingston
Area: 14 sq mi (35 sq km)
Population: 2,141
Major Language(s): English
Major Religion(s): Protestant Christianity
Currency: Australian dollar

Palau

Belau

Official Name: Republic of Palau
Capital: Koror
Area: 177 sq mi (458 sq km)
Population: 20,796
Major Language(s): Palauan, Tobi, English
Major Religion(s): Christianity
Currency: US dollar

Norway

Norge

Official Name: Kingdom of Norway
Capital: Oslo
Area: 125,021 sq mi (323,802 sq km)
Population: 4,660,539
Major Language(s): Norwegian
Major Religion(s): Protestant Christianity
Currency: Norwegian kroner

Panama

Official Name: Republic of Panama
Capital: Panama
Area: 30,193 sq mi (78,200 sq km)
Population: 3,360,474
Major Language(s): Spanish, English
Major Religion(s): Roman Catholic Christianity
Currency: balboa

Oman

Uman

Official Name: Sultanate of Oman
Capital: Muscat
Area: 82,031 sq mi (212,460 sq km)
Population: 3,418,085
Major Language(s): Arabic
Major Religion(s): Sunni Islam
Currency: Omani rial

Papua New Guinea

Papuaniugini

Official Name: Independent State of Papua New Guinea
Capital: Port Moresby
Area: 178,704 sq mi (462,840 sq km)
Population: 6,057,263
Major Language(s): Tok Pisin, English, Hiri Motu
Major Religion(s): Protestant Christianity
Currency: kina

Pakistan

Official Name: Islamic Republic of Pakistan
Capital: Islamabad
Area: 310,403 sq mi (803,940 sq km)
Population: 176,242,949
Major Language(s): Punjabi, Sindhi, Siraiki, Pashtu, Urdu
Major Religion(s): Sunni Islam
Currency: Pakistani rupee

Paraguay

Official Name: Republic of Paraguay
Capital: Asuncion
Area: 157,047 sq mi (406,750 sq km)
Population: 6,995,655
Major Language(s): Spanish, Guarani
Major Religion(s): Roman Catholic Christianity
Currency: guaraní

Peru

Official Name: Republic of Peru
Capital: Lima
Area: 496,226 sq mi (1,285,220 sq km)
Population: 29,546,963
Major Language(s): Spanish, Quechua, Aymara
Major Religion(s): Roman Catholic Christianity
Currency: nuevo sol

Portugal

Official Name: Portuguese Republic
Capital: Lisbon
Area: 35,672 sq mi (92,391 sq km)
Population: 10,707,924
Major Language(s): Portuguese
Major Religion(s): Roman Catholic Christianity
Currency: Euro

Philippines

Pilipinas

Official Name: Republic of the Philippines
Capital: Manila
Area: 115,831 sq mi (300,000 sq km)
Population: 97,976,603
Major Language(s): Filipino, Tagalog, English
Major Religion(s): Roman Catholic Christianity
Currency: Philippine peso

Puerto Rico

*(territory of the United States
with commonwealth status)*

Official Name: Commonwealth of Puerto Rico
Capital: San Juan
Area: 5,324 sq mi (13,790 sq km)
Population: 3,971,020
Major Language(s): Spanish, English
Major Religion(s): Roman Catholic Christianity
Currency: US dollar

Pitcairn Islands

*(overseas territory of the
United Kingdom)*

Official Name: Pitcairn, Henderson, Ducie, and Oeno Islands
Capital: Adamstown
Area: 18 sq mi (47 sq km)
Population: 48
Major Language(s): English
Major Religion(s): Protestant Christianity
Currency: New Zealand dollar

Qatar

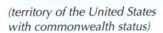

Official Name: State of Qatar
Capital: Doha
Area: 4,416 sq mi (11,437 sq km)
Population: 833,285
Major Language(s): Arabic
Major Religion(s): Sunni Islam
Currency: Qatari rial

Poland

Polska

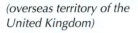

Official Name: Republic of Poland
Capital: Warsaw
Area: 120,726 sq mi (312,679 sq km)
Population: 38,482,919
Major Language(s): Polish
Major Religion(s): Roman Catholic Christianity
Currency: zloty

Romania

Official Name: Romania
Capital: Bucharest
Area: 91,699 sq mi (237,500 sq km)
Population: 22,215,421
Major Language(s): Romanian, Hungarian
Major Religion(s): Eastern Orthodox Christianity
Currency: lei

Country and Dependency Profiles

Russia

Rossiya

Official Name: Russian Federation
Capital: Moscow
Area: 6,592,772 sq mi (17,075,200 sq km)
Population: 140,041,247
Major Language(s): Russian, many minority languages
Major Religion(s): Russian Orthodox, Muslim
Currency: Russian ruble

Saint Lucia

Official Name: Saint Lucia
Capital: Castries
Area: 238 sq mi (616 sq km)
Population: 160,267
Major Language(s): English, French patois
Major Religion(s): Roman Catholic Christianity
Currency: East Caribbean dollar

Rwanda

Official Name: Republic of Rwanda
Capital: Kigali
Area: 10,169 sq mi (26,338 sq km)
Population: 10,746,311
Major Language(s): Kinyarwanda, French, English
Major Religion(s): Roman Catholic Christianity
Currency: Rwandan franc

Saint Pierre and Miquelon

Saint-Pierre et Miquelon
(territorial overseas collectivity of France)

Official Name: Territorial Collectivity of Saint Pierre and Miquelon
Capital: Saint-Pierre
Area: 93 sq mi (242 sq km)
Population: 7,051
Major Language(s): French
Major Religion(s): Roman Catholic Christianity
Currency: Euro

Saint Helena

(overseas territory of the United Kingdom)

Official Name: Saint Helena
Capital: Jamestown
Area: 159 sq mi (413 sq km)
Population: 7,637
Major Language(s): English
Major Religion(s): Protestant Christianity
Currency: Saint Helenanian pound

Saint Vincent and the Grenadines

Official Name: Saint Vincent and the Grenadines
Capital: Kingstown
Area: 150 sq mi (389 sq km)
Population: 104,574
Major Language(s): English, French patois
Major Religion(s): Protestant Christianity
Currency: East Caribbean dollar

Saint Kitts and Nevis

Official Name: Federation of Saint Kitts and Nevis
Capital: Basseterre
Area: 101 sq mi (261 sq km)
Population: 40,131
Major Language(s): English
Major Religion(s): Protestant Christianity
Currency: East Caribbean dollar

Samoa

Official Name: Independent State of Samoa
Capital: Apia
Area: 1,137 sq mi (2,944 sq km)
Population: 219,998
Major Language(s): Samoan, English
Major Religion(s): Protestant Christianity
Currency: tala

San Marino

Official Name: Republic of San Marino
Capital: San Marino
Area: 24 sq mi (61 sq km)
Population: 30,324
Major Language(s): Italian
Major Religion(s): Roman Catholic Christianity
Currency: Euro

Serbia

Srbija

Official Name: Republic of Serbia
Capital: Belgrade
Area: 29,913 sq mi (77,474 sq km)
Population: 7,379,339
Major Language(s): Serbian
Major Religion(s): Eastern Orthodox Christianity
Currency: Serbian dinal

São Tomé and Príncipe

São Tomé e Príncipe

Official Name: Democratic Republic of São Tomé and Príncipe
Capital: São Tomé
Area: 386 sq mi (1,001 sq km)
Population: 212,679
Major Language(s): Portuguese
Major Religion(s): Roman Catholic Christianity
Currency: dobra

Seychelles

Official Name: Republic of Seychelles
Capital: Victoria
Area: 176 sq mi (455 sq km)
Population: 87,476
Major Language(s): Creole
Major Religion(s): Roman Catholic Christianity
Currency: Seychelles rupee

Saudi Arabia

Al Arabiyah as Suudiyah

Official Name: Kingdom of Saudi Arabia
Capital: Riyadh
Area: 830,000 sq mi (2,149,690 sq km)
Population: 28,686,633
Major Language(s): Arabic
Major Religion(s): Sunni Islam
Currency: Saudi riyal

Sierra Leone

Official Name: Republic of Sierra Leone
Capital: Freetown
Area: 27,699 sq mi (71,740 sq km)
Population: 6,440,053
Major Language(s): Mende, Temne, English, Krio
Major Religion(s): Sunni Islam
Currency: leone

Senegal

Official Name: Republic of Senegal
Capital: Dakar
Area: 75,749 sq mi (196,190 sq km)
Population: 13,711,597
Major Language(s): French, Wolof, Pulaar, Jola, Mandinka
Major Religion(s): Sunni Islam
Currency: CFA franc

Singapore

Official Name: Republic of Singapore
Capital: Singapore
Area: 268 sq mi (693 sq km)
Population: 4,657,542
Major Language(s): Mandarin Chinese, English, Malay, Hokkien
Major Religion(s): Buddhism, Sunni Islam
Currency: Singapore dollar

Country and Dependency Profiles

Slovakia
Slovensko

Official Name: Slovak Republic
Capital: Bratislava
Area: 18,859 sq mi (48,845 sq km)
Population: 5,463,046
Major Language(s): Slovak, Hungarian
Major Religion(s): Roman Catholic Christianity
Currency: Euro

South Africa

Official Name: Republic of South Africa
Capital: Bloemfontein (judicial), Cape Town (legislative),
Area: 471,011 sq mi (1,219,912 sq km)
Population: 49,052,489
Major Language(s): Zulu, Xhosa, Afrikaans, Sepedi, English, Setswana, Sesotho
Major Religion(s): Protestant Christianity
Currency: rand

Slovenia
Slovenija

Official Name: Republic of Slovenia
Capital: Ljubljana
Area: 7,827 sq mi (20,273 sq km)
Population: 2,005,692
Major Language(s): Slovenian
Major Religion(s): Roman Catholic Christianity
Currency: Euro

Spain
España

Official Name: Kingdom of Spain
Capital: Madrid
Area: 194,897 sq mi (504,782 sq km)
Population: 40,525,002
Major Language(s): Spanish, Catalan, Galician, Basque
Major Religion(s): Roman Catholic Christianity
Currency: Euro

Solomon Islands

Official Name: Solomon Islands
Capital: Honiara
Area: 10,985 sq mi (28,450 sq km)
Population: 595,613
Major Language(s): English, many indigenous languages
Major Religion(s): Protestant Christianity
Currency: Solomon Islands dollar

Sri Lanka
Shri Lamka/Ilankai

Official Name: Democratic Socialist Republic of Sri Lanka
Capital: Colombo
Area: 25,332 sq mi (65,610 sq km)
Population: 21,324,791
Major Language(s): Sinhala, Tamil
Major Religion(s): Buddhism
Currency: Sri Lankan rupee

Somalia
Soomaaliya

Official Name: Somalia
Capital: Mogadishu
Area: 246,201 sq mi (637,657 sq km)
Population: 9,832,017
Major Language(s): Somali, Arabic, Italian
Major Religion(s): Sunni Islam
Currency: Somali shilling

Sudan
As-Sudan

Official Name: Republic of the Sudan
Capital: Khartoum
Area: 967,499 sq mi (2,505,810 sq km)
Population: 41,087,825
Major Language(s): Arabic, English
Major Religion(s): Sunni Islam
Currency: Sudanese pound

Suriname

Official Name: Republic of Suriname
Capital: Paramaribo
Area: 63,039 sq mi (163,270 sq km)
Population: 481,267
Major Language(s): Dutch, English
Major Religion(s): Hinduism, Christianity
Currency: Surinamese dollar

Swaziland

eSwatini

Official Name: Kingdom of Swaziland
Capital: Mbabane
Area: 6,704 sq mi (17,363 sq km)
Population: 1,337,186
Major Language(s): English, siSwati
Major Religion(s): Christianity, Indigenous beliefs
Currency: emalangeni

Sweden

Sverige

Official Name: Kingdom of Sweden
Capital: Stockholm
Area: 173,732 sq mi (449,964 sq km)
Population: 9,059,651
Major Language(s): Swedish
Major Religion(s): Protestant Christianity
Currency: Swedish kronor

Switzerland

Schweiz/Suisse/Svizzera/Svizra

Official Name: Swiss Confederation
Capital: Bern
Area: 15,942 sq mi (41,290 sq km)
Population: 7,604,467
Major Language(s): German, French, Italian, Romansch
Major Religion(s): Christianity
Currency: Swiss franc

Syria

Suriyah

Official Name: Syrian Arab Republic
Capital: Damascus
Area: 71,498 sq mi (185,180 sq km)
Population: 20,178,485
Major Language(s): Arabic
Major Religion(s): Sunni Islam
Currency: Syrian pound

Taiwan

*T'ai-wan
(unresolved status; has limited
international recognition as the
legitimate representative of China)*

Official Name: Taiwan
Capital: Taipei
Area: 13,892 sq mi (35,980 sq km)
Population: 22,974,347
Major Language(s): Mandarin Chinese, Taiwanese
Major Religion(s): Buddhism
Currency: New Taiwan dollar

Tajikistan

Tojikiston

Official Name: Republic of Tajikistan
Capital: Dushanbe
Area: 55,251 sq mi (143,100 sq km)
Population: 7,349,145
Major Language(s): Tajik, Russian
Major Religion(s): Sunni Islam
Currency: somoni

Tanzania

Official Name: United Republic of Tanzania
Capital: Dodoma
Area: 364,900 sq mi (945,087 sq km)
Population: 41,048,532
Major Language(s): Swahili, English
Major Religion(s): Sunni Islam, Christianity
Currency: Tanzanian shilling

Country and Dependency Profiles

Thailand

Prathet Thai

Official Name: Kingdom of Thailand
Capital: Bangkok
Area: 198,457 sq mi (514,000 sq km)
Population: 65,998,436
Major Language(s): Thai
Major Religion(s): Buddhism
Currency: baht

Trinidad and Tobago

Official Name: Republic of Trinidad and Tobago
Capital: Port-of-Spain
Area: 1,980 sq mi (5,128 sq km)
Population: 1,229,953
Major Language(s): English, French, Spanish
Major Religion(s): Roman Catholic Christianity, Hinduism
Currency: Trinidad and Tobago dollar

Timor-Leste

Timor Lorosa'e/Timor-Leste

Official Name: Democratic Republic of Timor-Leste
Capital: Dili
Area: 5,794 sq mi (15,007 sq km)
Population: 1,131,612
Major Language(s): Tetum, Portuguese, Indonesian
Major Religion(s): Roman Catholic Christianity
Currency: US dollar

Tunisia

Tunis

Official Name: Tunisian Republic
Capital: Tunis
Area: 63,170 sq mi (163,610 sq km)
Population: 10,486,339
Major Language(s): Arabic, French
Major Religion(s): Sunni Islam
Currency: Tunisian dinar

Togo

Official Name: Togolese Republic
Capital: Lome
Area: 21,925 sq mi (56,785 sq km)
Population: 6,031,808
Major Language(s): French, many African languages
Major Religion(s): Indigenous beliefs, Christianity
Currency: CFA franc

Turkey

Turkiye

Official Name: Republic of Turkey
Capital: Ankara
Area: 301,384 sq mi (780,580 sq km)
Population: 76,805,524
Major Language(s): Turkish, Kurdish
Major Religion(s): Sunni Islam
Currency: lira

Tonga

Official Name: Kingdom of Tonga
Capital: Nuku'alofa
Area: 289 sq mi (748 sq km)
Population: 120,898
Major Language(s): Tongan, English
Major Religion(s): Protestant Christianity
Currency: pa'anga

Turkmenistan

Official Name: Turkmenistan
Capital: Ashgabat
Area: 188,456 sq mi (488,100 sq km)
Population: 4,884,887
Major Language(s): Turkmen, Russian, Uzbek
Major Religion(s): Sunni Islam
Currency: manats

Turks and Caicos Islands

(overseas territory of the United Kingdom)

Official Name: Turks and Caicos Islands
Capital: Grand Turk
Area: 166 sq mi (430 sq km)
Population: 22,942
Major Language(s): English
Major Religion(s): Protestant Christianity
Currency: US dollar

United Arab Emirates

Al Imarat al Arabiyah al Muttahidah

Official Name: United Arab Emirates
Capital: Abu Dhabi
Area: 32,278 sq mi (83,600 sq km)
Population: 4,798,491
Major Language(s): Arabic
Major Religion(s): Sunni Islam
Currency: Emirati dirham

Tuvalu

Official Name: Tuvalu
Capital: Funafuti
Area: 10 sq mi (26 sq km)
Population: 12,373
Major Language(s): Tuvaluan, English, Samoan
Major Religion(s): Protestant Christianity
Currency: Tuvaluan dollar

United Kingdom

Official Name: United Kingdom of Great Britain and Northern Ireland
Capital: London
Area: 94,526 sq mi (244,820 sq km)
Population: 61,113,205
Major Language(s): English, Welsh, Scottish Gaelic
Major Religion(s): Christianity
Currency: British pound

Uganda

Official Name: Republic of Uganda
Capital: Kampala
Area: 91,136 sq mi (236,040 sq km)
Population: 32,369,558
Major Language(s): English, Ganda, Swahili, other African languages
Major Religion(s): Christianity
Currency: Ugandan shilling

United States

Official Name: United States of America
Capital: Washington, D.C.
Area: 3,794,083 sq mi (9,826,630 sq km)
Population: 307,212,123
Major Language(s): English, Spanish
Major Religion(s): Christianity
Currency: US dollar

Ukraine

Ukrayina

Official Name: Ukraine
Capital: Kyiv
Area: 233,090 sq mi (603,700 sq km)
Population: 45,700,395
Major Language(s): Ukrainian, Russian
Major Religion(s): Eastern Orthodox Christianity
Currency: hryvnia

Uruguay

Official Name: Oriental Republic of Uruguay
Capital: Montevideo
Area: 68,039 sq mi (176,220 sq km)
Population: 3,494,382
Major Language(s): Spanish
Major Religion(s): Christianity
Currency: Uruguayan peso

Uzbekistan

Ozbekiston

Official Name: Republic of Uzbekistan
Capital: Tashkent
Area: 172,742 sq mi (447,400 sq km)
Population: 27,606,007
Major Language(s): Uzbek, Russian
Major Religion(s): Sunni Islam
Currency: soum

Vietnam

Viet Nam

Official Name: Socialist Republic of Vietnam
Capital: Hanoi
Area: 127,244 sq mi (329,560 sq km)
Population: 86,967,524
Major Language(s): Vietnamese, French, Chinese
Major Religion(s): Non-religious beliefs dominate
Currency: dong

Vanuatu

Official Name: Republic of Vanuatu
Capital: Port-Vila
Area: 4,710 sq mi (12,200 sq km)
Population: 218,519
Major Language(s): Many local languages
Major Religion(s): Protestant Christianity
Currency: vatu

Virgin Islands

(territory of the United States)

Official Name: United States Virgin Islands
Capital: Charlotte Amalie
Area: 737 sq mi (1,910 sq km)
Population: 109,825
Major Language(s): English, Spanish
Major Religion(s): Christianity
Currency: US dollar

Venezuela

Official Name: Bolivarian Republic of Venezuela
Capital: Caracas
Area: 352,144 sq mi (912,050 sq km)
Population: 26,814,843
Major Language(s): Spanish, many Amerindian languages
Major Religion(s): Roman Catholic Christianity
Currency: bolivar

Wallis and Futuna

Wallis et Futuna
(overseas collectivity of France)

Official Name: Territory of the Wallis and Futuna Islands
Capital: Mata-Utu
Area: 106 sq mi (274 sq km)
Population: 15,289
Major Language(s): Wallisian, Futunian, French
Major Religion(s): Roman Catholic Christianity
Currency: CFP franc

Yemen

Al Yaman

Official Name: Republic of Yemen
Capital: Sanaa
Area: 203,850 sq mi (527,970 sq km)
Population: 23,822,783
Major Language(s): Arabic
Major Religion(s): Sunni Islam
Currency: Yemeni rial

Zimbabwe

Official Name: Republic of Zimbabwe
Capital: Harare
Area: 150,804 sq mi (390,580 sq km)
Population: 11,392,629
Major Language(s): English, many African (Bantu) languages
Major Religion(s): Christianity, Indigenous beliefs
Currency: Zimbabwean dollar

Zambia

Official Name: Republic of Zambia
Capital: Lusaka
Area: 290,586 sq mi (752,614 sq km)
Population: 11,862,740
Major Language(s): English, many African (Bantu) languages
Major Religion(s): Christianity, Islam
Currency: Zambian kwacha

Unit IX

Geographic Index

Unit IX

The geographic index contains approximately 1,500 names of cities, states, countries, rivers, lakes, mountain ranges, oceans, capes, bays, and other geographic features.

The name of each geographical feature in the index is accompanied by a geographical coordinate (latitude and longitude) in degrees and by the page number of the primary map on which the geographical feature appears. Where the geographical coordinates are for specific places or points, such as a city or a mountain peak, the latitude and longitude figures give the location of the map symbol denoting that point. Thus, Los Angeles, California, is at 34N and 118W and the location of Mt. Everest is 28N and 107E.

The coordinates for political features (countries or states) or physical features (oceans, deserts) that are areas rather than points are given according to the location of the name of the feature on the map, except in those cases where the name of the feature is separated from the feature (such as a country's name appearing over an adjacent ocean area because of space requirements). In such cases, the feature's coordinates will indicate the location of the center of the feature. The coordinates for the Sahara Desert will lead the reader to the place name "Sahara Desert" on the map; the coordinates for North Carolina will show the center location of the state since the name appears over the adjacent Atlantic Ocean. Finally, the coordinates for geographical features that are lines rather than points or areas will also appear near the center of the text identifying the geographical feature.

Alphabetizing follows general conventions; the names of physical features such as lakes, rivers, mountains are given as: proper name, followed by the generic name. Thus "Mount Everest" is listed as "Everest, Mt." Where an article such as "the," "le," "al" appears in a geographic name, the name is alphabetized according to the article. Hence, "La Paz" is found under "L" and not under "P."

GEOGRAPHIC INDEX

NAME/DESCRIPTION	LATITUDE & LONGITUDE	PAGE	NAME/DESCRIPTION	LATITUDE & LONGITUDE	PAGE
Aral Sea	45N, 60E	202	Baja California (st., Mex.)	30N, 110W	167
Archangel, Russia (city)	65N, 41E	202	Baja California Sur (st., Mex.)	25N, 110W	167
Arctic Bay, Canada (city)	73N, 85W	171	Baku, Azerbaijan (city, nat. cap.)	40N, 50E	181
Arctic Ocean	75N, 160W	211	Balearic Islands	29N, 3E	182
Arequipa, Peru (city)	16S, 71W	175	Balkan Peninsula	40N, 21E	182
Argentina (country)	39S, 67W	175	Balkhash, Lake (lake, Asia)	47N, 75E	202
Arizona (st., USA)	34N, 112W	167	Ballarat, Victoria, Australia (city)	38S, 144E	207
Arkansas (st., USA)	37N, 94W	167	Baltic Sea	56N, 18E	182
Arkhangelsk, Russia (city)	75N, 160W	181	Baltimore, Maryland, USA (city)	39N, 77W	167
Armenia (country)	40N, 45E	181	Bamako, Mali (city, nat. cap.)	13N, 8W	191
Arnhem, Cape	11S, 139E	207	Banda Sea	5N, 127E	206
Arno (riv., Europe)	44N, 11E	182	Bandar Seri Begawan, Brunei (city, nat. cap.)	5N, 115E	206
Aru Islands	6S, 134E	206	Bandjarmasin, Indonesia (city)	3S, 115E	206
Aruwimi (riv., Africa)	2N, 27E	191	Bandung, Indonesia (city)	7S, 108E	206
As Sudd	9N, 26E	191	Bangalore, India (city)	13N, 75E	204
Ascension (island)	9S, 13W	191	Banghazi, Libya (city)	32N, 20E	191
Ashburton (riv., Australasia)	23S, 115W	207	Bangkok, Thailand (city, nat. cap.)	14N, 100E	206
Ashgabat, Turkmenistan (city, nat. cap.)	38N, 58E	202	Bangladesh (country)	23N, 92E	204
Asmera, Eritrea (city, nat. cap.)	15N, 39E	191	Bangui, Central African Republic (city, nat. cap.)	4N, 19E	196
Astana, Kazakhstan (city, nat. cap.)	51N, 71E	202	Bani (riv., Africa)	14n, 4W	191
Astrakhan, Russia (city)	46N, 48E	181	Banjul, Gambia (city, nat. cap.)	13N, 17W	191
Asuncion, Paraguay (city, nat. cap.)	25S, 57W	175	Banks Island	73N, 125W	168
Aswan, Egypt (city)	24N, 33E	191	Baotou, China (city)	41N, 110E	205
Asyut, Egypt (city)	27N, 31E	191	Barbados (country)	13N, 60W	175
Atacama Desert	23S, 70W	176	Barcelona, Spain (city)	41N, 2E	181
Atar, Mauritania (city)	21N, 13W	196	Barents Sea	69N, 40E	202
Athabasca, Lake (lake, N.Am.)	60N, 109W	168	Barnaul, Russia (city)	53N, 84E	202
Athabaska (riv., N.Am.)	58N, 114W	168	Bass Strait	40S, 146E	207
Athens, Greece (city, nat. cap.)	38N, 24E	181	Baton Rouge, Louisiana, USA (city, st. cap.)	30N, 91W	167
Atlanta, Georgia, USA (city, st. cap.)	34N, 84W	167	Batumi, Georgia (city)	42N, 42E	202
Atlantic Ocean	30N, 40W	168	Beaufort Sea	72N, 135W	168
Atlas Mountains	31N, 6W	191	Beijing, China (city, nat. cap.)	40N, 116E	205
Auckland, New Zealand (city)	37S, 175E	207	Beira, Mozambique (city)	20S, 35E	197
Augusta, Maine, USA (city, st. cap.)	44N, 70W	167	Beirut, Lebanon (city, nat. cap.)	34N, 35E	203
Austin, Texas, USA (city, st. cap.)	30N, 98W	167	Belarus (country)	52N, 27E	181
Australia (country)	20S, 135W	207	Belém, Brazil (city)	1S, 48W	179
Austria (country)	47N, 14E	181	Belfast, Northern Ireland, UK (city)	55N, 6W	181
Azerbaijan (country)	38N, 48E	181	Belgium (country)	51N, 4E	181
Azov, Sea of	48N, 36E	181	Belgrade, Serbia (city, nat. cap.)	45N, 21E	181
Bab el Mandeb (strait)	13N, 42E	191	Belize (country)	18S, 102W	167
Baffin Bay	74N, 65W	168	Bellinghausen Sea	72S, 85W	212
Baffin Island	70N, 72W	168	Belmopan, Belize (city, nat. cap.)	18S, 89W	167
Baghdad, Iraq (city, nat. cap.)	33N, 44E	203	Belo Horizonte, M.G. (city, st. cap., Braz.)	20S, 43W	175
Bahamas (island)	25N, 75W	168	Belyando (riv., Australasia)	22S, 147W	207
Bahia (st., Brazil)	13S, 42W	175	Bengal, Bay of	15N, 90E	204
Bahia Blanca, Argentina (city)	39S, 62W	175	Benguela, Angola (city)	13S, 13E	191
Bahrain (country)	26N, 50E	203	Beni, Rio (riv., S.Am.)	14S, 67W	176
Baikal, Lake (lake, Asia)	52N, 105E	202	Benin (country)	10N, 4E	191

GEOGRAPHIC INDEX

GEOGRAPHIC INDEX

GEOGRAPHIC INDEX

GEOGRAPHIC INDEX

NAME/DESCRIPTION	LATITUDE & LONGITUDE	PAGE	NAME/DESCRIPTION	LATITUDE & LONGITUDE	PAGE
Faeroe Islands	62N, 11W	182	Garonne (riv., Europe)	45N, 1E	182
Fairbanks, Alaska, USA (city)	65N, 148W	167	Gascoyne (riv., Australasia)	25S, 115E	207
Falkland Islands (Islas Malvinas)	52S, 60W	176	Gaza, Palestinian Terr. (city)	32N, 34E	203
Farewell, Cape (NZ)	40S, 170E	207	Gdansk, Poland (city)	54N, 19E	181
Fargo, North Dakota, USA (city)	47N, 97W	167	Geelong, Aust. (city)	38S, 144E	207
Farquhar, Cape	24S, 114E	207	Gees Gwardafuy (island)	15N, 50E	191
Feira de Santana, Brazil (city)	12S, 39W	179	Geneva, Switzerland (city)	46N, 6E	182
Fez, Morocco (city)	34N, 5W	196	Geographe Bay	35S, 115E	207
Fiji (country)	17S, 178E	208	Georgetown, Guyana (city, nat. cap.)	8N, 58W	175
Finland (country)	62N, 28E	181	Georgia (country)	42N, 44E	181
Finland, Gulf of	60N, 20E	182	Georgia (st., USA)	30N, 82W	167
Fitzroy (riv., Australasia)	17S, 125E	207	Germany (country)	50N, 12E	181
Flinders (island)	40S, 148E	207	Ghana (country)	8N, 3W	191
Flinders Ranges	31S, 139E	207	Ghardaia, Algeria (city)	32N, 4E	196
Florianopolis, Brazil (city)	28S, 49W	180	Gibraltar, Gibraltar, UK	36N, 5W	182
Florida (st., USA)	28N, 83W	167	Gibraltar, Strait of	37N, 6W	182
Florida Peninsula	26N, 82W	168	Gibson Desert	24S, 124E	209
Florida, Straits of	28N, 80W	168	Gilbert (riv., Australasia)	8S, 142E	207
Fly (riv., Australasia)	8S, 143E	207	Glama (riv., Europe)	61N, 11E	182
Formosa (st., Argentina)	23S, 60W	175	Glasgow, Scotland, UK (city)	56N, 6W	181
Formosa, Formosa (city, st. cap., Argen.)	27S, 58W	175	Gobi Desert	48N, 105E	205
Fort Good Hope, Canada (city)	66N, 129W	171	Goiania, Brazil (city)	17S, 49W	179
Fort Severn, Canada (city)	56N, 88W	171	Goiás (st., Brazil)	15S, 50W	175
Fort Worth, Texas, USA (city)	33N, 97W	167	Gonder, Ethiopia (city)	13N, 37E	196
Fortaleza, Brazil (city)	4S, 39W	179	Good Hope, Cape of	33S, 18E	191
Foxe Basin	66N, 77W	168	Goteborg, Sweden (city)	58N, 12E	181
France (country)	46N, 4E	181	Gotland (island)	57N, 20E	190
Frankfort, Kentucky, USA (city, st. cap.)	38N, 85W	167	Gran Chaco	23S, 70W	176
Frankfurt, Germany (city)	50N, 9E	181	Grand Erg Occidental	29N, 0	191
Franz Josef Land (island)	80N, 40E	202	Grand Erg Oriental	30N, 7E	191
Fraser (riv., N.Am.)	52N, 122W	168	Great Artesian Basin	25S, 145E	209
Fredericton, N.B., Canada (city, prov. cap.)	46N, 67W	167	Great Australian Bight	33S, 130E	209
Freetown, Sierra Leone (city, nat. cap.)	8N, 13W	191	Great Barrier Reef	15S, 145E	209
Fremantle, Australia (city)	33S, 116E	207	Great Basin	39N, 117W	168
French Guiana (dept., France)	4N, 52W	175	Great Bear Lake (lake, N.Am.)	67N, 120W	168
Fria, Cape	18S, 12E	191	Great Dividing Range	20S, 145E	209
Fuzhou, China (city)	26N, 119E	205	Great Indian Desert	25N, 72E	305
Gabes, Gulf of	33N, 12E	191	Great Plains	40N, 105W	168
Gabes, Tunisia (city)	34N, 10E	191	Great Salt Lake (lake, N.Am.)	40N, 113W	168
Gabon (country)	2S, 12E	191	Great Sandy Desert	23S, 125E	209
Gaborone, Botswana (city, nat. cap.)	25S, 25E	191	Great Slave Lake (lake, N.Am.)	62N, 110W	168
Gairdiner, Lake	32S, 136E	207	Great Victoria Desert	30S, 125E	209
Galveston, Texas, USA (city)	29N, 95W	167	Greater Khingan Range	50N, 120E	205
Gambia (country)	13N, 15W	191	Greece (country)	39N, 21E	181
Gambia (riv., Africa)	13N, 15W	191	Greenland (island)	78N, 40W	168
Ganges (riv., Asia)	27N, 85E	204	Gregory Range	18S, 145E	207
Ganges Plain	27N, 82E	204	Grey Range	26S, 145E	207

GEOGRAPHIC INDEX

NAME/DESCRIPTION	LATITUDE & LONGITUDE	PAGE	NAME/DESCRIPTION	LATITUDE & LONGITUDE	PAGE
Grootfontein, Namibia (city)	20S, 18E	197	Hidalgo (st., Mex.)	20N, 98W	167
Guadalajara, Jalisco, Mexico (city, st. cap.)	21N, 103W	173	Himalayas	26N, 80E	204
Guadalcanal (island)	9S, 160E	209	Hindu Kush	30N, 70E	204
Guadalquivir (riv., Europe)	37N, 5W	182	Ho Chi Minh City, Vietnam (city)	11N, 107E	206
Guanajuato (st., Mex.)	22N, 100W	167	Hobart, Tasmania, Australia (city, st. cap.)	43S, 147E	209
Guanajuato, Guan., Mexico (city, st. cap.)	21N, 101W	167	Hokkaido (island)	43N, 142E	205
Guangzhou, China (city)	23N, 113E	205	Honduras (country)	16N, 87W	167
Guatemala (country)	14N, 90W	167	Hong Kong, China (city)	22N, 114E	205
Guatemala, Guatemala (city, nat. cap.)	15N, 91W	167	Honiara, Solomon Islands (city, nat. cap.)	9S, 160E	207
Guayaquil, Ecuador (city)	2S, 80W	175	Honolulu, Hawaii, USA (city, st. cap.)	21N, 158W	167
Guayaquil, Gulf of	3S, 83W	176	Honshu (island)	38N, 140E	205
Guerrero (st., Mex.)	18N, 102W	167	Horn of Africa	10N, 50E	191
Guiana Highlands	5N, 60W	176	Horn, Cape	55S, 70W	176
Guinea (country)	10N, 10W	191	Houston, Texas, USA (city)	30N, 95W	167
Guinea, Gulf of	3N, 0	191	Howe, Cape	37S, 150E	207
Guinea-Bissau (country)	12N, 15W	191	Huambo, Angola (city)	13S, 16E	197
Guiyang, China (city)	27N, 107E	205	Huang (riv., Asia)	30N, 105E	205
Gulf of Riga	58N, 24E	182	Hudson Bay	60N, 90W	168
Guyana (country)	6N, 57W	175	Hudson Strait	63N, 70W	168
Hadejia (riv., Africa)	12N, 10E	191	Hue, Vietnam (city)	15N, 110E	198
Hainan (island)	19N, 110E	205	Hughes, Aust. (city)	30S, 130E	207
Haiti (country)	18N, 72W	167	Hungary (country)	48N, 20E	181
Hakodate, Japan (city)	42N, 140E	205	Huron, Lake (lake, N.Am.)	45N, 85W	168
Halifax Bay	18S, 146E	207	Hyderabad, India (city)	17N, 79E	204
Halifax, Nova Scotia, Canada (city, prov. cap.)	45N, 64W	171	Hyderabad, Pakistan (city)	25N, 68E	204
Halmahera (island)	1N, 128E	206	Ibadan, Nigeria (city)	7N, 4E	191
Hamburg, Germany (city)	54N, 10E	181	Iberian Peninsula	42N, 4W	182
Hamilton, Bermuda, UK	32N, 65W	182	Ica, Peru (city)	14S, 76W	179
Hammersley Range	23S, 116W	207	Iceland (country)	64N, 20W	181
Hangzhou, China (city)	30N, 120E	205	Idaho (st., USA)	43N, 113W	167
Hanoi, Vietnam (city, nat. cap.)	21N, 106E	206	Illinois (st., USA)	44N, 90W	167
Harare, Zimbabwe (city, nat. cap.)	18S, 31E	197	Iloilo, Philippines (city)	11N, 123E	206
Harbin, China (city)	46N, 127E	205	India (country)	23N, 80E	204
Harer, Ethiopia (city)	10N, 42E	191	Indiana (st., USA)	46N, 88W	167
Hargeysa, Somalia (city)	9N, 44E	191	Indianapolis, Indiana, USA (city, st. cap.)	40N, 86W	167
Harrisburg, Pennsylvania, USA (city, st. cap.)	40N, 77W	167	Indonesia (country)	2S, 120E	206
Hartford, Connecticut, USA (city, st. cap.)	42N, 73W	167	Indus (riv., Asia)	25N, 70E	204
Hatteras, Cape	32N, 73W	168	Ionian Sea	38N, 19E	182
Havana, Cuba (city, nat. cap.)	23N, 82W	167	Iowa (st., USA)	43N, 95W	167
Hawaii (st., USA)	21N, 156W	167	Iqaluit, Nunavut, Canada (city, terr. cap.)	63N, 68W	167
Hawaiian Islands	23N, 152W	172	Iquique, Chile (city)	20S, 70W	180
Heart, Afghanistan (city)	34N, 62E	198	Iquitos, Peru (city)	4S, 74W	175
Hebrides Islands	58N, 8W	182	Iran (country)	30N, 55E	203
Hejaz Mountains	30N, 40E	203	Iraq (country)	30N, 50E	203
Helena, Montana, USA (city, st. cap.)	47N, 112W	167	Ireland (country)	54N, 8W	181
Helsinki, Finland (city, nat. cap.)	60N, 25E	181	Irish Sea	51N, 7W	182
Hermosillo, Sonora, Mexico (city, st. cap.)	29N, 111W	173	Irkutsk, Russia (city)	52N, 104E	202

GEOGRAPHIC INDEX

NAME/DESCRIPTION	LATITUDE & LONGITUDE	PAGE	NAME/DESCRIPTION	LATITUDE & LONGITUDE	PAGE
Irrawaddy (riv., Asia)	25N, 95E	198	Karakorum Range	32N, 78E	204
Irtysh (riv., Asia)	50N, 70E	202	Kasai (riv., Africa)	5S, 18E	191
Isfahan, Iran (city)	33N, 52E	203	Kashi, China (city)	39N, 76E	205
Islamabad, Pakistan (city, nat. cap.)	34N, 73E	204	Kassala, Sudan (city)	15N, 36E	196
Israel (country)	31N, 36E	181	Katanga Plateau	11S, 26E	191
Istanbul, Turkey (city)	41N, 29E	203	Katherine, Aust. (city)	14S, 132E	207
Italy (country)	42N, 12E	181	Kathmandu, Nepal (city, nat. cap.)	28N, 85E	204
Jackson, Mississippi, USA (city, st. cap.)	32N, 90W	167	Katowice, Poland (city)	50N, 19E	181
Jacksonville, Florida, USA (city)	30N, 82W	167	Kattegat, Strait of	57N, 11E	182
Jaipur, India (city)	27N, 76E	204	Kazakh Uplands	49N, 72E	202
Jakarta, Indonesia (city, nat. cap.)	6S, 107E	206	Kazakhstan (country)	50N, 70E	202
Jalisco (st., Mex.)	20N, 105W	167	Kazan, Russia (city)	56N, 49E	202
Jamaica (country)	18N, 78W	167	Kentucky (st., USA)	37N, 88W	167
James Bay	54N, 81W	168	Kenya (country)	0, 35E	191
Japan (country)	35N, 138E	205	Kenya, Mt. 17,058	0, 37E	191
Japura, Rio (riv., S.Am.)	3S, 65W	176	Khabarovsk, Russia (city)	48N, 135E	202
Java (island)	6N, 110E	206	Khambhat, Gulf of	20N, 73E	198
Jayapura, Indonesia (city)	3S, 141E	207	Kharkiv, Ukraine (city)	50N, 36E	181
Jebel Toubkal 13,671	31N, 8W	191	Khartoum, Sudan (city, nat. cap.)	16N, 33E	196
Jeddah, Saudi Arabia (city)	22N, 39E	203	Kiev, Ukraine (city, nat. cap.)	50N, 31E	189
Jefferson City, Missouri, USA (city, st. cap.)	39N, 92W	167	Kigali, Rwanda (city, nat. cap.)	2S, 30E	191
Jerusalem, Israel (city, nat. cap.)	32N, 35E	203	Kikwit, Dem. Rep. of the Congo (city)	5S, 19E	197
Jinan, China (city)	37N, 117E	205	Kilimanjaro, Mt. 19,340	4N, 35E	191
Ji-Parana, Brazil (city)	11S, 62W	179	Kimberly, South Africa (city)	29S, 25E	191
João Pessoa, Paraiba (city, st. cap., Braz.)	7S, 35W	175	King Leopold Ranges	16S, 125E	207
Johannesburg, South Africa (city)	26S, 27E	191	Kingston, Jamaica (city, nat. cap.)	18N, 77W	167
Jordan (country)	32N, 36E	181	Kinshasa, Dem. Rep. Congo (city, nat. cap.)	4S, 15E	191
Juan Fernandez Islands	33S, 80W	176	Kirghiz Steppe	40N, 65E	202
Jubba (riv., Africa)	3N, 43E	191	Kisangani, Dem. Rep. Congo (city)	1N, 25E	191
Jujuy (st., Argentina)	23S, 67W	175	Kitakyushu, Japan (city)	34N, 130E	205
Juneau, Alaska, USA (city, st. cap.)	58N, 134W	167	Kjollen Range	65N, 12E	182
Jutland Peninsula	56N, 9E	182	Kobe, Japan (city)	34N, 135E	205
K2, Mt. 28,250	30N, 70E	204	Kodiak Island	58N, 152W	168
Kabul, Afghanistan (city, nat. cap.)	35N, 69E	203	Kolkata, India (city)	23N, 88E	204
Kalahari Desert	25S, 20E	191	Kolwezi, Dem. Rep. of the Congo (city)	11S, 25E	197
Kalgourie-Boulder, Australia (city)	31S, 121E	207	Kolyma (riv., Asia)	70N, 160E	202
Kaliningrad, Russia (city)	55N, 21E	181	Kolyma Range	65N, 165E	202
Kamchatka Peninsula	55N, 159E	202	Komsomolsk, Russia (city)	51N, 137E	202
Kampala, Uganda (city, nat. cap.)	0, 33E	196	Korea Strait	32N, 130W	205
Kandahar, Afghanistan (city)	32N, 66E	203	Korea, North (country)	40N, 128E	205
Kano, Nigeria (city)	12N, 9E	196	Korea, South (country)	3S, 130W	205
Kanpur, India (city)	26N, 80E	204	Kosovo (country)	42N, 21E	181
Kansas (st., USA)	40N, 98W	167	Kotto (riv., Africa)	7N, 23E	191
Kansas City, Missouri, USA (city)	39N, 95W	167	Krasnoyarsk, Russia (city)	56N, 93E	202
Kaohsiung, Taiwan (city)	23N, 120E	205	Krishna (riv., Asia)	15N, 76E	204
Kara Sea	69N, 65E	202	Kuala Lumpur, Malaysia (city, nat. cap.)	3N, 107E	206
Karachi, Pakistan (city)	25N, 67E	204	Kunlun Mountains	36N, 90E	205

GEOGRAPHIC INDEX

NAME/DESCRIPTION	LATITUDE & LONGITUDE	PAGE	NAME/DESCRIPTION	LATITUDE & LONGITUDE	PAGE
Kunming, China (city)	25N, 106E	205	Lincoln, Nebraska, USA (city, st. cap.)	41N, 97W	167
Kuril Islands	46N, 147E	202	Lisbon, Portugal (city, nat. cap.)	39N, 9W	182
Kutch, Gulf of	23N, 70E	198	Lithuania (country)	56N, 24E	181
Kuwait (country)	29N, 48E	203	Little Rock, Arkansas, USA (city, st. cap.)	35N, 92W	167
Kuwait, Kuwait (city, nat. cap.)	29N, 48E	203	Liverpool, England, UK (city)	53N, 3W	181
Kyoto, Japan (city)	35N, 136E	205	Ljubljana, Slovenia (city, nat. cap.)	46N, 15E	189
Kyrgyzstan (country)	40N, 75E	202	Llanos	33N, 103W	176
Kyushu (island)	30N, 130W	205	Logone (riv., Africa)	10N, 14E	191
La Pampa (st., Argentina)	36S, 70W	175	Loire (riv., Europe)	47N, 1E	182
La Paz, Baja California Sur, Mexico (city, st. cap.)	24N, 110W	173	Lome, Togo (city, nat. cap.)	6N, 1E	196
La Paz, Bolivia (city, nat. cap.)	17S, 68W	175	London, UK (city, nat. cap.)	51N, 0	182
La Plata, Argentina (city)	35S, 58W	175	Londonderry, Cape	14S, 125E	207
La Rioja (st., Argentina)	30S, 70W	175	Long Island	40N, 73W	168
La Rioja, La Rioja (city, st. cap., Argen.)	29S, 67W	175	Lopez, Cape	1S, 8E	191
Laayoune, Morocco (city)	27N, 13W	196	Los Angeles, California, USA (city)	34N, 118W	167
Labrador Peninsula	52N, 60W	168	Louisiana (st., USA)	30N, 90W	167
Labrador Sea	60N, 55W	168	Lower Hutt, New Zealand (city)	45S, 175E	207
Lachlan (riv., Australasia)	34S, 145E	207	Luanda, Angola (city, nat. cap.)	9S, 13E	197
Ladoga, Lake	61N, 31E	182	Lubumbashi, Dem. Rep. of the Congo (city)	12S, 27E	197
Lagos, Nigeria (city)	6N, 3E	196	Lusaka, Zambia (city, nat. cap.)	15S, 28E	191
Lahore, Pakistan (city)	32N, 74E	204	Luxembourg (country)	50N, 6E	181
Lake Erie (lake, N.Am.)	42N, 85W	168	Luxembourg, Luxembourg (city, nat. cap.)	50N, 6E	182
Lansing, Michigan, USA (city, st. cap.)	43N, 85W	167	Luxor, Egypt (city)	26N, 33E	196
Lanzhou, China (city)	36N, 104E	205	Luzon (island)	17N, 121E	206
Laos (country)	20N, 105E	206	Luzon Strait	20N, 121E	206
Laptev Sea	73N, 120E	202	Lyon, France (city)	46N, 5E	181
Las Vegas, Nevada, USA (city)	36N, 115W	167	Macapa, Brazil (city)	0, 51W	179
Latvia (country)	56N, 24E	181	MacDonnell Ranges	23S, 135E	207
Le Havre, France (city)	50N, 0	181	Macedonia (country)	41N, 21E	181
Lebanon (country)	34N, 35E	181	Maceió, Alagoas (city, st. cap., Braz.)	10S, 36W	175
Leeds, England, UK (city)	54N, 2W	181	Mackenzie (riv., N.Am.)	68N, 130W	168
Lena (riv., Asia)	70N, 125E	202	Mackenzie Mountains	63N, 130W	168
Lesotho (country)	30S, 27E	191	Macquarie (riv., Australasia)	33S, 146E	207
Leticia, Brazil (city)	4S, 70W	179	Madagascar (country)	20S, 46E	191
Leveque, Cape	16S, 123E	207	Madeira, Rio (riv., S.Am.)	5S, 60W	176
Leyte (island)	12N, 130E	206	Madison, Wisconsin, USA (city, st. cap.)	43N, 89W	167
Lhasa, Tibet (China) (city)	30N, 91E	198	Madrid, Spain (city, nat. cap.)	40N, 4W	182
Liberia (country)	6N, 10W	191	Mafia (island)	7S, 39E	191
Libreville, Gabon (city, nat. cap.)	0, 9E	196	Magadan, Russia (city)	60N, 151E	202
Libya (country)	27N, 17E	191	Magdalena, Rio (riv., S.Am.)	8N, 74W	176
Libyan Desert	27N, 25E	191	Magellan, Strait of	54S, 68W	176
Liechtenstein (country)	47N, 9E	181	Main (riv., Europe)	50N, 9E	182
Ligurian Sea	43N, 9E	182	Maine (st., USA)	46N, 70W	167
Lille, France (city)	51N, 3E	181	Makkah (Mecca), Saudi Arabia (city)	21N, 40E	203
Lilongwe, Malawi (city, nat. cap.)	14S, 34E	197	Malabo, Equatorial Guinea (city, nat. cap.)	4N, 9E	191
Lima, Peru (city, nat. cap.)	12S, 77W	179	Malacca, Strait of	3N, 100E	206
Limpopo (riv., Africa)	22S, 30E	191	Malawi (country)	13S, 35E	191

GEOGRAPHIC INDEX

NAME/DESCRIPTION	LATITUDE & LONGITUDE	PAGE	NAME/DESCRIPTION	LATITUDE & LONGITUDE	PAGE
Malaysia (country)	3N, 110E	206	Melbourne, Victoria, Australia (city, st. cap.)	38S, 145E	209
Male, Maldives (city, nat. cap.)	4N, 73E	197	Melville, Cape	15S, 145E	207
Malekula (island)	16S, 166E	207	Memphis, Tennessee, USA (city)	35N, 90W	167
Mali (country)	17N, 5W	191	Mendoza (st., Argentina)	35S, 70W	175
Malta (country)	36N, 16E	181	Mendoza, Mendoza (city, st. cap., Argen.)	33S, 69W	175
Managua, Nicaragua (city, nat. cap.)	12N, 86W	167	Menongue, Angola (city)	15S, 18E	197
Manama, Bahrain (city, nat. cap.)	26N, 51E	203	Merauke, Indonesia (city)	9S, 140E	207
Manaus, Brazil (city)	3S, 60W	179	Merida, Yucatán, Mexico (city, st. cap.)	21N, 90W	173
Manchester, England, UK (city)	53N, 2W	181	Meuse (riv., Europe)	51N, 4E	182
Manchurian Plain	45N, 125E	205	Mexicali, Baja California, Mexico (city, st. cap.)	33N, 115W	173
Mandalay, Myanmar (city)	22N, 96E	206	Mexico (country)	30N, 110W	167
Mangoky (riv., Africa)	22S, 46E	191	Mexico (st., Mex.)	18N, 100W	167
Manila, Philippines (city, nat. cap.)	15N, 121E	206	Mexico City, Mexico (city, nat. cap.)	19N, 99W	173
Manitoba (prov., Can.)	52N, 93W	167	Mexico, Gulf of	26N, 90W	168
Maputo, Mozambique (city, nat. cap.)	26S, 33E	191	Miami, Florida, USA (city)	26N, 80W	167
Mar del Plata, Argentina (city)	38S, 58W	180	Michigan (st., USA)	45N, 82W	167
Maraba, Brazil (city)	5S, 49W	179	Lake Michigan (lake, N.Am.)	45N, 90W	168
Maracaibo, Lake	10N, 72W	176	Michoacan (st., Mex.)	17N, 107W	167
Maracaibo, Venezuela (city)	11N, 72W	175	Milan, Italy (city)	45N, 9E	181
Marajó Island	1S, 49W	176	Milwaukee, Wisconsin, USA (city)	43N, 88W	167
Maranhao (st., Brazil)	4S, 45W	175	Minas Gerais (st., Brazil)	17S, 45W	175
Maranon, Rio (riv., S.Am.)	5S, 75W	176	Mindanao (island)	8N, 125E	206
Marie Byrd Land	78S, 125W	212	Mindoro (island)	13N, 120E	206
Marne (riv., Europe)	49N, 4E	182	Minneapolis, Minnesota, USA (city)	45N, 93W	167
Maroua, Cameroon (city)	11N, 14E	196	Minnesota (st., USA)	45N, 90W	167
Marrakesh, Morocco (city)	32N, 8W	196	Minsk, Belarus (city, nat. cap.)	54N, 28E	181
Marseille, France (city)	43N, 5E	181	Misiones (st., Argentina)	25S, 55W	175
Marshall Islands	8N, 171E	208	Mississippi (riv., N.Am.)	28N, 90W	168
Maryland (st., USA)	37N, 76W	167	Mississippi (st., USA)	30N, 90W	167
Masai Steppe	5S, 35E	191	Missouri (riv., N.Am.)	41N, 96W	168
Maseru, Lesotho (city, nat. cap.)	29S, 27E	191	Missouri (st., USA)	35N, 92W	167
Mashad, Iran (city)	36N, 59E	203	Mitchell (riv., Australasia)	16S, 143E	207
Massachusetts (st., USA)	42N, 70W	167	Mitu, Colombia (city)	1N, 70W	179
Matadi, Dem. Rep. of the Congo (city)	6S, 13E	197	Mobile, AL (city)	31N, 88W	167
Mato Grosso (st., Brazil)	15S, 55W	175	Moçambique, Mozambique (city)	15S, 40E	191
Mato Grosso do Sul (st., Brazil)	20S, 55W	175	Mogadishu, Somalia (city, nat. cap.)	2N, 45E	191
Mato Grosso Plateau	16S, 52W	176	Moldova (country)	49N, 28E	181
Mauritania (country)	20N, 10W	191	Mombasa, Kenya (city)	4S, 40E	191
Mazar-e Sharif, Afghanistan (city)	37N, 67E	203	Monaco (country)	44N, 8E	181
Mazatlan, Mexico (city)	23N, 106W	173	Monaco, Monaco (city)	44N, 8E	181
Mbandaka, Dem. Rep. Congo (city)	0, 18E	191	Mongolia (country)	45N, 100E	205
Mbeya, Tanzania (city)	9S, 33E	197	Mongolian Plateau	47N, 110E	205
McKinley, Mt. 20,320	62N, 150W	168	Monrovia, Liberia (city, nat. cap.)	6N, 11W	191
Medan, Indonesia (city)	4N, 99E	206	Montana (st., USA)	50N, 110W	167
Medellin, Colombia (city)	6N, 76W	175	Montenegro (country)	42N, 19E	181
Mediterranean Sea	36N, 16E	182	Monterrey, Nuevo León, Mexico (city, st. cap.)	26N, 100W	173
Mekong (riv., Asia)	15N, 108E	206	Montevideo, Uruguay (city, nat. cap.)	35S, 56W	175

GEOGRAPHIC INDEX

GEOGRAPHIC INDEX

NAME/DESCRIPTION	LATITUDE & LONGITUDE	PAGE	NAME/DESCRIPTION	LATITUDE & LONGITUDE	PAGE
North West Cape	22S, 115W	207	Otway, Cape	40S, 142W	207
Northern Territory (st., Aust.)	20S, 134W	207	Ouachita Mountains	34N, 95W	168
Northwest Territories (terr., Can.)	65N, 125W	167	Ougadougou, Burkina Faso (city, nat. cap.)	12N, 2W	191
Norway (country)	62N, 8E	181	Owen Stanley Range	9S, 148E	207
Norwegian Sea	65N, 5E	182	Ozark Plateau	37N, 93W	168
Nouakchott, Mauritania (city, nat. cap.)	18N, 16W	191	Pachuca, Hidalgo, Mexico (city, st. cap.)	20N, 99W	173
Noumea, New Caledonia (city)	22S, 167E	207	Pacific Ocean	20N, 115W	168
Nova Scotia (prov., Can.)	46N, 67W	167	Padang, Indonesia (city)	1S, 100E	206
Novaya Zemlya (island)	72N, 55E	202	Pakistan (country)	25N, 72E	204
Novosibirsk, Russia (city)	55N, 83E	202	Palana, Russia (city)	59N, 160E	202
Nubian Desert	20N, 30E	191	Palawan (island)	10N, 119E	206
Nuevo Leon (st., Mex.)	25N, 100W	167	Palembang, Indonesia (city)	3S, 105E	206
Nukualofa, Tonga (city, nat. cap.)	21S, 175W	208	Palmas, Cape	8N, 8W	191
Nullarbor Plain	34S, 125W	209	Palmas, Tocantins (city, st. cap., Braz.)	10S, 49W	175
Nunavut (terr., Can)	70N, 95w	167	Pamir Mountains	32N, 70E	204
Nyasa, Lake	10S, 35E	191	Pampas	36S, 73W	176
Oakland, CA (city)	38N, 122W	167	Panama (country)	10N, 80W	167
Oaxaca (st., Mex.)	17N, 97W	167	Panama City, Panama (city, nat. cap.)	9N, 80W	173
Oaxaca, Oaxaca, Mexico (city, st. cap.)	17N, 97W	173	Gulf of Panama	10N, 80W	176
Ob (riv., Asia)	60N, 78E	202	Panay (island)	11N, 122E	206
Obo, Central African Republic (city)	5N, 27E	196	Pantanal	20S, 58W	176
Oder (riv., Europe)	51N, 16E	182	Papua New Guinea (country)	6S, 144E	207
Ohio (riv., N.Am.)	38N, 85W	168	Papua, Gulf of	8S, 144E	207
Ohio (st., USA)	42N, 85W	167	Pará (st., Brazil)	4S, 54W	175
Okavango Delta	21S, 23E	191	Paraguay (country)	23S, 60W	175
Okhotsk, Russia (city)	59N, 140E	202	Paraíba (st., Brazil)	6S, 35W	175
Okhotsk, Sea of	57N, 150E	202	Paramaribo, Suriname (city, nat. cap.)	5N, 55W	175
Oklahoma (st., USA)	36N, 95W	167	Paraná (st., Brazil)	25S, 55W	175
Oklahoma City, Oklahoma, USA (city, st. cap.)	35N, 98W	167	Paraná, Entre Rios (city, st. cap., Argen.)	32S, 60W	175
Olympia, Washington, USA (city, st. cap.)	47N, 123W	167	Parana, Rio (riv., S.Am.)	20S, 50W	176
Omaha, Nebraska, USA (city)	41N, 96W	167	Paranaiba, Rio (riv., S.Am.)	20S, 55W	176
Oman (country)	20N, 55E	203	Paris, France (city, nat. cap.)	49N, 2E	181
Gulf of Oman	23N, 55E	203	Pasadas, Misiones (city, st. cap., Argen.)	27S, 56W	175
Omdurman, Sudan (city)	16N, 32E	191	Pasay City, Philippines (city)	15N, 121E	206
Omsk, Russia (city)	55N, 73E	202	Patagonia	43S, 70W	176
Lake Onega	62N, 35E	202	Patna, India (city)	26N, 85E	204
Ontario (prov., Can.)	50N, 90W	168	Peace (riv., N.Am.)	55N, 120W	168
Lake Ontario (lake, N.Am.)	45N, 77W	168	Pechora Basin	65N, 58E	202
Oodnadatta, Aust. (city)	28S, 135E	207	Pempa (island)	5S, 39E	191
Oran, Algeria (city)	36N, 1W	191	Pennsylvania (st., USA)	43N, 80W	167
Orange (riv., Africa)	27S, 18E	191	Perm, Russia (city)	58N, 56E	202
Oregon (st., USA)	46N, 120W	167	Pernambuco (st., Brazil)	7S, 36W	175
Orinoco, Rio (riv., S.Am.)	8N, 65W	176	Persian Gulf	28N, 50E	203
Orkney Islands	60N, 0	182	Perth, Western Australia, Australia (city, st. cap.)	32S, 116E	209
Osaka, Japan (city)	35N, 135E	205	Peru (country)	10S, 75W	175
Oslo, Norway (city, nat. cap.)	60N, 11W	181	Peshawar, Pakistan (city)	34N, 72E	204
Ottawa, Canada (city, nat. cap.)	45N, 76W	167	Petropavlovsk Kamchatskiy, Russia (city)	53N, 159E	202

GEOGRAPHIC INDEX

NAME/DESCRIPTION	LATITUDE & LONGITUDE	PAGE	NAME/DESCRIPTION	LATITUDE & LONGITUDE	PAGE
Philadelphia, Pennsylvania, USA (city)	40N, 75W	167	Pune, India (city)	19N, 74E	204
Philippine Sea	15N, 125E	206	Punta Arenas, Chile (city)	53S, 71W	180
Philippines (country)	15N, 120E	206	Punta Negra	6S, 81W	176
Phnom Penh, Cambodia (city, nat. cap.)	12N, 105E	206	Purus, Rio (riv., S.Am.)	5S, 68W	176
Phoenix, Arizona, USA (city, st. cap.)	34N, 112W	167	Putrajaya, Malaysia (city)	3N, 102E	206
Piauí (st., Brazil)	7S, 44W	175	Putumayo, Rio (riv., S.Am.)	3S, 74W	176
Pierre, South Dakota, USA (city, st. cap.)	44N, 100W	167	Pyongyang, North Korea (city, nat. cap.)	39N, 126E	205
Pietermaritzburg, South Africa (city)	30S, 30E	191	Pyrenees Mountains	43N, 2E	182
Pittsburgh, Pennsylvania, USA (city)	40N, 80W	167	Qaraghandy, Kazakhstan (city)	50N, 73E	202
Plateau of Iran	26N, 60E	198	Qatar	25N, 51E	203
Plateau of Tibet	26N, 85E	205	Qilian Shan	39N, 98E	205
Platte (riv., N.Am.)	41N, 105W	168	Qingdao, China (city)	36N, 120E	205
Po (riv., Europe)	45N, 12E	182	Qinling Mountains	33N, 107E	205
Podgorica, Montenegro (city, nat. cap.)	42N, 19E	189	Quebec (prov., Can.)	52N, 70W	167
Point Barrow	70N, 156W	168	Quebec, Quebec (city, prov. cap., Can.)	47N, 71W	167
Poland (country)	54N, 20E	181	Queen Charlotte Islands	50N, 130W	168
Port Elizabeth, South Africa (city)	34S, 26E	191	Queen Elizabeth Islands	75N, 110W	168
Port Lincoln, Aust. (city)	35S, 135E	207	Queen Maud Land	72S, 10E	212
Port Louis, Mauritius (city, nat. cap.)	20S, 57E	196	Queensland (st., Aust.)	24S, 145E	207
Port Moresby, Papua New Guinea (city, nat. cap.)	9S, 147E	206	Querataro (st., Mex.)	22N, 96W	167
Port Sudan, Sudan (city)	20N, 37E	196	Queretaro, Querétaro, Mexico (city, st. cap.)	21N, 100W	173
Port Vila, Vanuatu (city, nat. cap.)	17S, 169E	207	Quezon City, Philippines (city)	15N, 121E	206
Port-au-Prince, Haiti (city, nat. cap.)	19N, 72W	167	Quintana Roo (st., Mex.)	18N, 88W	167
Port-Gentil, Gabon (city)	1S, 9E	196	Quito, Ecuador (city, nat. cap.)	0, 79W	179
Portland, Oregon, USA (city)	46N, 123W	167	Québec, Québec, Canada (city, prov. cap.)	47N, 71W	171
Porto Alegre, Brazil (city)	30S, 51W	180	Rabat, Morocco (city, nat. cap.)	34N, 7W	191
Porto Novo, Benin (city, nat. cap.)	7N, 3E	191	Rainier, Mt. 14,410	48N, 120W	168
Porto Velho, Rondonia (city, st. cap., Braz.)	9S, 64W	175	Raleigh, North Carolina, USA (city, st. cap.)	36N, 79W	167
Port-of-Spain, Trinidad and Tobago (city, nat. cap.)	11N, 62W	173	Rangoon, Myanmar (Burma) (city, nat. cap.)	17N, 96E	198
Porto-Novo, Benin (city)	6N, 3E	196	Rapid City, South Dakota, USA (city)	44N, 103W	167
Portugal (country)	38N, 8W	181	Rawalpindi, Pakistan (city)	34N, 73E	204
Port-Vila, Vanuatu (city, nat. cap.)	18S, 168E	208	Rawson, Chubuy (city, st. cap., Argen.)	43S, 65W	175
Posadas, Argentina (city)	27S, 56W	180	Recife, Brazil (city)	8S, 35W	179
Potosí, Bolivia (city)	20S, 66W	175	Recife, Pernambuco (city, st. cap., Braz.)	8S, 35W	175
Prague, Czechia (city, nat. cap.)	50N, 14E	181	Red River (of the North) (riv., N.Am.)	50N, 98W	175
Pretoria, South Africa (city, nat. cap.)	26S, 28E	191	Red Sea	20N, 35E	191
Prince Edward Island (island)	50N, 67W	168	Regina, Saskatchewan, Canada (city, prov. cap.)	50N, 105W	171
Prince Edward Island (prov., Can.)	50N, 67W	167	Reindeer Lake (lake, N.Am.)	57N, 100W	168
Prince of Wales Island	73N, 97W	168	Repulse Bay, Canada (city)	67N, 86W	171
Prince Rupert, Canada (city)	54N, 130W	171	Resistencia, Chaco (city, st. cap., Argen.)	27S, 59W	175
Pripyat (riv., Europe)	52N, 29E	182	Resolute, Canada (city)	75N, 95W	171
Pristina, Kosovo (city, nat. cap.)	43N, 21E	189	Resolution Island	61N, 65W	168
Providence, Rhode Island, USA (city, st. cap.,)	42N, 71W	167	Réunion (island)	21S, 55E	191
Puebla, Mexico (st.)	18N, 96W	167	Reykjavik, Iceland (city, nat. cap.)	64N, 22W	181
Puebla, Puebla, Mexico (city, st. cap.)	19N, 98W	173	Rhine (riv., Europe)	50N, 10E	182
Puerto Deseado, Argentina (city)	48S, 66W	180	Rhode Island (st., USA)	42N, 70W	167
Puerto Monte, Chile (city)	42S, 74W	175	Rhone (riv., Europe)	42N, 8E	182

GEOGRAPHIC INDEX

GEOGRAPHIC INDEX

NAME/DESCRIPTION	LATITUDE & LONGITUDE	PAGE	NAME/DESCRIPTION	LATITUDE & LONGITUDE	PAGE
Saudi Arabia (country)	25N, 50E	203	South Cape	8S, 150E	207
Savannah, Georgia, USA (city)	32N, 81W	167	South Carolina (st., USA)	33N, 79W	167
Sayan Range	45N, 90E	198	South China Sea	15N, 115E	206
Scandanavia (peninsula)	63N, 15E	182	South Dakota (st., USA)	45N, 100W	167
Sea of Japan (East Sea)	40N, 135E	205	South Island	45S, 170E	209
Sea of Marmara	41N, 28E	182	South Magnetic Pole	64S, 138E	212
Seattle, Washington, USA (city)	48N, 122W	167	South Pole	90S, 0	212
Seine (riv., Europe)	49N, 3E	182	Southampton Island	68N, 86W	168
Semarang, Indonesia (city)	7S, 110E	206	Southern Alps	45S, 170E	207
Senegal (country)	15N, 15W	191	Southwest Cape	47S, 167E	207
Senegal (riv., Africa)	15N, 15W	191	Spain (country)	38N, 4W	181
Seoul, South Korea (city, nat. cap.)	38N, 127E	205	Spokane, Washington, USA (city)	48N, 117W	167
Serbia and Montenegro	44N, 20E	181	Springfield, Illinois, USA (city, st. cap.)	40N, 90W	167
Sergipe (st., Brazil)	12S, 36W	175	Sri Jawewardenepura Kotte, Sri Lanka (city, nat. cap.)	7N, 80E	204
Sev Dvina (riv., Asia)	60N, 50E	202	Sri Lanka (country)	8N, 80E	204
Sevastapol, Ukraine (city)	45N, 33E	189	Srinagar, India (city)	34N, 75E	204
Severnaya Zemlya (island)	80N, 88E	202	St. Helena (island)	16S, 5W	191
Seville, Spain (city)	37N, 6W	181	St. John's, Newfoundland, Canada (city, prov. cap.)	48N, 53W	171
Shanghai, China (city)	31N, 121E	205	St. Lawrence (riv., N.Am.)	50N, 65W	168
Shelikov Gulf	60N, 158E	202	St. Lawrence, Gulf of	50N, 65W	168
Shenyeng, China (city)	42N, 123E	205	St. Louis, Missouri, USA (city)	39N, 90W	167
Shenzhen, China (city)	23N, 114E	205	St. Marie, Cape	25S, 45E	191
Shetland Islands	60N, 5W	182	St. Paul, Minnesota, USA (city, st. cap.)	45N, 93W	167
Shikoku (island)	34N, 130E	205	St. Petersburg, Russia (city)	60N, 30E	181
Shiraz, Iran (city)	30N, 52E	203	Stanley, Falkland Islands, UK	52S, 58W	182
Sicily (island)	38N, 14E	182	Stanovoy Range	55N, 125E	202
Sierra Leone (country)	6N, 14W	191	Stavanger, Norway (city)	59N, 6E	181
Sierra Madre Occidental	27N, 108W	168	Steep Point	25S, 115E	207
Sierra Madre Oriental	27N, 100W	168	Stewart Island	47S, 167E	207
Sierra Nevada	38N, 120W	168	Stockholm, Sweden (city, nat. cap.)	59N, 18E	181
Sikhote Range	45N, 135E	202	Stuart Range	32S, 135E	207
Simpson Desert	25S, 136E	209	Stuttgart, Germany (city)	49N, 9E	181
Sinai Peninsula	28N, 33E	191	Sucre, Bolivia (city, nat. cap.)	19S, 65W	179
Sinaloa (st., Mex.)	25N, 110W	167	Sudan (country)	10N, 30E	191
Singapore (city, nat. cap.)	1N, 104E	206	Sulaiman Range	28N, 70E	198
Singapore (country)	1N, 104E	206	Sulu Islands	8N, 120E	206
Singapore, Singapore (city, nat. cap.)	1N, 104E	206	Sulu Sea	10N, 120E	206
Skagerrak, Strait of	58N, 8E	182	Sumatra (island)	0, 100E	206
Skopje, Macedonia (city, nat. cap.)	42N, 21E	181	Sumba (island)	10S, 120E	206
Slovenia (country)	47N, 14E	181	Sumbawa (island)	8S, 116E	206
Socotra (island)	12N, 54E	191	Lake Superior (lake, N.Am.)	50N, 90W	168
Sofia, Bulgaria (city, nat. cap.)	43N, 23E	181	Surabaya, Indonesia (city)	7S, 113E	206
Solomon Islands (country)	7S, 160E	207	Surat, India (city)	21N, 73E	204
Somalia (country)	5N, 45E	191	Surgut, Russia (city)	61N, 73E	202
Sonora (st., Mex.)	30N, 110W	167	Suriname (country)	5N, 55W	175
South Africa (country)	30S, 25E	191	Suva, Fiji (city, nat. cap.)	18S, 178E	208
South Australia (st., Aust.)	30S, 125E	207			

GEOGRAPHIC INDEX

GEOGRAPHIC INDEX

GEOGRAPHIC INDEX

Sources

Maps

Map 21	World Christian Database
Map 24	Living Tongues Institute for Endangered Languages
Map 27	Population Reference Bureau
Map 36	Food and Agriculture Organization, United Nations
Map 39	World Health Organization
Map 70	Deutsche Gesellschaft für Technische Zusammenarbeit
Map 75	World Development Indicators, World Bank
Map 76	World Development Indicators, World Bank
Map 77	World Development Indicators, World Bank
Map 82	Oak Ridge National Laboratory Landscan 2004; Climatic Research Unit, University of East Anglia; Fourth Assessment Report of the Intergovernmental Panel on Climate Change Working Group II
Map 83	Fourth Assessment Report of the Intergovernmental Panel on Climate Change Working Group II
Map 87	Water Systems Analysis Group, University of New Hampshire
Map 89	Beverage Marketing Corporation
Map 93	Studies of Jack R. Harlan, Professor of Plant Genetics/University of Illinois; John Bartholomew & Son, Ltd./National Library of Scotland Archive
Map 99	Compiled by Donald L. Carrick, NGS using data from: IUCN/NYZS/WWF African Rhino Survey, Andrew Laurie/Cambridge University, and Thomas Foose/American Association of Zoological Parks and Aquariums
Map 101	World Wildlife Fund, IUCN
Map 102	World Bank, World Development Indicators
Map 104	USDA Natural Resources Conservation Service; Gerhard Bechtold; Johannes Lehmann, Cornell University
Map 106	Yale University, Columbia University, and LinkedByAir (compiled by *Newsweek* Magazine)
Map 119	Freedom House; Interparliamentary Union; CIA World Factbook
Map 124	Julian Burger, United Nations

References

Allen, John L. (1997). *Student Atlas of Environmental Issues.* Guilford, CT: Dushkin/McGraw-Hill.

American Association for the Advancement of Science. (2000). *AAAS Atlas of Population and Environment.* Berkeley-Los Angeles-London: The University of California Press.

An atmosphere of uncertainty. (1987, April). *National Geographic,* 171.

Cincotta, Richard P., Robert Engelman and Daniele Anastasion. (2003). *The Security Demographic.* Washington, DC: Population Action International.

Conservation International. (2002). *Global Hotspots of Diversity.* Washington, DC.

Crabb, C. (1993, January). Soiling the planet. *Discover, 14* (1), 74–75.

DeBlij, H. J., & Muller, P. (2006). *Geography: Realms, Regions and Concepts* (12th ed., revised). New York: John Wiley & Sons.

Department of Geography, Pennsylvania State University. (1996). Unpublished computer model output. State College, PA: Pennsylvania State University.

Deutsche Gesellschaft für Technische Zusammenarbeit (GTZ). (2010). *International Fuel Prices 2009, 6th Edition* [Online]. Available: www.gtz.de/fuelprices.

Domke, K. (1988). *War and the Changing Global System.* New Haven, CT: Yale University Press.

Economic consequences of the accident at Chernobyl nuclear plant. (1987). PlanEcon Reports, 3.

Environmental Protection Agency. (1996). Unpublished data [Online]. Available: http://www.epa.gov.

Fagan, B. M. (1998). *People of the Earth* (9th ed.). New York: Longman.

Fellman, J., Getis, A., & Getis, J. (1995). *Human Geography: Landscapes of Human Activities* (4th ed.). Dubuque, IA: Wm. C. Brown Publishers.

Fuller, Harold, ed. (1971). *World Patterns: The Aldine College Atlas.* Chicago: Aldine Publishing Co.

Hoebel, E. A. (1966). *Anthropology: the Study of Man* (3rd ed.). New York: McGraw-Hill.

Johnson, D. (1977). *Population, Society, and Desertification.* New York: United Nations Conference on Desertification, United Nations Environment Programme.

Köppen, W., & Geiger, R. (1954). *Klima der Erde* [Climate of the Earth]. Darmstadt, Germany: Justus Perthes.

Kuchler, A. W. (1949). Natural vegetation. *Annals of the Association of American Geographers, 39.*

Lindeman, M. (1990). *The United States and the Soviet Union: Choices for the 21st Century.* Guilford, CT: McGraw-Hill/Dushkin.

Mather, J. R. (1974). *Climatology: Fundamentals and Applications.* New York: McGraw-Hill.

McGuire, V. L. (2009). *Water-level changes in the High Plains aquifer, predevelopment to 2007, 2005–06, and 2006–07: U.S. Geological Survey Scientific Investigations Report 2009–5019* [Online]. Available: http://pubs.usgs.gov/sir/2009/5019/.

Mexico under siege. *Los Angeles Times* [Online]. Available: http://projects.latimes.com/mexico-drug-war/.

Miller, G. T. (1992). *Living in the Environment* (7th ed.). Belmont, CA: Wadsworth.

Murphy, R. E. (1968). Landforms of the world [Map supplement No. 91]. *Annals of the Association of American Geographers, 58* (1), 198–200.

National Aeronautics and Space Administration. (2003–2005). Unpublished data and images [Online]. Available: http://www.nasa.gov.

National Geographic Society. (1999). *Atlas of the World,* (7th ed.). Washington, DC: National Geographic Society.

National Oceanic and Atmospheric Administration. (2010). Unpublished data [Online]. Available: http://www.noaa.gov.

Population Action International. (2003). *The Security Demographic: Population and Civil Conflict After the Cold War.* Washington, DC.

Population Reference Bureau. (2010). *2010 World Population Data Sheet* [Online]. Available: http://www.prb.org/.

Rand McNally. (1996). *Goode's World Atlas* (19th ed.). Chicago: Rand McNally and Co.

Rand McNally Answer Atlas. (1996). Chicago: Rand McNally and Co.

Rand McNally. (2010). *Goode's Atlas of Human Geography* (22nd ed.). New York: John Wiley & Sons, Inc.

Rourke, J. T. (2003). *International Politics on the World Stage* (9th ed). Guilford, CT: McGraw-Hill/Dushkin.

Scupin, R., and Decorse, C. R. (2001). *Anthropology: A Global Perspective* (4th ed.). Upper Saddle River, NJ: Prentice Hall.

Shelley, F., and Clarke, A. (1994). *Human and Cultural Geography: A Global Perspective*. Dubuque, IA: Wm. C. Brown Publishers.

Smith, Dan. (1997). *The State of War and Peace Atlas* (3rd ed.). Penguin Books: New York.

Spector, L. S., and Smith, J. R. (1990). *Nuclear Ambitions: The Spread of Nuclear Weapons*. Boulder, CO: Westview Press.

This Fragile Earth [map]. (1988, December). *National Geographic*, 174.

Thornthwaite, C. W., & Mather, J. R. (1955). *The Water Balance* [Publications in Climatology No. 8]. Centerton, NJ: Drexel Institute of Technology, Laboratory of Climatology.

Times Atlas of World History. (1978). Maplewood, NJ: Hammond.

United Nations Department of Economic and Social Affairs, Population Division. (2009). *International Migration 2009*. [Online]. Available: http://www.un.org/esa/population/publications/2009Migration_Chart/2009IttMig_chart.htm.

United Nations Department of Economic and Social Affairs, Population Division. (2009). *World Urbanization Prospects* [Online]. Available: http://esa.un.org/unpd/wup/.

United Nations Food and Agriculture Organization. (2009). *FAO Statistical Yearbook* [Online]. Available: http://www.fao.org/economic/ess/publications-studies/statistical-yearbook/en/.

United Nations Food and Agriculture Organization (FAO). (1995). *Forest Resources Assessment 1990: Global Synthesis* [FAO Forestry Paper NO. 124]. Rome: FAO.

United Nations High Commissioner for Refugees. (2010). *Global Trends: Refugees, Asylum-seekers, Returnees, Internally Displaced and Stateless Persons* [Online]. Available: http://www.unhcr.org/4c11f0be9.html.

U.S. Census Bureau. (1998). *World Population Profile*. Washington, DC: U.S. Government Printing Office.

U.S. Central Intelligence Agency. (2010). *World Factbook* [Online]. Available: https://www.cia.gov/library/publications/the-world-factbook/.

U.S. Department of Energy. (1996). *U.S.–Canada Memorandum of Intent on Transboundary Air Pollution*. Washington, DC: U.S. Government Printing Office.

U.S. Department of State. (2005). *Stateman's Year-Book 2005*. Washington, DC: U.S. Government Printing Office.

U.S. Department of State. (2010). *Country Reports on Human Rights Practices* [Online]. Available: http://www.state.gov/g/drl/rls/hrrpt/.

U.S. Department of Agriculture, Forest Service. (1989). *Ecoregions of the Continents*. Washington, DC: U.S. Government Printing Office.

U.S. Department of Agriculture, Natural Resources Conservation Service. (2010). *World Soil Resources Map* [Online]. Available: http://soils.usda.gov/use/worldsoils/mapindex/.

The World Bank. (2010). *World Development Indicators* [Online]. Available: http://data.worldbank.org/.

World Conservation Monitoring Centre. (1996). Unpublished data. Cambridge, England: World Conservation Monitoring Centre.

World Health Organization. (2010). *Global Health Observatory* [Online]. Available: http://www.who.int/gho/.

World Resources Institute. *World Resources 2002–2004: A Guide to the Global Environment*. New York: Oxford University Press.

World Resources Institute. (2005). *World Resources 2005—The Wealth of the Poor. Managing Ecosystems to Fight Poverty*. Washington, DC: Oxford University Press.